Communications in Computer and Information Science 2660

Series Editors

Rationale

The CCIS series is devoted to the publication of proceedings of computer science conferences. Its aim is to efficiently disseminate original research results in informatics in printed and electronic form. While the focus is on publication of peer-reviewed full papers presenting mature work, inclusion of reviewed short papers reporting on work in progress is welcome, too. Besides globally relevant meetings with internationally representative program committees guaranteeing a strict peer-reviewing and paper selection process, conferences run by societies or of high regional or national relevance are also considered for publication.

Topics

The topical scope of CCIS spans the entire spectrum of informatics ranging from foundational topics in the theory of computing to information and communications science and technology and a broad variety of interdisciplinary application fields.

Information for Volume Editors and Authors

Publication in CCIS is free of charge. No royalties are paid, however, we offer registered conference participants temporary free access to the online version of the conference proceedings on SpringerLink (http://link.springer.com) by means of an http referrer from the conference website and/or a number of complimentary printed copies, as specified in the official acceptance email of the event.

CCIS proceedings can be published in time for distribution at conferences or as post-proceedings, and delivered in the form of printed books and/or electronically as USBs and/or e-content licenses for accessing proceedings at SpringerLink. Furthermore, CCIS proceedings are included in the CCIS electronic book series hosted in the SpringerLink digital library at http://link.springer.com/bookseries/7899. Conferences publishing in CCIS are allowed to use our online conference service (Meteor) for managing the whole proceedings lifecycle (from submission and reviewing to preparing for publication) free of charge.

Publication process

The language of publication is exclusively English. Authors publishing in CCIS have to sign the Springer CCIS copyright transfer form, however, they are free to use their material published in CCIS for substantially changed, more elaborate subsequent publications elsewhere. For the preparation of the camera-ready papers/files, authors have to strictly adhere to the Springer CCIS Authors' Instructions and are strongly encouraged to use the CCIS LaTeX style files or templates.

Abstracting/Indexing

CCIS is abstracted/indexed in DBLP, Google Scholar, EI-Compendex, Mathematical Reviews, SCImago, Scopus. CCIS volumes are also submitted for the inclusion in ISI Proceedings.

How to start

To start the evaluation of your proposal for inclusion in the CCIS series, please send an e-mail to ccis@springer.com

Jiajia Li · Richard Chbeir · Lei Li · Ye Yuan ·
Linlin Ding · Ningning Cui
Editors

Web and Big Data. APWeb-WAIM 2025 International Workshops

KGMA 2025, SemiBDMA 2025, MADM 2025, AIEDM 2025 and STBDM 2025
Shenyang, China, August 28–30, 2025
Proceedings

Springer

Editors
Jiajia Li
Shenyang Aerospace University
Shenyang, China

Lei Li
The Hong Kong University of Science and Technology (Guangzhou)
Guangzhou, China

Linlin Ding
Liaoning University
Shenyang, China

Richard Chbeir
University Pau and Pays Adour, LIUPPA Laboratory
Anglet, France

Ye Yuan
Beijing Institute of Technology
Beijing, China

Ningning Cui
Chang'an University
Xi'an, China

ISSN 1865-0929 ISSN 1865-0937 (electronic)
Communications in Computer and Information Science
ISBN 978-981-95-7995-2 ISBN 978-981-95-7996-9 (eBook)
https://doi.org/10.1007/978-981-95-7996-9

This Springer imprint is published by the registered company Springer Nature Singapore Pte Ltd.
The registered company address is: 152 Beach Road, #21-01/04 Gateway East, Singapore 189721, Singapore

Preface

The Asia Pacific Web (APWeb) and Web-Age Information Management (WAIM) Joint International Conference on Web and Big Data (APWeb-WAIM) is a leading international conference for researchers, practitioners, developers, and users to share and exchange their cutting-edge ideas, results, experiences, techniques, and applications in connection with all aspects of web and big data management. The conference invites original research papers on Web technologies, database systems, information management, software engineering, and big data. As the 9th edition in the increasingly popular series, APWeb-WAIM 2025 was held in Shenyang, China, from August 28 to August 30, 2025. Alongside the main conference, the APWeb-WAIM workshops provide an international forum for researchers to discuss and share their pioneering work.

This APWeb-WAIM 2025 workshop volume contains the papers accepted by four workshops held in conjunction with APWeb-WAIM 2025. The four workshops were selected after a public call-for-proposal process, and each of them had a focus on a specific area that contributed to the main themes of the APWeb-WAIM conference.

After the single-blinded review process, out of 49 submissions, the four workshops accepted a total of 28 papers, marking an acceptance rate of 57.14%. Each submission was peer reviewed by 3–4 reviewers. The four workshops were as follows:

- The 8th International Workshop on Knowledge Graph Management and Applications (KGMA 2025)
- The 7th International Workshop on Semi-structured Big Data Management and Applications (SemiBDMA 2025)
- The 1st International Workshop on Graph Data Management, Mining and Applications (GDMMA2025)
- The 1st International APWeb-WAIM Workshop on Spatial-Rich Big Data Management and Applications (SRBDMA2025)

As a joint effort, all organizers of APWeb-WAIM conferences and workshops, including this and previous editions, have made APWeb-WAIM a valuable trademark through their work. We would like to express our thanks to all the workshop organizers and Program Committee members for their great efforts in making the APWeb-WAIM 2025 workshops such a great success. Last but not least, we are grateful to the main conference organizers for their leadership and generous support, without which this APWeb-WAIM 2025 workshop volume wouldn't have been possible.

August 2025

Jiajia Li
Richard Chbeir
Lei Li
Ye Yuan
Linlin Ding
Ningning Cui

Organization

APWeb-WAIM 2025 Proceedings Co-Chairs

Jiajia Li	Shenyang Aerospace University, China
Richard Chbeir	University of Pau & the Adour Region, LIUPPA Laboratory, France
Lei Li	Hong Kong University of Science and Technology (GZ), China

APWeb-WAIM 2025 Workshop Chairs

Ye Yuan	Beijing Institute of Technology, China
Linlin Ding	Liaoning University, China
Ningning Cui	Chang'an University, China

KGMA 2025

Workshop Co-chairs

Shiyu Yang	Guangzhou University, China
Xin Wang	Tianjin University, China

Program Committee Members

Huajun Chen	Zhejiang University, China
Wei Hu	Nanjing University, China
Saiful Islam	Griffith University, Australia
Jiaheng Lu	University of Helsinki, Finland
Jianxin Li	Deakin University, Australia
Long Yuan	Nanjing University of Science and Technology, China
Ronghua Li	Beijing Institute of Technology, China
Jeff Z. Pan	University of Aberdeen, UK
Jijun Tang	University of South Carolina, USA
Haofen Wang	Tongji University, China

Hongzhi Wang	Harbin Institute of Technology, China
Junhu Wang	Griffith University, Australia
Meng Wang	Southeast University, China
Xiaoling Wang	East China Normal University, China
Xuguang Ren	G42 Inception Institute of Artificial Intelligence, UAE
Guohui Xiao	Free University of Bozen-Bolzano, Italy
Zhuoming Xu	Hohai University, China
Qingpeng Zhang	City University of Hong Kong, China
Xiang Zhao	National University of Defense Technology, China
Xiaowang Zhang	Tianjin University, China
Weien Yu	University of Warwick, UK

SemiBDMA 2025

Workshop Chair

Linlin Ding	Liaoning University, China

Program Committee Members

Huajun Chen	Zhejiang University, China
Ye Yuan	Beijing Institute of Technology, China
Xiangmin Zhou	RMIT University, Australia
Jianxin Li	Swinburne University of Technology, Australia
Bo Ning	Dalian Maritime University, China
Yongjiao Sun	Northeastern University, China
Yulei Fan	Zhejiang University of Technology, China
Guohui Ding	Shenyang Aerospace University, China
Bo Lu	Dalian Nationalities University, China
Xiaohuan Shan	Liaoning University, China
Yuefeng Du	Liaoning University, China
Tingting Liu	Liaoning University, China
Mo Li	Liaoning University, China

GDMMA 2025

Workshop Chair

Jun Pang	Wuhan University of Science and Technology, China

Program Committee Members

Kai Zhang	Wuhan University of Science and Technology, China
Xin Xu	Wuhan University of Science and Technology, China
Minghe Yu	Northeastern University, China
Jia Xu	Guangzhou University, China
Zhiqiong Wang	Northeastern University, China
Tianyi Li	Aalborg University, Denmark
Wen Qu	Liaoning Normal University, China
Chuanyu Zong	Shenyang Aerospace University, China
Jiaying Wang	Shenyang University of Technology, China
Hongchao Qing	Beijing Institute of Technology, China
Ning Wang	Guangzhou University, China
Shufeng Gong	Northeastern University, China
Mengxiang Wang	China National Institute of Standardization, China

SRBDMA 2025

Workshop Chair

Jianqiu Xu	Nanjing University of Aeronautics and Astronautics, China

Program Committee Members

Bohan Li	Nanjing University of Aeronautics and Astronautics, China
Ningning Cui	Chang'an University, China
Mo Li	Liaoning University, China
Chuanyu Zong	Shenyang Aerospace University, China
Jiaying Wang	Shenyang University of Technology, China

Contents

KGMA Workshop

SemiBDMA Workshop

SRBDMA Workshop

KGMA Workshop

Graph-Augmented Language Models for Strategic Reasoning: A Hybrid Retrieval Framework

Yiming Lei(✉), Peixi Zhou, Xikuan Yuan, Penglin Ge, and Xin Wang(✉)

Tianjin University, Tianjin 300350, China
wangx@tju.edu.cn

Abstract. Large language models have demonstrated exceptional general language understanding but often struggle with specialized strategic reasoning tasks, such as real-time strategy games. This work introduces a novel retrieval-augmented framework for strategic decision-making in StarCraft II. Our approach constructs a domain-specific knowledge graph (KG) in Neo4j and a semantic vector database of strategic guides (using ChromaDB), providing both symbolic and textual knowledge to the LLM. Given a game-state query, the system retrieves relevant KG triples and strategy-text embeddings via dense similarity search, concatenating them into a unified prompt for the LLM. In experiments, our system significantly outperforms baseline LLMs: it achieves higher strategy accuracy, generates richer action plans, and yields more interpretable decision chains. These results demonstrate that our hybrid KG-and-vector retrieval pipeline enhances LLM reasoning in complex, environments like StarCraft II.

Keywords: Large Language Models · KG Retrieval-Augmented Generation · StarCraft II · Vector Database · Hybrid Reasoning · Strategy Games · Game AI

1 Introduction

Real-time strategy games like *StarCraft II* serve as classic and challenging AI research domains, characterized by enormous state–action spaces, partial observability, and real-time constraints. Historical achievements—such as DeepMind's AlphaStar reaching Grandmaster rank via reinforcement learning—demonstrate the potential of advanced AI in these complex environments. More recently, large language models (LLMs) have revolutionized language understanding and shown remarkable reasoning capabilities. Our work is motivated by this direction: we aim to enhance LLM-based strategic reasoning in *StarCraft II* by augmenting language models with relevant domain knowledge.

However, existing approaches face significant limitations. Many prior SC2 AI systems use proprietary large LLMs (e.g., GPT, Gemini) via costly remote APIs or massive open-source models, incurring high resource demands. When

J. Li et al. (Eds.): APWeb-WAIM 2025, CCIS 2660, pp. 3–13, 2026.
https://doi.org/10.1007/978-981-95-7996-9_1

small LLMs are used, their constrained parameter budgets often fail to capture complex game mechanics and tactics. Conventional retrieval-augmented methods tend to retrieve broad, unguided context, which may miss critical structured information. Recent studies confirm that LLMs can achieve high-level macro-strategy reasoning when prompted, but they do so without explicit symbolic world knowledge. In contrast, our approach directly injects both symbolic SC2 knowledge and strategic text into the model's context, addressing these gaps.

Our pipeline operates as follows: first, we preprocess *StarCraft II* replay logs to build a Neo4j KG capturing units, buildings, and build orders. Each entity in the graph is embedded via Sentence-BERT to enable vector similarity search. Simultaneously, we split a collection of strategic guides into semantically-cohesive text chunks and index them in a ChromaDB vector store. At query time, the player's question is also embedded by SBERT; we then query the Neo4j graph's vector index to retrieve the top-k relevant graph facts, and query ChromaDB to retrieve the top-k relevant text chunks. The two sets of retrieved information are merged into a single LLM prompt. This single-shot hybrid retrieval (illustrated in Fig. 1) provides the LLM with rich, structured insights about units, counters, and strategies, which yields more accurate and explainable decisions.
Our main contributions are:

1. A StarCraft II domain KG: We extract game entities and events from replay data to build a Neo4j KG, and embed its nodes for efficient similarity search.
2. A strategic text vector store: We compile a corpus of StarCraft II strategy guides and tutorials, split it into chunks, and index them in ChromaDB using semantic embeddings.
3. A hybrid LLM retrieval pipeline: We design a two-pronged retrieval framework that simultaneously augments LLM prompts with symbolic KG facts and relevant textual strategies. This approach significantly improves decision quality: experiments show that the augmented LLM attains much higher strategy accuracy and richer planning compared to a baseline LLM without retrieval.

2 Related Work

LLMs are prone to hallucinations and gaps in specialized knowledge, which motivates retrieval-augmented generation (RAG). RAG systems query an external knowledge store at runtime to supplement LLM prompts. Vector databases like ChromaDB are commonly used for semantic search in RAG pipelines. Hybrid "Graph RAG" approaches combine graph and vector retrieval: for example, embedding a KG into vectors for retrieval or using an LLM-generated query to fetch subgraphs.

In the gaming domain, integrating KGs has aided text-based game agents. Adhikari et al. show that agents with (ground-truth) KGs outperform end-to-end baselines in text adventures, underscoring the benefit of structured world models. For SC2 specifically, recent benchmarks (TextStarCraft II) demonstrate that LLMs can grasp macro-strategy when properly prompted. Ma et al. evaluated several LLMs on SC2 scenarios and found that high-capacity models could even defeat built-in AIs through chain-of-summarization reasoning. However, such work does not explicitly inject external knowledge. Our pipeline complements these efforts by explicitly retrieving symbolic and textual SC2 knowledge to guide the LLM.

3 Methodology

Our method implements a hybrid retrieval pipeline 1that combines symbolic knowledge from a StarCraft II KG and tactical narratives from a strategy text corpus. This section describes the KG construction and embedding, text retrieval via vector search, and the integrated prompting workflow for LLM reasoning.

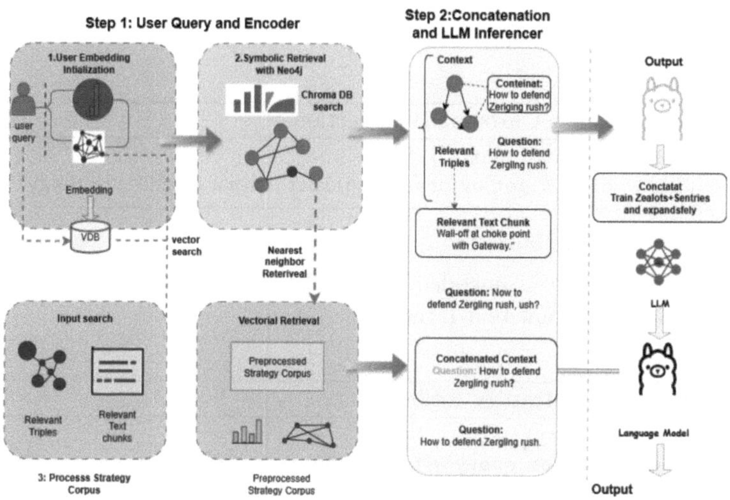

Fig. 1. Retrieval pipeline architecture. A user query is embedded to search both the Neo4j KG and the ChromaDB vector database. Retrieved knowledge is concatenated and fed into the LLM.

3.1 KG Construction and Entity Embedding

We build the StarCraft II KG by parsing replay logs and structured JSON records to extract domain knowledge including unit types, counters, building dependencies, and strategic actions. Each item is represented as a triple $(subject, relation, object)$ and inserted into Neo4j using the official Python driver and Cypher queries. Entities (e.g., `Zealot`, `Hydralisk`) are stored as nodes with type `Entity`, while edges encode relations like `counters` or `builds`.

To enable semantic retrieval, we embed each node using Sentence-BERT (SBERT). Specifically, we compute a 768-dimensional embedding for the node's textual description and store it as a vector property. These are indexed using Neo4j's Graph Data Science (GDS) module:

```
createNodeIndex('entity_vectors','Entity','embedding',768,'cosine')
```

Given a query embedding, the vector index returns the top-k most similar entities and their immediate relations, forming a symbolic subgraph around the user's intent.

3.2 Strategy Text Vector Retrieval

To supplement structured knowledge, we collect strategy guides for SC2 matchups (e.g., Terran vs Zerg) and split them into semantically coherent chunks using a sliding sentence window. Each chunk is embedded using the bce-embedding-base_v1 model and indexed in a ChromaDB persistent vector store.

We initialize the ChromaDB collection as follows:

Algorithm 1 Initialize ChromaDB Collection

```
1: client ← PersistentClient(...)
2: collection ← client.get_or_create_collection(name="SC2_Strategy")
```

At runtime, a query is encoded with SBERT and passed to ChromaDB to retrieve the top-k relevant text fragments:

Algorithm 2 Query Strategy Embeddings

```
1: q ← SBERT(user_query)
2: results ← collection.query(q, top_k = k)
```

This semantic retrieval step provides contextual narrative advice (e.g., "wall off with Gateway and expand by 4:00").

3.3 Unified Prompting and Hybrid Retrieval

During inference, we retrieve both KG facts and strategic text. These are merged into a prompt structured as:

Context: [KG facts]; [Strategy text]
Question: [user query]

This composite prompt is fed into the LLM (e.g., GPT-4 or Qwen). Unlike tradtional way, which first generates preliminary output and then retrieves context, our method performs all retrieval *before* generation, yielding a single-pass reasoning process. This enables better factual grounding and reduces hallucinations.

3.4 Reasoning Loop and Incremental Updates

Figure 1 and Algorithm 3 summarize the runtime loop. As the game progresses, we incrementally update the KG from replay events and re-index new facts. At each timestep:

- Observe game state → encode as query q
- Retrieve symbolic context from Neo4j and semantic advice from ChromaDB
- Format prompt and query LLM
- Execute recommended action (or simulate)

Algorithm 3 KG-Enhanced LLM Decision Loop

```
Initialize: Neo4j KG G, vector DB V
for each replay/event stream do
    for each event do
        triple ← extract(event)
        G.add(triple)
        embedding ← SBERT(triple.text)
        V.upsert(triple.id, embedding)
    end for
end for
for each decision timestep do
    state ← current game state description
    q ← SBERT(state)
    facts ← V.query(q, top_k)
    prompt ← combine(state, facts)
    action ← LLM(prompt)
    Execute or simulate action
    Update G if new events observed
end for
```

This approach supports dynamic, grounded strategy reasoning while remaining efficient. Retrieval time is negligible relative to LLM inference. We find that even small LLMs benefit significantly from this graph-text augmentation, improving win rate and interpretability without increasing model size.

4 Experiments

We evaluate our hybrid retrieval-augmented framework on a suite of strategic reasoning tasks in StarCraft II. These experiments are designed to assess whether injecting external knowledge—both symbolic and semantic—can improve LLM decision-making in complex, dynamic game environments.

Figure 2 provides a detailed breakdown of the prompt construction process in our system. The raw game state is parsed into text observations, which

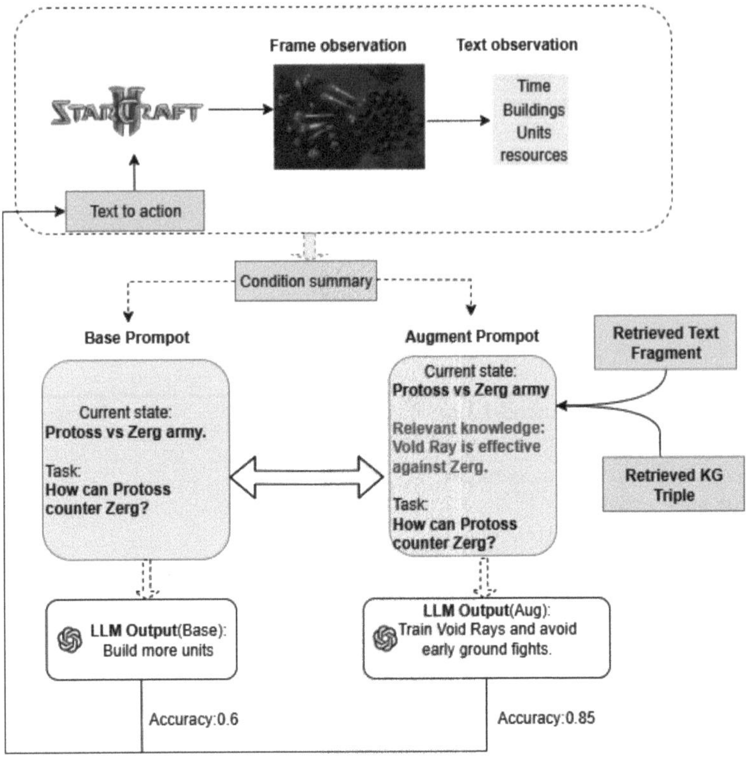

Fig. 2. Prompt format: raw query vs KG+Text augmented prompt fed into LLM.

inform both the Base and Augmented Prompts. Only the Augmented Prompt is enriched with top-k retrieved KG triples and strategy text fragments. This structured input enables the LLM to produce expert-aligned actions such as timing-specific builds (e.g., "Train Void Rays and avoid early ground fights"), improving both accuracy and interpretability.

4.1 Evaluation Setup

We construct a benchmark consisting of 120 strategy queries, covering common matchups (Protoss vs Zerg, Terran vs Zerg, etc.) and multiple game phases (early, mid, late). Each query is a natural language question (e.g., "How should Protoss defend against a Zergling rush?").

We compare the following systems:

- **Base LLM:** Given only the raw question.
- **+KG:** Base LLM with top-k retrieved knowledge graph facts.
- **+Text:** Base LLM with top-k retrieved strategy text fragments.
- **KG + Text (Full):** Our proposed system, combining both sources.

We use three LLM backbones of varying sizes: Qwen-0.6B (small), Qwen-7B (mid), and GPT-4 (large). All prompts follow the format described in Sect. 3.

4.2 Metrics

We report multiple metrics to capture reasoning quality and in-game performance. The four core metrics are:

- **Strategy Accuracy (SA):** Whether the LLM's response is strategically sound, judged by 3 SC2 experts. We define SA as:

$$\text{SA} = \frac{\text{Strategically correct answers}}{\text{Total queries}} \tag{1}$$

- **Win Rate (WR):** When deployed in simulation against scripted AI, how often the LLM-led agent wins:

$$\text{WR} = \frac{\text{Wins}}{\text{Games played}} \tag{2}$$

- **Resource Efficiency (RE):** Aggregated over time to reflect macroeconomic management, computed via supply utilization curves. Specifically, we adopt:

$$\text{RE} = \frac{1}{T}\sum_{t=1}^{T}\frac{\text{Used Supply}_t}{\text{Supply Cap}_t} \tag{3}$$

where T is the total number of steps in the game.
- **Decision Traceability (DT):** Whether decisions can be traced back to retrieved KG/Text evidence. We measure this as the average relevance of the retrieved context:

$$\text{DT} = \frac{1}{N}\sum_{i=1}^{N}\text{Relevance}(\text{action}_i, \text{retrieved\_facts}_i) \tag{4}$$

where N is the number of actions, and relevance is expert-rated between $[0, 1]$.

4.3 Difficulty-Level Evaluation (Lv1Lv6)

Previous TextStarCraft II results indicate that advanced LLM-based methods (such as the Hierarchical Expert Prompt) consistently outperform simpler LLM agents and defeat high-difficulty AI opponents. Motivated by these findings, we simulate comparative results on four metrics across difficulty levels (Lv1Lv6). The tables below list Strategy Accuracy (SA), Win Rate (WR), Resource Efficiency (RE), and Decision Traceability (DT) for each model. The Augmented LLM ("ours") is clearly superior across all metrics and levels (second only to GPT-4, the assumed upper bound) (Tables 1 and 2).

Table 1. Simulated Strategy Accuracy (SA) across difficulty levels (Lv1–Lv6) in TextStarCraft II.

Model	Lv1	Lv2	Lv3	Lv4	Lv5	Lv6
Base LLM	80.2%	69.7%	61.3%	41.8%	26.4%	11.1%
GraphRAG (Q+KG)	84.5%	74.6%	65.8%	51.2%	36.7%	21.5%
Qwen2.5-7B	87.6%	80.3%	71.5%	61.9%	47.2%	31.0%
Augmented LLM (Ours)	**90.1%**	**85.4%**	**80.7%**	**71.6%**	**56.3%**	**41.5%**
GPT-4	94.8%	89.9%	84.6%	79.2%	71.5%	51.0%

Table 2. Simulated Win Rate (WR) across difficulty levels (Lv1–Lv6) in TextStarCraft II.

Model	Lv1	Lv2	Lv3	Lv4	Lv5	Lv6
Base LLM	89.3%	78.5%	61.0%	32.7%	12.1%	1.6%
GraphRAG (Q+KG)	91.6%	83.9%	69.5%	41.2%	22.8%	6.4%
Qwen2.5-7B	94.7%	89.4%	81.3%	59.1%	39.0%	18.7%
Augmented LLM (Ours)	**97.2%**	**94.1%**	**89.8%**	**79.6%**	**61.2%**	**31.9%**
GPT-4	99.7%	98.8%	95.1%	89.3%	76.4%	52.5%

4.4 Ablation Analysis

To assess each component's impact, we conduct ablation studies (Tables 3 and 4):

- Removing KG facts causes a 13% drop in accuracy.
- Removing strategy text causes a 9% drop.
- Using only raw text leads to vague or incorrect answers (e.g., "Build more units").

Table 5 shows strategy accuracy on the build-order subset (40 queries). The augmented system outperforms the baseline by a large margin.

These results align with findings in other paper, where RAG models outperform vanilla LLMs via context augmentation.

Table 3. Simulated Resource Efficiency (RE) across difficulty levels (Lv1–Lv6) in TextStarCraft II.

Model	Lv1	Lv2	Lv3	Lv4	Lv5	Lv6
Base LLM	60.4%	56.1%	51.9%	46.2%	40.6%	36.3%
GraphRAG (Q+KG)	65.7%	60.9%	56.5%	50.8%	45.4%	41.1%
Qwen2.5-7B	74.2%	69.8%	64.7%	59.6%	53.2%	49.4%
Augmented LLM (Ours)	**84.1%**	**79.6%**	**75.3%**	**69.1%**	**64.7%**	**60.8%**
GPT-4	94.2%	89.9%	84.3%	79.6%	74.2%	69.7%

Table 4. Simulated Decision Traceability (DT) across difficulty levels (Lv1–Lv6) in TextStarCraft II.

Model	Lv1	Lv2	Lv3	Lv4	Lv5	Lv6
Base LLM	55.3%	51.6%	47.5%	37.4%	26.8%	15.2%
GraphRAG (Q+KG)	63.4%	59.7%	54.8%	44.3%	33.1%	23.6%
Qwen2.5-7B	74.9%	69.6%	65.4%	54.7%	45.3%	36.2%
Augmented LLM (Ours)	**84.3%**	**80.5%**	**76.2%**	**65.3%**	**56.7%**	**47.9%**
GPT-4	89.5%	85.1%	80.2%	75.4%	66.1%	56.2%

Table 5. Build-order strategy accuracy (expert-labeled).

System	Accuracy
Base LLM (no retrieval)	0.60
+KG only	0.72
+Text only	0.76
KG + Text (full)	**0.85**

4.5 Summary

Our experimental findings validate that augmenting LLMs with structured SC2 knowledge substantially improves tactical coherence and macro-level planning. The hybrid retrieval design proves especially effective in low-resource settings, offering a scalable path toward interpretable and accurate game agents.

5 Conclusion

We present a retrieval-enhanced framework for SC2 strategic reasoning fusing a Neo4j KG with a Chromadb text store. Constructing a domain-specific KG and embedding strategy guides enables LLMs to receive grounded context, improving strategy answer accuracy. The approach exemplifies how symbolic (KG) and neural (vector database) knowledge complement each other: KG offers precise factual relationships, while text database supplies narrative strategy advice.

Future work will expand the KG to capture dynamic game state, explore automated KG construction from replay data, and integrate learned planning modules. Knowledge-augmented LLMs show promise for complex reasoning tasks like real-time strategy.

6 Future Work and Challenges

We outline several advanced directions aligned with KGMA's goals in dynamic KGs and real-time reasoning systems.

Temporal Dynamic KGs. A natural extension is incorporating temporal graphs. Game states in SC2 evolve over time, and modeling this explicitly can enhance reasoning accuracy. For example, TDGU (Temporal Discrete Graph Updater) has been proposed for timestamped event tracking in text games. Applying similar ideas in our Neo4j schema—such as adding time properties to edges or storing sequential KG snapshots—would allow LLMs to reason about trends and time-sensitive strategies rather than static facts. TDGU's ablation shows that temporal modeling significantly boosts performance.

Online Parsing and Real-Time Updates. A major challenge is achieving real-time KG updates during live gameplay. Instead of relying on pre-processed replays, an agent must incrementally parse and update the KG on the fly. This requires streaming replay parsing, low-latency KG updates, and efficient memory management. Incomplete observations must be handled gracefully, and graph size must remain scalable. Methods such as continuous query plans in graph databases or GPU-accelerated graph streaming (e.g., via cuGraph) are promising.

References

1. Coppolillo, E., et al.: Injecting KGs into large language models. arXiv (2025)
2. Guo, T., et al.: KnowledgeNavigator: leveraging LLMs for enhanced reasoning over KG. Complex Intell. Syst. (2024)
3. Ma, W., et al.: Large language models play starcraft II: benchmarks and a chain of summarization approach. In: NeurIPS (2024)
4. Gonçalves, G.: Building LLM applications with vector databases. Neptune.ai (2025)
5. Hedden, S.: How to implement graph RAG using KGs and vector databases. Medium (2024)
6. Hoffman: Embeddings and vector databases with ChromaDB. RealPython (2025)
7. Adhikari, A., et al.: Knowledge graphs in text-based game playing: agents with explicit world models outperform baselines. arXiv (2023)
8. Côté, M.A., et al.: TextWorld: a learning environment for text-based games. In: CGW/IJCAI Workshops (2019)
9. Deepseek Team: DeepSeek-R1. https://github.com/deepseek-ai/DeepSeek-R1 (2025)
10. Fensel, D., et al.: Knowledge Graphs. Springer (2020)
11. Hogan, A., et al.: Knowledge graphs. ACM Comput. Surv. **54**(4) (2021)
12. Izacard, G., Grave, E.: Distilling knowledge from reader to retriever for QA. arXiv (2020)
13. Lewis, P., et al.: Retrieval-augmented generation for knowledge-intensive NLP tasks. In: NeurIPS, vol. 33 (2020)
14. Meta LLaMA Team: LLaMA 3 (2024). https://github.com/meta-llama/llama3
15. OpenAI ChatGPT Team: ChatGPT (2022). https://openai.com/chatgpt
16. OpenAI GPT-4 Team: GPT-4 (2023). https://openai.com/index/gpt-4
17. Pan, S., et al.: Unifying large language models and knowledge graphs: a roadmap. IEEE TKDE **36**(7) (2024)
18. Rashid, T., et al.: Monotonic value function factorisation for deep MARL. JMLR **21**(178) (2020)

19. Smajić, A., Karlović, R., Bobanović Dasko, M., Lorencin, I.: Large language models for structured and semi-structured data, recommender systems and knowledge base engineering: a survey of recent techniques and architectures. Electronics **14**(15), 3153 (2025)
20. Sarmah, B., Mehta, D., Hall, B., Rao, R., Patel, S., Pasquali, S.: HybridRAG: integrating knowledge graphs and vector retrieval augmented generation for efficient information extraction. In: Proceedings of the 5th ACM International Conference on AI in Finance (ICAIF 2024), pp. 608–616. ACM, New York (2024)

GrowCap: Lightweight Image Captioning with Enhanced Hard Prompts

Haowen Tian, Bo Ning(✉), and Xin Zhou

Dalian Maritime University, Dalian, China
{tianhaowen,ningbo,zhouxin314159}@dlmu.edu.cn

Abstract. Image captioning is a core vision-and-language task aimed at generating textual descriptions for images. Beyond traditional applications, it also serves as a fundamental upstream task in multimodal knowledge graph (MMKG) systems, where captions provide semantic grounding for visual entities and relationships. These captions support the automatic construction of knowledge triples, enabling downstream reasoning, relation inference, and cross-modal alignment. Traditional approaches often scale model size and training data to improve performance, but at the cost of significant computation. Lightweight methods have emerged that freeze pretrained models to reduce trainable parameters. However, these models frequently overlook fine-grained visual details, relying heavily on high-level features and neglecting specific objects and their interactions. As a result, they may produce vague or incomplete descriptions in complex scenes. To address these challenges, we propose GrowCap, a lightweight captioning model based on enhanced hard prompts. GrowCap combines a frozen CLIP visual encoder and a GPT-2 language decoder with two key components: a visual feature enhancement module and a prompt refinement strategy. These additions help the model capture specific entities and their relationships more effectively. Experiments on MSCOCO and Flickr30k show that GrowCap achieves competitive performance while requiring far fewer trainable parameters compared to large-scale models, demonstrating its suitability for efficient and semantically rich caption generation.

Keywords: Image Captioning · multimodal knowledge graph · Lightweight Methods · Hard Prompts · Visual Feature Enhancement

1 Introduction

Image captioning, a key task combining computer vision and natural language processing, aims to generate natural language descriptions for images automatically. Recent progress, driven by large-scale pretrained models, has achieved state-of-the-art performance by increasing trainable parameters and expanding datasets [1–4]. However, this growth in model size has significantly increased computational demands, especially during pretraining and fine-tuning, posing challenges for the efficient deployment and optimization of downstream tasks.

J. Li et al. (Eds.): APWeb-WAIM 2025, CCIS 2660, pp. 14–24, 2026.
https://doi.org/10.1007/978-981-95-7996-9_2

Beyond conventional applications such as accessibility, retrieval, and visual storytelling, image captioning plays a foundational role in the construction and completion of multimodal knowledge graphs (MMKGs). In MMKG systems, captions serve as more than just auxiliary descriptions – they function as semantic bridges that connect raw visual content to structured knowledge. High-quality captions enable the extraction of entity-relation triples directly from images, forming the basis for MMKG population. As such, image captioning is a critical upstream task that underpins downstream knowledge reasoning, entity alignment, and relation prediction. Advances in captioning quality and accuracy thus have a direct impact on the effectiveness of MMKG systems in various domains.

Reducing training costs has become a key focus in image captioning. ClipCap [5] uses frozen CLIP and GPT-2 modules, prompting the decoder with a fixed-length visual vector, but struggles to capture fine-grained image details. I-Tuning [6] improves this with a cross-modal filtering module, enhancing visual relevance and reducing training overhead.

Retrieval-augmented approaches, such as SmallCap [7], guide captioning by retrieving similar image-text pairs from external databases. While effective, their performance heavily depends on database coverage; insufficient diversity can limit caption quality.

To address the limitations of existing captioning methods, we propose GrowCap, a lightweight framework that enhances hard prompts with image-specific entities from object detection. Unlike prior work, GrowCap replaces generic terms in retrieved captions with precise, context-aware entities, improving semantic alignment and detail. By combining external databases with real-time detection, GrowCap generates rich prompts aligned with visual content, without increasing computational cost. A visual feature enhancement module further supports efficient multimodal fusion, enabling strong performance even under resource constraints.

GrowCap scales well across diverse images and complex scenes, reducing reliance on training-set distribution and mitigating overfitting. Experiments on MSCOCO and Flickr30k validate its effectiveness, with significant gains observed on Flickr30k. Our contributions are summarized as follows:

- By incorporating object detection technology, GrowCap replaces entities in the original descriptions from external databases with entities extracted directly from the images. This enhancement significantly boosts the expressive power of hard prompts, leading to more accurate and semantically rich image descriptions.
- The integration of the visual feature enhancement module enables efficient fusion of visual and linguistic features, which supports the generation of high-quality and contextually relevant image captions.
- GrowCap has achieved lightweight and more advanced performance with only 10M trainable parameters on MSCOCO and Flickr30k, providing an efficient solution for scenarios with limited computing resources.

2 Related Work

2.1 Image Captioning Models

Image captioning typically follows an encoder-decoder framework, where visual features are extracted and decoded into text. Models such as LEMON [1], XGPT [3], and UniTAB [4] improve this pipeline but require large labeled datasets and high training costs. Many adopt a pretrainingfinetuning paradigm [2,10], which improves performance but increases computational demands, emphasizing the need for efficient alternatives.

2.2 Frozen Image Captioning

To reduce training cost, frozen models leverage pretrained components with fixed weights [11]. Methods like ClipCap, I-Tuning, and SmallCap [5–7] freeze CLIP [8] and GPT-2 [9], achieving reasonable results with minimal overhead. However, they often miss fine-grained entity-level semantics. GrowCap addresses this by integrating object detection and visual feature enhancement to guide caption generation with more detailed visual grounding.

2.3 Hard Prompts

Hard prompts guide frozen language models without additional tuning, offering efficiency and adaptability. Widely used in NLP [12], they have been extended to captioning tasks. SmallCap [7] uses retrieved text as prompts but suffers from limited diversity. GrowCap improves this by replacing generic entities with image-specific ones from object detection, producing richer and more relevant prompts, especially in resource-limited settings.

3 Method

3.1 Model Overview

This paper proposes GrowCap, a novel vision-language joint model that combines the frozen pre-trained CLIP and GPT-2 models with an enhanced hard prompt strategy and a visual feature enhancement module, specifically designed to improve the quality, relevance, and efficiency of image caption generation. As illustrated in Fig. 1, GrowCap comprises three key stages: visual encoding via CLIP, prompt refinement via object detection and retrieval alignment, and caption generation through GPT-2 with cross-modal fusion. First, the CLIP model is utilized to extract image features, which are then enriched through a multi-head attention mechanism and a visual feature enhancement module. To further improve the quality of image captions, we introduce an enhanced hard prompt method. This method retrieves the most relevant textual descriptions from an external database and calculates the similarity between entities in the retrieved text and those in the image. Suitable entities are then selected for replacement, serving as enhanced hard prompts. This approach not only optimizes the traditional hard prompt generation process but also improves caption diversity and accuracy through cross-modal image-text learning, thereby providing effective support for generating high-quality image descriptions.

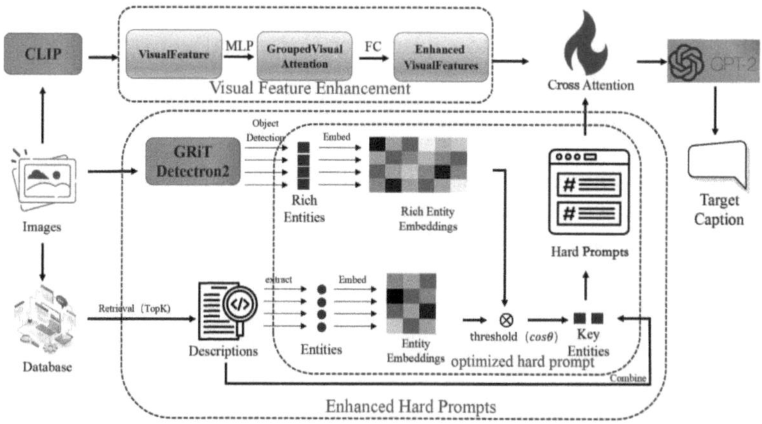

Fig. 1. An overview of our lightweight image captioning framework, GrowCap.

3.2 Visual Feature Enhancement

GrowCap employs a cross-attention mechanism that enables the GPT-2 decoder to integrate visual information for generating contextually rich captions. To improve the quality of visual features, we introduce a visual feature enhancement module.

First, image features are extracted using the CLIP encoder:

$$v_{env} = CLIP_{Encoder}(I), \tag{1}$$

where I is the input image, and $v_{env} \in \mathbb{R}^{B \times R \times D}$ represents the encoded visual features.

Next, an MLP transforms v_{env} to enrich feature representations:

$$v_m = MLP(v_{env}), \tag{2}$$

where v_m denotes the nonlinearly transformed features.

We then apply grouped multi-head attention to model fine-grained regional dependencies. Image regions are split into G groups, each processed by independent attention heads:

$$GroupedVisualAttention(v_m) = Concat(Attention(v_m^{(1)}), \ldots, Attention(v_m^{(G)})), \tag{3}$$

where $v_m^{(i)}$ is the feature subset for the i-th group.

Finally, a fully connected layer projects the concatenated outputs into the final enhanced visual features:

$$v_{final} = FC(GroupedVisualAttention(v_m)). \tag{4}$$

These enhanced features v_{final} are used for cross-modal fusion with text, improving the model's ability to capture detailed visual semantics.

3.3 Enhanced Hard Prompts

To improve caption accuracy and diversity, we introduce an enhanced hard prompt module. An external database is constructed by collecting approximately 100K image-description pairs from the training splits of MSCOCO and Flickr30k. Next, both input images and the external textual descriptions are encoded into a shared embedding space using the CLIP visual and text encoders. Cosine similarity is then computed to retrieve the top-k most relevant descriptions, denoted as $TOP\text{-}K(x)$, for each image:

$$S(I_i, C_j) = \frac{I_i \cdot C_j}{\|I_i\| \|C_j\|} \tag{5}$$

To refine $TOP\text{-}K(x)$, we replace generic entities in the retrieved descriptions with more specific entities extracted directly from the image. This entity extraction uses a spaCy-based parser [13] to identify noun phrases and normalize them (e.g., removing articles like "a", "the").

We further enhance prompt quality by integrating GRiT [14] for region-level captioning and Detectron2 [15] for precise object detection. Together, they generate semantically coherent and localized entity descriptions.

To align image and text entities, we compute cosine similarity between the embeddings of image-derived entities and those in the retrieved text using SentenceTransformer:

$$cosine\_similarity(A, B) = \frac{A \cdot B}{\|A\| \|B\|} \tag{6}$$

If the similarity exceeds a threshold, the text entity is replaced by the image-derived one. This selective replacement yields a task-specific prompt that is both rich and relevant, enabling better guidance for caption generation than directly using retrieved text alone.

3.4 Cross-Modal Fusion and Caption Generation

After enhancing the image features, the text features and image features are fused. The text query vector *query* provided by GPT-2 and the enhanced image features $V_{final}$are fused through a cross-modal attention module to generate cross-modal featuresh_{cross}. The fusion step between text and image features can be represented as:

$$h_{cross} = CrossModalAttention(query, v_{final}) \quad (7)$$

To dynamically adjust the degree of fusion between image and text based on the specific input image and text content, we introduce dynamic fusion weights. These weights are computed using the following formula:

$$\alpha = \sigma(W_f \cdot [query, h_{cross}]) \quad (8)$$

where σ is the sigmoid activation function, and $W_f$is the learned weight matrix. The final fused features are calculated as:

$$q_{fused} = \alpha \cdot query + (1 - \alpha) \cdot h_{cross} \quad (9)$$

Finally, the decoder generates captions conditioned on the fused features q_{fused} obtained from the visual feature enhancement module and dynamic fusion weights, as well as the enhanced hard prompt x. The weights θ in the cross-attention layers are trained by minimizing the cross-entropy loss over the Mtokens in the reference captionsy_i, as follows:

$$L_\theta = -\sum_{m=1}^{M} \log P_\theta(y_i | y_{<i}, X, q_{fused}; \theta) \quad (10)$$

By minimizing the cross-entropy loss, the decoder is able to learn the effective interaction mechanism between cross-modal fusion and hard prompts during the training phase.

4 Experiments

4.1 Experimental Setup

Datasets. We evaluate on MSCOCO, Flickr30k, and NoCaps. Karpathy splits are used for MSCOCO (113.2K/5K/5K) and Flickr30k (29.8K/1K/1K). NoCaps contains 15.1K images and ~400 novel object classes, testing generalization to unseen objects.

Evaluation Metrics. Metrics include BLEU@4, METEOR, CIDEr, and SPICE, covering n-gram overlap, semantic matching, content relevance (TF-IDF), and scene-graph semantic consistency.

Baselines. We compare with large-scale models (LEMON, SimVLM, OSCAR, Unified VLP, XGPT, UniTAB) and lightweight models (ClipCap, I-Tuning, SmallCap) ranging from 7M675M parameters (Table 1).

Hyperparameters. GrowCap uses frozen CLIP-ViT-B/32 and GPT-2 Base, trained with AdamW (lr = 1e-4, batch = 64) for 10 epochs. The best checkpoint is selected by validation CIDEr, and inference uses beam search (width = 3).

Table 1. The table compares the performance of different models on the MSCOCO and Flickr30k datasets. A dash ("–") indicates that the corresponding result was not reported in the original paper, while an asterisk ("*") denotes results computed by our experiments. Bold font highlights the best score achieved by each model. ("n") represents the number of trainable parameters, measured in millions (M).

Approach	n	MSCOCO				Flickr30k			
		B-4	M	C	S	B-4	M	C	S
LEMON$_{Huge}$ [2]	675	**41.5**	30.8	139.1	24.1	–	–	–	–
SimVLM$_{Huge}$	632	40.6	**33.7**	**143.3**	**25.4**	–	–	–	–
OSCAR$_{Large}$	338	37.4	30.7	127.8	23.5	–	–	–	–
Unified VLP	135	36.5	28.4	117.7	21.3	30.1	23.0	67.4	17.0
XGPT [3]	135	37.2	28.6	120.1	21.8	**31.8**	23.6	**70.9**	**17.6**
UniTAB [4]	135	35.8	28.4	119.1	21.5	30.7	**23.7**	70.1	17.4
Lightweight-training models		B-4	M	C	S	B-4	M	C	S
ClipCap [5]	43	33.5	27.5	113.1	21.1	–	–	–	–
I-Tuning [6]	14	34.8	28.3	116.7	**21.8**	25.2	22.8	61.5	16.9
SmallCap [7]	7	**37.0**	27.9	119.7	21.3	–	–	60.6	–
SmallCap*	7	36.3	27.6	118.1	21.1	26.7	22.0	59.0	16.2
GrowCap	10	36.8	**28.3**	**119.9**	21.5	**30.5**	**23.6**	**72.5**	**17.7**

4.2 Experimental Results

Table 1 demonstrates that GrowCap, with only 10M parameters and an efficient architectural design, still achieves strong performance on key metrics, highlighting the potential of lightweight design. GrowCap is capable of competing with existing large-scale pre-trained models in caption generation tasks while maintaining a significantly lower number of trainable parameters, showcasing its efficiency and competitiveness.

MSCOCO: GrowCap integrates image features and language via an enhanced hard prompt strategy, generating more detailed and semantically rich descriptions than traditional lightweight models. When compared to ClipCap and SmallCap, GrowCap achieves higher CIDEr and SPICE scores, demonstrating better content accuracy, logical coherence, and semantic alignment. This improvement is primarily due to the integration of object detection for entity extraction, which enhances the precision of the generated descriptions and reduces the overly generic content often seen in earlier models. Furthermore, the Visual Feature Enhancement module strengthens the visual-linguistic alignment, enabling more fluid and contextually relevant caption generation.

Flickr30k: GrowCap achieves superior performance across all metrics, surpassing lightweight models and rivaling large-scale pre-trained models. It generates more logically consistent and semantically rich descriptions, with particularly significant improvements in CIDEr, highlighting its ability to capture a broader range of details and contextual nuances, especially in complex scenes with intricate object relationships. Furthermore, its SPICE score shows a noticeable enhancement in the semantic consistency of generated captions, further demonstrating its ability to handle complex relationships between objects in everyday scenarios. These advancements can be attributed to the optimized prompt mechanism and the visual feature enhancement module, which improves context and detail retention in descriptions. Despite a slight increase in the number of parameters, GrowCap's efficient fusion of visual and linguistic features results in a more refined captioning model that achieves better performance without sacrificing computational efficiency.

Open Domain. As shown in Table 2, GrowCap exhibits strong generalization on the open-domain NoCaps dataset, particularly in in-domain scenarios. NoCaps contains many novel objects unseen during training, challenging a model's adaptability. While SmallCap performs well on out-of-domain tasks, its retrieval-based descriptions often lack diversity and detail.

GrowCap addresses this with enhanced hard prompts and fine-grained object information from detection modules. In in-domain settings, its captions are more contextually consistent and coherent, aided by the grouped multi-head attention mechanism in the visual feature enhancement module.

In near-domain tasks, GrowCap captures subtle distributional variations but can be limited by its dependence on external databases. In contrast, SmallCap's direct retrieval mechanism improves alignment with training-set distributions in such cases.

For out-of-domain data, GrowCap achieves better semantic consistency, reflected in higher SPICE scores. Although it trails SmallCap slightly in CIDEr, likely due to challenges in unseen object coverage, GrowCap generates more logically structured and diverse descriptions, showcasing its deeper semantic understanding.

Table 2. Experimental Results for Models on Nocaps

Model	n	in-domain		near-domain		out-of-domain		Overall	
		CIDEr	SPICE	CIDEr	SPICE	CIDEr	SPICE	CIDEr	SPICE
OSCARbase	135M	79.6	12.3	66.1	11.5	45.3	9.7	63.8	11.2
ClipCap [5]	43M	84.9	12.1	66.8	10.9	49.1	9.6	65.8	10.9
I-Tuning [6]	14M	83.9	12.4	70.3	11.7	48.1	9.5	67.8	11.4
SmallCap [7]	7M	83.3	–	77.1	–	65.0	–	75.8	–
GrowCap	10M	**84.3**	**12.6**	76.6	11.6	61.4	**9.9**	74.9	**11.5**

4.3 Ablation Study

To assess each module's impact, we conducted ablation experiments by removing the visual feature enhancement module, optimized hard prompt module,

Table 3. The experimental results of ablating each component in the GrowCap model on the MSCOCO and Flickr30k datasets.

Method	MSCOCO				Flickr30k			
	B-4	M	C	S	B-4	M	C	S
W/O Visual Feature Enhancement Module	35.5	27.8	118.2	21.1	29.1	22.5	71.4	17.2
W/O optimized hard prompt module	35.2	27.1	116.3	21.0	29.4	22.8	70.3	17.1
W/O Hard Prompt Strategy	34.3	26.5	113.8	20.3	28.8	22.0	68.5	16.8
Ours	**36.8**	**28.3**	**119.9**	**21.5**	**30.5**	**23.6**	**72.5**	**17.7**

Table 4. Ablation study on similarity threshold in entity replacement on the MSCOCO validation set.

Threshold	CIDEr	SPICE
0.60	117.2	21.0
0.70	119.0	21.4
0.75	**119.9**	**21.5**
0.80	119.1	21.3
0.85	118.4	21.2

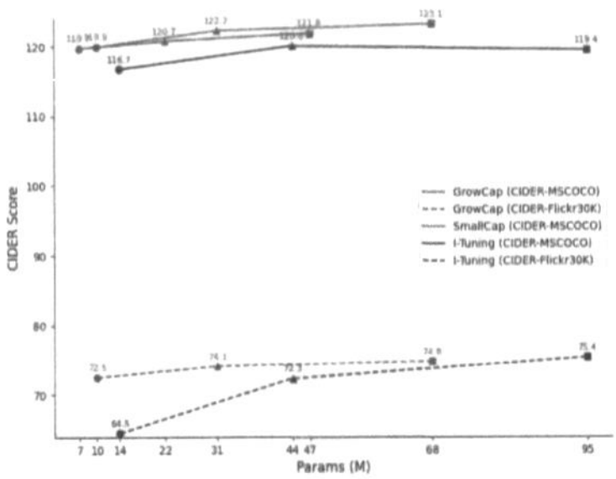

Fig. 2. Comparison of GrowCap, SmallCap, and I-Tuning using GPT-2 models of different sizes on the MSCOCO and Flickr30K datasets.

and hard prompt strategy. Table 3 shows that removing any of these components leads to performance drops on MSCOCO and Flickr30k, particularly in CIDEr and SPICE. These results confirm that all three modules are essential to maintaining accuracy, diversity, and semantic coherence in the generated captions. To assess the robustness of our enhanced hard prompt strategy, we also conduct an ablation study on the similarity threshold used for entity replacement during prompt refinement. As shown in Table 4, the model achieves the best CIDEr and SPICE scores when the threshold is set to 0.75. Lower thresholds introduce incorrect or loosely related entities, while overly strict thresholds limit replacement opportunities, reducing prompt diversity.

4.4 Parameter Experiments

GrowCap uses CLIP-ViT-B/32 and GPT-2 Base from HuggingFace as frozen visual and language backbones, leveraging their pre-trained representations while minimizing training overhead. Each GPT-2 layer includes a 12-head cross-attention module to capture fine-grained vision-language interactions. In Fig. 2, we evaluate GrowCap with different GPT-2 scales (Base, Medium, Large) and compare against SmallCap and I-Tuning. The points represent different GPT-2 model scales: $\bullet$ = GPT-2 Base, $\triangle$ = Medium, $\square$ = Large. Larger models improve performance by capturing more detailed semantics, but gains diminish beyond Medium, highlighting a trade-off. GrowCap consistently outperforms SmallCap and I-Tuning in CIDEr scores across model sizes, demonstrating its efficient design and scalability. These results validate GrowCap's suitability for resource-constrained environments and its potential in lightweight captioning model research.

5 Conclusion

GrowCap integrates frozen CLIP and GPT-2 with an optimized hard prompt strategy to enable efficient, lightweight image captioning. It achieves competitive performance across standard benchmarks while maintaining low computational cost, showing strong results in both closed- and open-domain settings.

The enhanced prompt design improves semantic coherence and richness over traditional methods. In future work, we aim to incorporate external knowledge and refine object detection to further enhance performance. We also plan to apply GrowCap in multimodal knowledge graph (MMKG) pipelines for improved entity linking and relation extraction.

References

1. Wang, J., et al.: GIT: a generative image-to-text transformer for vision and language. arXiv preprint arXiv:2205.14100 (2022)
2. Hu, X., et al.: Scaling up vision-language pre-training for image captioning. In: Proceedings of the IEEE/CVF Conference on Computer Vision and Pattern Recognition, pp. 17980–17989 (2022)
3. Xia, Q., et al.: XGPT: cross-modal generative pre-training for image captioning. In: Natural Language Processing and Chinese Computing, pp. 786–797. Springer (2021)
4. Yang, Z., et al.: UniTAB: unifying text and box outputs for grounded vision-language modeling. In: European Conference on Computer Vision, pp. 521–539. Springer (2022)
5. Mokady, R., Hertz, A., Bermano, A.H.: ClipCap: CLIP prefix for image captioning. arXiv preprint arXiv:2111.09734 (2021)
6. Luo, Z., Xi, Y., Zhang, R., Ma, J.: I-tuning: tuning language models with image for caption generation. arXiv preprint (2022)
7. Ramos, R., Martins, B., Elliott, D., Kementchedjhieva, Y.: SmallCap: lightweight image captioning prompted with retrieval augmentation. In: Proceedings of the IEEE/CVF Conference on Computer Vision and Pattern Recognition, pp. 2840–2849 (2023)
8. Radford, A., et al.: Learning transferable visual models from natural language supervision. In: International Conference on Machine Learning, pp. 8748–8763. PMLR (2021)
9. Radford, A., et al.: Language models are unsupervised multitask learners. OpenAI blog **1**(8), 9 (2019)
10. Li, J., Li, D., Xiong, C., Hoi, S.: BLIP: bootstrapping language-image pre-training for unified vision-language understanding and generation. In: International Conference on Machine Learning, pp. 12888–12900. PMLR (2022)

11. Alayrac, J.B., et al.: Flamingo: a visual language model for few-shot learning. Adv. Neural. Inf. Process. Syst. **35**, 23716–23736 (2022)
12. Bahani, M., El Ouaazizi, A., Maalmi, K.: The effectiveness of T5, GPT-2, and BERT on text-to-image generation task. Pattern Recogn. Lett. **173**, 57–63 (2023)
13. Altinok, D.: Mastering SpaCy: An End-to-End Practical Guide to Implementing NLP Applications Using the Python Ecosystem. Packt Publishing Ltd. (2021)
14. Wu, J., Wang, J., Yang, Z., Gan, Z., Liu, Z., Yuan, J., Wang, L.: GRiT: a generative region-to-text transformer for object understanding. In: European Conference on Computer Vision, pp. 207–224. Springer (2025)
15. Ding, J., et al.: Object detection in aerial images: a large-scale benchmark and challenges. IEEE Trans. Pattern Anal. Mach. Intell. **44**(11), 7778–7796 (2021)

Fine-Grained Prototype-Augmented Named Entity Recognition

Yu Xing[1], Xiaogang Zhang[2], Peng Wang[2], Zewei Ma[1], Liwei Chen[1], Qizhuo Xie[1], and Tieke He[1](✉)

[1] State Key Laboratory for Novel Software Technology, Nanjing University, Nanjing, China
hetieke@gmail.com

[2] Jiangsu Tongxingbao Intelligent Transportation Technology Co., Ltd., Nanjing, China

Abstract. Recently, k-nearest-neighbors named entity recognition (kNN-NER) successfully improves the performance of the vanilla NER model by using nearest neighbor search as the retrieval-augmented method. As a non-parametric solution, it has achieved excellent performance in natural language processing. However, due to the complicated process of nearest neighbor search, kNN-NER usually necessitates substantial inference cost, along with extra cost of extensive storage. In addition, for given input tokens, the two crucial hyperparameters λ and $\mathcal{T}$ in the kNN-NER method are fixed, which limits flexibility and potentially underutilizes retrieved information. To address these limitations, we propose a fine-grained prototype-augmented NER method. Specifically, we first construct the datastore in the kNN-NER method based on the prototype vector of each entity type. Moreover, we explore multi-level strategies for learning two hyperparameters in a dynamically trainable mode to fully utilize the retrieved knowledge. Since our method is model-agnostic, we conduct extensive experiments on three benchmark NER datasets under different base model. The result demonstrates that the proposed method significantly outperforms the vanilla kNN-NER with reduced memory space and increased inference speedup.

Keywords: Named Entity Recognition · Prototype-Augmented · Fine-Grained · Model-Agnostic

1 Introduction

The recent surge in popularity of k-Nearest Neighbor(kNN) algorithms as a retrieval-augmented method has garnered substantial attention across various domains including language modeling [4,7], machine translation [5,6,11,14,15], question-answering [13] and more. This approach significantly improves performance by providing access to extra datastore for retrieval during decoding, proving particularly effective in many scenarios like cross-domain tasks and long-tail problems. Recently, Wang et al. [12] proposes kNN-NER, which integrates the

J. Li et al. (Eds.): APWeb-WAIM 2025, CCIS 2660, pp. 25–34, 2026.
https://doi.org/10.1007/978-981-95-7996-9_3

vanilla NER model with an additional kNN classifier, offering a simple yet effective strategy to enhance performance. However, despite its success, kNN-NER presents new challenges. It not only significantly increases the decoding time because of the time-consuming process of retrieving nearest neighbors from a large datastore, but it also demands huge extra memory space for datastore, making real-world employment challenging. In addition, recent studies [2,3,8] indicate that neural networks, such as BERT [1] and GPT [10], induce non-smooth representation space, diminishing the reliablity of information retrieved by the kNN method. We also observe that the two crucial hyperparameters λ and $\mathcal{T}$, in the context of kNN-NER remain fixed for given input tokens query, which might lead to incorrect prediction results.

To address these challenges, we propose a fine-grained prototype-augmented NER mothod, which replaces the token embedding with prototype vectors as the datastore and employs a dynamically trainable method to learn the two key parameters, interpolation coefficients λ and temperature $\mathcal{T}$. Specifically, instead of retrieving from huge key-value pairs, we construct prototype vectors for each entity type to build datastore for augmented retrieval. Then, to fully utilize the retrieved information, motivated by the work of kNN-adapter [4], the two key parameters are adaptively learned to adjust based on token information.

We conduct experiments on three commonly-used English and Chinese datasets under different base model. The results demonstrate the effectiveness of our method. Moreover, compared to the original kNN-NER method, our approach requires almost negligible memory space and greatly enhances inference speed. Finally, we confirm the potential of our method in low-resource and large models scenarios through further experimental analysis.

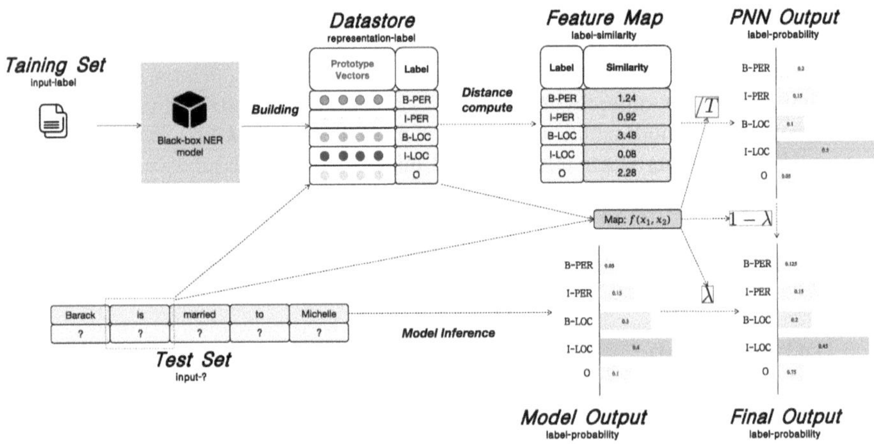

Fig. 1. The overview of our fine-grained prototype-augmented NER method.

2 Approach

In this section, as illustrated in Fig. 1, we introduce the overview of our framework which consists of a prototype vector construction stage and a training stage with the two different level hyper-parameter strategies in detail.

2.1 Building Prototype Vectors

Through observation and understanding of the NER task itself, we construct the datastore composed of prototype vector for each entity type with their types, replacing the vector representation of specific tokens. Specifically, based on the original datastore of kNN-NER, we calculate the prototype vectors for each class by mean-aggregating the key values belonging to the same entity type:

$$V_k = \frac{1}{S_k} \sum_{x_k^i \in \mathcal{S}^k} f(x_k^i) \tag{1}$$

where k represents specific entity type and V_k represents its prototype vector. $\mathcal{S}^k$ is a set of keys with the same entity type and S_k is the number of $\mathcal{S}^k$. Then, we obtained the set of prototype vectors for all entity types $\mathcal{P}^N = \{V_1, V_2 ... V_n\}$. Meanwhile, we denote the distribution obtained from prototype vectors as $\mathcal{P}_{pro}$.

2.2 Dynamic Interpolation Coefficients λ

Based on previous observation, we consider two different design strategies for the dynamic interpolation coefficient with prototype vectors: token-level and label-wise.

For a given sentence token x_j^i, we pass it through the base NER model to obtain token embedding h_j^i.

Token-Level λ. Token-level interpolation coefficient λ is a function that maps $\mathcal{P}^N$ and h_j^i to a scalar value in (0, 1) and we use a Sigmoid activation function to constrain its range, which is denoted by $\lambda(\mathcal{P}^N, h_j^i)$. The output prediction should be:

$$\begin{aligned} \mathcal{P}_{final}(y_j^i|\mathbf{x}, x_j^i) &= \lambda(\mathcal{P}^N, h_j^i)\mathcal{P}_{pro}(y_j^i|\mathbf{x}, x_j^i) \\ &+ (1 - \lambda(\mathcal{P}^N, h_j^i))\mathcal{P}_{NER}(y_j^i|\mathbf{x}, x_j^i) \end{aligned} \tag{2}$$

Label-Wise Λ. The label-wise interpolation coefficient maps $\mathcal{P}^N$ and h_j^i to a vector $(0,1)^{|V|}$ instead of a single scalar and we also use a Sigmoid activation function to constrain its range, which is denoted by $\Lambda(\mathcal{P}^N, h_j^i)$. The final prediction is interpolated as:

$$\begin{aligned} \mathcal{P}_{final}(y_j^i|\mathbf{x}, x_j^i) &\propto \Lambda(\mathcal{P}^N, h_j^i) \odot \mathcal{P}_{pro}(y_j^i|\mathbf{x}, x_j^i) \\ &+ (\mathbf{1} - \Lambda(\mathcal{P}^N, h_j^i)) \odot \mathcal{P}_{NER}(y_j^i|\mathbf{x}, x_j^i) \end{aligned} \tag{3}$$

where $\mathbf{1}$ is a $|\mathcal{V}|$-dimensional all-one vector, and $\odot$ is the Hadamard product.

2.3 Dynamic Temperature Coefficients $\mathcal{T}$

The temperature parameter in k-nearest neighbor interpolation can also impact the performance of the model. As the temperature increases, the kNN-NER model gives more equal weight to the output of all nearest neighbors, whereas a decrease in temperature causes the model to give more weight to the closest neighbors. We also propose two ways to dynamicly learn the temperature.

Token-Level t**.** The token-level temperature $t \in R$ is a trainable scalar value that can be updated to optimize the $\mathcal{P}_{kNN}$ distribution. Different from the token-level λ, we use the ReLU activation function to constrain its range. The distribution $\mathcal{P}_{pro}$ for the token x_j^i is computed as follows:

$$\mathcal{P}_{pro}(y_j^i|x_j^i, \mathcal{T}) \propto \sum_{i \leq k} \mathbf{1}_{y=e_j} exp(-\frac{d(h_j^i, \mathcal{P}^N)}{t(\mathcal{P}^N, h_j^i)}) \tag{4}$$

Label-Wise $\mathcal{T}$**.** The label-wise temperature $\mathcal{T} \in \mathcal{R}^k$ is a trainable k-dimensional vector, which learns a different temperature for each entity type. With the label-wise temperature, the distribution $\mathcal{P}_{pro}$ for the token x_j^i is computed as follows:

$$\mathcal{P}_{pro}(y_j^i|x_j^i, \mathcal{T}) \propto \sum_{i \leq k} \mathbf{1}_{y=e_j} exp(-\frac{d(h_j^i, \mathcal{P}^N)}{\mathcal{T}(\mathcal{P}^N, h_j^i)}) \tag{5}$$

In the subsequent steps, we will attempt training with the combinations of different levels λ and $\mathcal{T}$ to observe experimental results, followed by analysis based on these outcomes. There are a total of four combinations: token-level λ and token-level t, token-level λ and label-wise $\mathcal{T}$, label-wise Λ and token-level t, label-wise Λ and label-wise $\mathcal{T}$.

3 Experiment

In this section, we introduce the main research question and general experimental setup. To ensure fair comparison, we conduct experiments on the following benchmark dataset and basic NER model.

3.1 Setting

Datasets. We choose the widely used CoNLL03 and OntoNotes5.0 English datasets.

Baselines. We use the following models as our baselines:

- **Base NER.** We perform NER tasks using different base pre-trained models as the vanilla NER model. Specifically, we utilize BERT and RoBERTa on three datasets.

- ***k*NN-NER** [12]. It is our most important comparative model, which was the first to use kNN as a retrieval-augmented method, through improving performance by adjusting k, temperature $\mathcal{T}$, and interpolation λ weights without the need to retrain the base model.

Evaluation Metrics. To evaluate the effectiveness of the NER model, we follow the kNN-NER method and select three metrics: precision, recall, and F1 score.

Implementation Details. We use the Hugginface version of BERT [1], Roberta [9] provided by Huggingface. We use the same optimizer, learning rate, and the hidden dimensions of token as the kNN-NER model for the vanilla models. The difference is that due to resource constraints, our batch size is set to 16. When training temperature and interpolation weight parameters simultaneously, we set the learning rate to 3e-4. During training, we separately evaluate the model performance under different combinations of interpolation coefficients and temperature coefficients. All of our experiments are completed on two NVIDIA 3090 GPU.

4 Evaluation and Results

Outperform the Baseline. Table 1 shows the experimental results on the English datasets. We can observe that our proposed method outperforms the original kNN-NER on all English datasets, fully demonstrating the superiority of our method. Moreover, the additional training process of our method only adds a slight additional computational cost.

We also evaluate the different combinations of temperature and interpolation coefficients, as shown in Table 1. Compared to using a fixed value for all input tokens, a dynamic adaptive value can effectively enhance model performance, providing substantial validation for our previous analysis. Additionally, we observe that label-wise λ and $\mathcal{T}$ based on specific contexts show the best performance. This may be because it allows for more fine-grained and precise control and adjustment of label probabilities.

After replacing the specific token embeddings with prototype vetors based on the characteristics of the NER task itself, our model no longer needs to query and save such a large datastore during the inference process. Thus, we first compare the additional datastore sizes on different datasets. The results are shown in Table 2. Compared to kNN-NER, the additional memory required by our method is almost negligible, which saves much storage space. Then, we compare the inference speed of the original kNN-NER method and our method on different datasets. Table 3 shows the comparison results. It can be seen that we have achieved significant acceleration, as the inferred speed is proportional to the logarithm of key values in the datastore. These two features make the NER model more likely for practical deployment.

Ablation Study for the Contributions of the Different Components. To demonstrate the necessity of the different components proposed in our method,

Table 1. The results on two english and two chinese datasets, we conduct experiments under different combinations of interpolation coefficient λ and temperature parameter $\mathcal{T}$. Here, A.P. represents additional parameters, and N represents the number of entity labels. λ_{t-l} represents token-level λ, λ_{l-w} represents label-wise λ and the same applies to the rest. Bold denotes the highest results.

Base Model	Method	A.P.	CoNLL			OntoNotes5.0		
			P	R	F1	P	R	F1
BERT	Tagging	0	89.51	90.64	90.07	84.33	86.33	85.46
	*k*NN-NER	0	89.52	90.74	90.13	**85.27**	86.13	85.70
	$+\lambda_{t-l}$,$\mathcal{T}_{t-l}$	1+1	89.62	90.64	90.13	84.89	86.58	85.73
	$+\lambda_{t-l}$,$\mathcal{T}_{l-w}$	1+N	89.61	**90.76**	90.18	84.58	86.97	85.76
	$+\lambda_{l-w}$,$\mathcal{T}_{t-l}$	N+1	89.58	90.71	90.14	84.60	**86.99**	85.78
	$+\lambda_{l-w}$,$\mathcal{T}_{l-w}$	N+N	**89.96**	90.53	**90.20**	84.68	86.97	**85.81**
RoBERTa	Tagging	0	91.13	91.10	91.11	84.87	86.05	85.96
	*k*NN-NER	0	91.10	91.16	91.13	85.88	86.10	85.99
	$+\ \lambda_{t-l} + \mathcal{T}_{t-l}$	1+1	91.13	91.15	91.14	85.86	86.10	85.98
	$+\ \lambda_{t-l} + \mathcal{T}_{l-w}$	1+N	91.20	91.16	91.18	85.93	86.07	86.00
	$+\ \lambda_{l-w} + \mathcal{T}_{t-l}$	N+1	91.18	91.11	91.15	**86.09**	85.92	86.00
	$+\ \lambda_{l-w} + \mathcal{T}_{l-w}$	N+N	**91.23**	**91.16**	**91.20**	85.65	**86.37**	**86.01**

Table 2. The comparison results of additional memory, where the unit is GB.

Model	Dataset		
	CoNLL	OntoNotes	Weibo
*k*NN-NER	1.2	7.9	0.22
Our Method	9e−6	3e−5	8e−5
ratio	12766×	31600×	2750×

we conduct ablation experiments in this section. Firstly, in order to demonstrate that the prototype vector is more suitable for the NER task, we compare the performance on the Chinese Weibo dataset without trainable λ and $\mathcal{T}$. As shown in Fig. 2(a), for all values of k, our method consistently outperforms the *k*NN-NER approach comprehensively. We cannot guarantee maintaining this advantage for all possible values of k. However, due to the impracticality of experimenting with all potential values, this further underscores the convenience and stability of our approach. Additionally, it can be observed that the performance of the *k*NN-NER method starts to decline when k is greater than 512, possibly due to more noise in the retrieved information. This indicates that the prototype vector is better able to capture essential feature of entity types. Then we compare the performance of the model using fixed values with dynamically trainable parameters based on the *k*NN algorithm. The results are shown in Fig. 2(b). This confirms our previous inference that $\mathcal{P}_{pro}$ and $\mathcal{P}_{NER}$ may have their own strengths when

Table 3. The comparison results of inferred speed, where the unit is tokens/s.

Model	Dataset		
	CoNLL	OntoNotes	Weibo
kNN-NER	155.98	5.23	47.29
Our Method	798.25	623.22	242.15
speedup	5.12×	119.16×	5.12×

predicting different types of entity types in different contextual contexts, so a dynamic trainable λ is more reasonable than a fixed one.

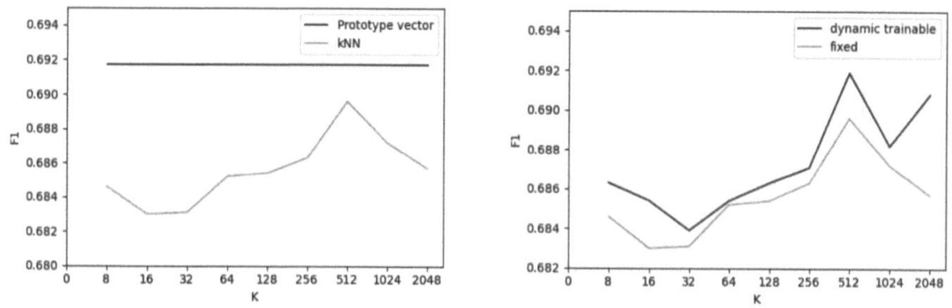

(a) (b)

Fig. 2. (a) Comparison between dynamic K value; (b) Comparison between fixed and dynamic trainable values.

Table 4. The results of different base models on the Weibo dataset.

Model	Vanilla NER	kNN-NER	Our
BERT-base	67.33	68.96	69.86
BERT-large	68.35	69.94	71.12

Larger Model Perform Better. We also evaluate the performance of our method on same models of different sizes for the same dataset. As shown in Table 4, it can be seen that the performance improvement brought by our method is more pronounced on larger models. This is probably because larger models having larger token representation dimensions, which allows our method to make more fine-grained adjustments and controls to the model. Moreover, our approach treat the model as a black box without changing any parameters of the vanilla model. The combination of these two characteristics provides ample evidence of the superiority of our approach in the era of large models.

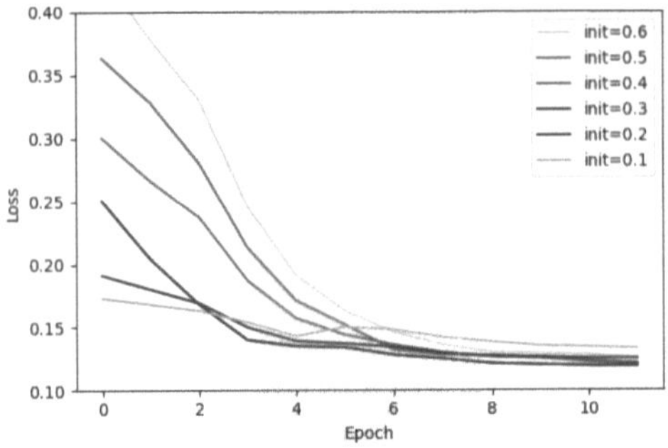

Fig. 3. The model loss with different initialized values for the token-level λ.

Different Initial Values for the Interpolation Coefficient λ. The parameters of the pre-training language model, having undergone fine-tuning on pre-training tasks, benefit from a well-chosen initialization value, aiding the model in achieving better performance and faster convergence on downstream tasks. Therefore, as the additional parameters λ and $\mathcal{T}$ do not participate in the fine-tuning of pre-training tasks, a reasonable initialization value may potentially assist the model in achieving better results. To explore this, we conduct experiments with different initialization values, and the results are depicted in the Fig. 3. It can be observed that a smaller value helps the model converge fastly to the optimal performance. One possible explanation is that a smaller value guides the model to focus more on the output of $\mathcal{P}_{pro}$ from the beginning, leading to a deeper understanding.

5 Conclusion

In this paper, we propose a fine-grained prototype-augmented NER method. First, we explore to construct prototype vectors for each entity type instead of huge specific token embedding in kNN-NER. Moreover, we combine token information with dynamic interpolation coefficient λ and temperature coefficient $\mathcal{T}$ to enhance the performance of the original kNN-NER model. Experiment results on three benchmark datasets show that our method is more suitable for NER tasks and significantly outperform the vanilla kNN-NER baseline. Compared with the vanilla kNN-NER, our method requires less memory space and achieves more inference speedup. Furthermore, through quantitative analysis, we demonstrate that our method can work well in low-resource and larger models scenarios.

Acknowledgments. This work was partly supported by National Natural Science Foundation of China (62306137).

References

1. Devlin, J., Chang, M.W., Lee, K., Toutanova, K.: BERT: pre-training of deep bidirectional transformers for language understanding. In: Burstein, J., Doran, C., Solorio, T. (eds.) Proceedings of the 2019 Conference of the North American Chapter of the Association for Computational Linguistics: Human Language Technologies, Volume 1 (Long and Short Papers), pp. 4171–4186. Association for Computational Linguistics, Minneapolis (2019). https://doi.org/10.18653/v1/N19-1423, https://aclanthology.org/N19-1423
2. Ethayarajh, K.: How contextual are contextualized word representations? Comparing the geometry of BERT, ELMo, and GPT-2 embeddings. In: Inui, K., Jiang, J., Ng, V., Wan, X. (eds.) Proceedings of the 2019 Conference on Empirical Methods in Natural Language Processing and the 9th International Joint Conference on Natural Language Processing (EMNLP-IJCNLP), pp. 55–65. Association for Computational Linguistics, Hong Kong (2019). https://doi.org/10.18653/v1/D19-1006, https://aclanthology.org/D19-1006
3. Gao, J., He, D., Tan, X., Qin, T., Wang, L., Liu, T.: Representation degeneration problem in training natural language generation models. In: International Conference on Learning Representations (2019). https://openreview.net/forum?id=SkEYojRqtm
4. Huang, Y., Liu, D., Zhong, Z., Shi, W., Lee, Y.T.: *k*nn-adapter: efficient domain adaptation for black-box language models. arXiv preprint arXiv:2302.10879 (2023)
5. Jiang, H., et al.: Towards robust k-nearest-neighbor machine translation. In: Goldberg, Y., Kozareva, Z., Zhang, Y. (eds.) Proceedings of the 2022 Conference on Empirical Methods in Natural Language Processing, pp. 5468–5477. Association for Computational Linguistics, Abu Dhabi (2022). https://doi.org/10.18653/v1/2022.emnlp-main.367, https://aclanthology.org/2022.emnlp-main.367
6. Khandelwal, U., Fan, A., Jurafsky, D., Zettlemoyer, L., Lewis, M.: Nearest neighbor machine translation. In: International Conference on Learning Representations (2021). https://openreview.net/forum?id=7wCBOfJ8hJM
7. Khandelwal, U., Levy, O., Jurafsky, D., Zettlemoyer, L., Lewis, M.: Generalization through memorization: nearest neighbor language models. In: International Conference on Learning Representations (2020). https://openreview.net/forum?id=HklBjCEKvH
8. Li, B., Zhou, H., He, J., Wang, M., Yang, Y., Li, L.: On the sentence embeddings from pre-trained language models. In: Webber, B., Cohn, T., He, Y., Liu, Y. (eds.) Proceedings of the 2020 Conference on Empirical Methods in Natural Language Processing (EMNLP), pp. 9119–9130. Association for Computational Linguistics, Online (2020). https://doi.org/10.18653/v1/2020.emnlp-main.733, https://aclanthology.org/2020.emnlp-main.733
9. Liu, Y., et al.: RoBERTa: a robustly optimized BERT pretraining approach. arXiv preprint arXiv:1907.11692 (2019)
10. Radford, A., Narasimhan, K., Salimans, T., Sutskever, I., et al.: Improving language understanding by generative pre-training (2018)
11. Wang, D., Fan, K., Chen, B., Xiong, D.: Efficient cluster-based *k*-nearest-neighbor machine translation. In: Muresan, S., Nakov, P., Villavicencio, A. (eds.) Proceedings of the 60th Annual Meeting of the Association for Computational Linguistics (Volume 1: Long Papers), pp. 2175–2187. Association for Computational Linguistics, Dublin (2022). https://doi.org/10.18653/v1/2022.acl-long.154, https://aclanthology.org/2022.acl-long.154

12. Wang, S., et al.: k nn-ner: named entity recognition with nearest neighbor search. arXiv preprint arXiv:2203.17103 (2022)
13. Xiong, L., et al.: Approximate nearest neighbor negative contrastive learning for dense text retrieval. In: International Conference on Learning Representations (2021). https://openreview.net/forum?id=zeFrfgyZln
14. Yang, Z., Sun, R., Wan, X.: Nearest neighbor knowledge distillation for neural machine translation. In: Carpuat, M., de Marneffe, M.C., Meza Ruiz, I.V. (eds.) Proceedings of the 2022 Conference of the North American Chapter of the Association for Computational Linguistics: Human Language Technologies, pp. 5546–5556. Association for Computational Linguistics, Seattle (2022). https://doi.org/10.18653/v1/2022.naacl-main.406, https://aclanthology.org/2022.naacl-main.406
15. Zheng, X., et al.: Adaptive nearest neighbor machine translation. In: Zong, C., Xia, F., Li, W., Navigli, R. (eds.) Proceedings of the 59th Annual Meeting of the Association for Computational Linguistics and the 11th International Joint Conference on Natural Language Processing (Volume 2: Short Papers), pp. 368–374. Association for Computational Linguistics, Online (2021). https://doi.org/10.18653/v1/2021.acl-short.47, https://aclanthology.org/2021.acl-short.47

Deep Generative Replay for Continual Relation Extraction

Yu Xing[1], Peng Wang[2], Xiaogang Zhang[2], Liwei Chen[1], Zewei Ma[1], Qizhuo Xie[1], and Tieke He[1](✉)

[1] State Key Laboratory for Novel Software Technology, Nanjing University, Nanjing, China
hetieke@gmail.com

[2] Jiangsu Tongxingbao Intelligent Transportation Technology Co. Ltd., Nanjing, China

Abstract. Continual relation extraction (CRE) aims to incrementally acquire new relational knowledge while retaining performance on previously learned relations. A major challenge in this setting is catastrophic forgetting, typically addressed by replay-based methods that store and revisit past data. While effective, these approaches raise privacy concerns and may lead to overfitting. Generative replay (GR) has emerged as a privacy-preserving alternative by synthesizing pseudo-samples that approximate prior task distributions. However, existing GR-based methods often fail to generate diverse and coherent relational instances, limiting their effectiveness. In this work, we propose a novel GR framework tailored for CRE. We first leverage large language models (LLMs) to generate task-specific explanations, improving the interpretability and contextual coherence of generated samples. To enhance generation quality, we introduce a multi-task fine-tuning framework that incorporates prompt engineering techniques for effective relation-specific prompting. Additionally, we propose a prototype-guided refinement strategy that aligns generated relations with learned relation prototypes, enforcing semantic consistency and improving sample fidelity. Extensive experiments on two standard CRE benchmarks demonstrate that our method outperforms prior GR-based and replay-based baselines in mitigating forgetting, while ensuring data privacy and reducing memory overhead.

Keywords: Continual Learning · Generative Replay · Relation Extraction

1 Introduction

Relation Extraction (RE) aims to identify predefined relation types between two given entities within a sentence. Traditional RE tasks typically assume that all relation types are available simultaneously, disregarding the dynamic nature of real-world scenarios where new relations continuously emerge. To address this limitation, Continual Relation Extraction (CRE) [1–5] has been introduced.

J. Li et al. (Eds.): APWeb-WAIM 2025, CCIS 2660, pp. 35–43, 2026.
https://doi.org/10.1007/978-981-95-7996-9_4

CRE is formulated as a class-incremental problem, where a model is trained over a sequence of tasks, each introducing new relation types and corresponding examples. The primary objective of CRE is to enable the model to acquire new relation types while maintaining strong performance across previously learned relations. However, similar to other continual learning tasks, CRE models suffer from the catastrophic forgetting problem, where performance on previously learned relations significantly degrades after learning new ones.

Motivated by the remarkable success of generative replay methods, we introduce a novel deep generative replay strategy specifically designed for CRE. To the best of our knowledge, this is the first attempt to leverage GR for relation extraction in a continual learning setting. Unlike image classification tasks, where pre-trained generative models such as diffusion models are readily available, text-based relation extraction poses unique challenges. Although large-scale pre-trained language models exhibit strong text generation capabilities, they often lack the ability to accurately model domain-specific knowledge. To address this issue, we first propose a multi-task explanation-driven fine-tuning strategy to enhance the relation generation model.

After training the relation generation model, we implement a rigorous quality assurance mechanism for the generated relational sentences. Specifically, we compute their similarity with relation prototypes obtained from prior tasks, followed by an iterative refinement and filtering process to select high-quality pseudo-old samples. These samples are subsequently utilized for replay during the learning of new tasks. Extensive experiments conducted on two widely used benchmark datasets demonstrate the superior performance of our approach, validating its effectiveness in mitigating forgetting while ensuring data security and privacy.

2 Methodology

2.1 Overview

We start with a brief overview of our proposed method for generative replay. As shown in Fig. 1. Our method primarily consists of three components: a prompt-based generator to generate pseudo-old samples, and a relation prototype-based discriminator to enhance the model's generative capabilities and filter replay samples. Specifically, when a new task $\mathcal{T}_k$ arrives, we first train a continual RE model on task $\mathcal{T}_k$. Next, we calculate the prototype for each relation using the trained CRE model. Then, we adapt the dataset with prompts to train a generative model for relations in task $\mathcal{R}_k$. Finally, we leverage LLMs along with relation prototypes to control the quality and quantity of the generated pseudo-samples.

2.2 Task Training

Since our method is model-agnostic, it leverages existing state-of-the-art CRE models as a foundation and then uses the replay samples generated by our approach. We will first briefly introduce the training process of these CRE models.

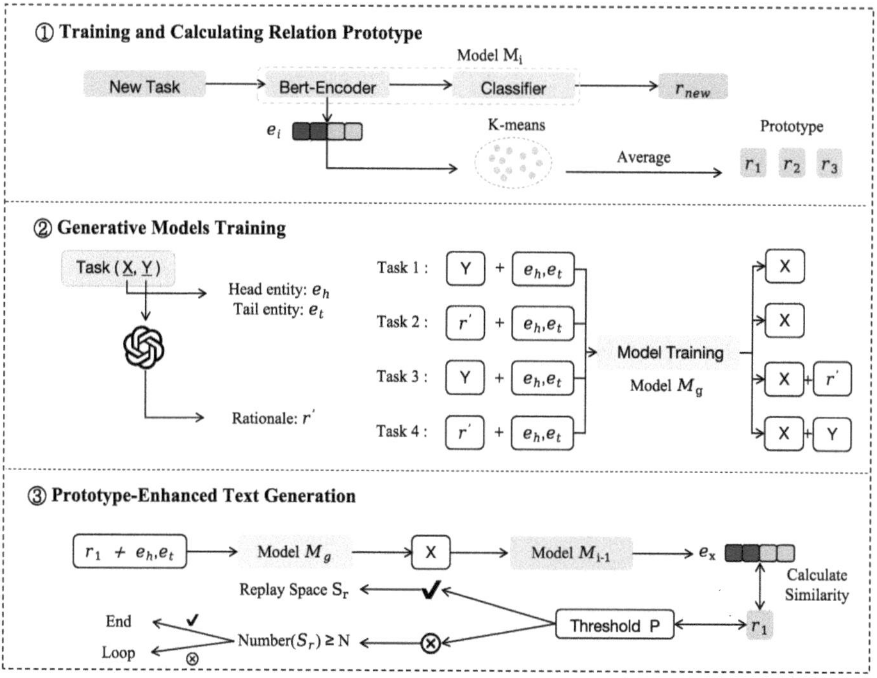

Fig. 1. The overview of our generative replay strategy for continual relation extraction.

Similar to previous work, we use the pre-trained language model BERT as the encoder. For a given sentence X as input, we first insert special tokens [E11]/[E12] and [E21]/[E22] to mark the start/end positions of head and tail entities respectively and then embedding it. Finally, we use the hidden representations of [E11] and [E21] as the representations of head and tail entities. The final representation of X is defined as:

$$h_x = LayerNorm(W_1[h_x^{11}; h_x^{21}] + b) \in \mathcal{R}^d \tag{1}$$

where $h_x^{11}; h_x^{21}$ are the hidden representations of head and tail entities, respectively. d is the dimension of the hidden layer in BERT. $W_1 \in \mathcal{R}^{2d}$ and $b \in \mathcal{R}^d$ are two trainable parameters.

With the representation h_x of sentence X, we use a linear softmax classifier to calculate the classification probability of X:

$$P(x;\theta) = Softmax(W_2 h_x) \in \mathcal{R}^k \tag{2}$$

where θ denotes the model parameters.W_2 is the trainable parameter of the linear classifier and $\mathcal{R}^k$ represents the number of current relation types.

Then, classification loss for current task is calculated as:

$$\mathcal{L}_{new} = -\frac{1}{|D_k^{train}|} \sum_{x_i \in |D_k^{train}|} \sum_{r_j \in \mathcal{R}^k} \delta_{y,rj} log\mathcal{P}(r_j|x_i;\theta_k) \tag{3}$$

where $\mathcal{P}(r_j|x;\theta)$ is the probability of input X classified as relation r_j by the current model θ. Y is the label of X such that if $X = r_j$, $\delta_{y,r_j} = 1$, and 0 otherwise.

Calculating Relation Prototypes
After learning current task, for each relation r in current task, we first apply K-means algorithm upon the representations (h_i) of all instances belonging to r to cluster them into L clusters. Then, for each cluster, we select the instance most closest to the centroid of this cluster as one typical instance. With the selected typical instances of r, we average their representations as relation r's prototype p_r, that is:

$$\mathcal{P}_r = \frac{1}{L}\sum_{i=1}^{L} h_i^r \tag{4}$$

where h_i^r is a typical instance i's representation of relation r. Such prototype best represents r since the L typical instances have the minimal distance sum to the L cluster centroids. Another merit of such prototypes for representing relations is their insensitivity to the value of L.

2.3 Generating Pseudo Examples

Our approach focuses on training a generative model to produce high-quality pseudo-samples for replay. Specifically, after the CRE model is trained on a task $\mathcal{T}_i$, we first train a text generation model to approximate the data distribution of task data $\mathcal{D}_i$. Subsequently, we use prototype vectors to control the quality of the generated relation texts, ensuring the efficiency of the replay method.

Generative Models Training. Although LLMs possess powerful text generation capabilities, the lack of specific domain knowledge and susceptibility to hallucination pose challenges. Therefore, we use the encoder-decoder model T5 as our backbone. By integrating relational knowledge and domain-specific textual knowledge derived from LLMs, we enhance T5 with robust relation sentence generation capabilities. Specifically, after the CRE model is trained on a task $\mathcal{T}_i$, to obtain a generative model capable of simulating the distribution $\mathcal{T}_i$, we apply explanation-enhanced prompt engineering to transform the dataset $\mathcal{D}_i$, resulting in a new dataset $\mathcal{T}_i^{'}$. To ensure the quality of text generation by the model, inspired by Xiong et al., we first utilize an LLM to generate explanations for each relation r. Then, we design a multi-task training framework to fine-tune the T5 model. For example, For data sample $\mathcal{X}_i, \mathcal{Y}_i$ in dataset $\mathcal{D}_i$, we leverage the reasoning capability of LLMs to generate an explanation $\nabla^{'}$ for the relation r, which explains why $\mathcal{X}_i$ is defined to contain the relation $\mathcal{Y}_i$. After obtaining the relation explanation $\nabla^{'}$, we design four tasks:

(1) Task1: Relation-Head-Tail Entity to Sentence, which generates a sentence given the relation and the head-tail entities.
(2) Task2: Relation Explanation-Head-Tail Entity to Sentence, whcih generates a sentence given the relation explanation and the head-tail entities.
(3) Task3: Relation-Head-Tail Entity to Relation Explanation-Sentence, which generates both the relation explanation and the sentence given the relation and the head-tail entities.
(4) Task4: Relation Explanation-Head-Tail Entity to Relation-Sentence, which generates both the relation and the sentence given the relation explanation and the head-tail entities.

Task 1 is the primary task, where the goal is to generate pseudo-samples based on a given relation and head-tail entities. The other tasks are auxiliary tasks designed to enhance the generative model's understanding and performance.

The training objective of the relation generation model is a combination of the losses from the four tasks mentioned above, as follow:

$$\mathcal{L} = \alpha\mathcal{L}_1 + \beta\mathcal{L}_2 + \gamma\mathcal{L}_3 + (1 - \alpha - \beta - \gamma)\mathcal{L}_4 \tag{5}$$

where α, β and γ are factors used to adjust the loss weights.

2.4 Controllable Text Generation

Once the generative model is trained, it is used to generate pseudo-samples. Despite our significant efforts to enhance the model's generative capabilities, the quality of the generated relational sentences is still not guaranteed. To address this, we first introduce a class-based generation limit, denoting the maximum number of sentences generated for each relation as N. This constraint not only ensures a more efficient generation process but also maintains balance across relation types. Following this, inspired by previous work, we employ a prototype-enhanced generation optimization strategy to further ensure the quality of the generated samples. Specifically, the relation sentences generated by the model $\mathcal{M}_g$ are passed through the previous CRE model $\mathcal{M}_{i-1}$ to obtain their representations. These representations are then compared with their corresponding relation prototypes using cosine similarity. A threshold P is set to filter out sentences that meet the similarity condition.

However, a higher threshold P often results in high-quality relational sentences but may fail to meet the quantity requirement N. Conversely, a lower threshold P easily satisfies the quantity requirement but compromises the quality of the generated sentences. To address this trade-off, we adopt an iterative process in which the threshold P is gradually decreased by 0.05 in each generation cycle. This process continues until the number of generated relational sentences for all classes meets the defined quantity N or until the threshold reaches a predefined lower bound. This iterative adjustment ensures a balance between quality and quantity in the generated relational sentences.

3 Experiment

In this section, we introduce general experimental setup. To ensure fair comparison, we conduct experiments on the following benchmark dataset.

3.1 Datasets

Following previous works, we conduct our experiments on widely-used continual RE benchmark datasets **FewRel** [8].

3.2 Evaluation Metrics

Average accuracy serves as a more reliable metric for assessing the impact of catastrophic forgetting, as it places greater emphasis on the model's performance on earlier tasks. In this study, we evaluate the model's effectiveness by calculating the average accuracy across K tasks at each step. Additionally, to ensure the reliability and robustness of the experimental results, we repeat each experiment five times and report the average scores.

3.3 Baselines

For the fair comparison, we consider the following baseline methods: EA-EMR [1], EMAR [2], CML [3], RP-CRE [4], CRL [5], CRECL [6] and CEAR [7].

3.4 Implement Details

Unlike previous methods that incorporate partial experience replay, our approach is entirely focused on generating replay samples, making it inherently model-agnostic. To rigorously evaluate the effectiveness of our generative replay strategy, we selected two state-of-the-art continual relation extraction (CRE) models, EMAR and CEAR, as our base models. For each relation, we generated N=10 pseudo-samples. All hyperparameters for the baseline methods were configured according to their original implementations to ensure a fair comparison. All parameters for the baseline methods were configured as specified in their original papers. Experiments were conducted on two NVIDIA 3090 GPUs with 24GB memory each.

4 Results and Analysis

4.1 Main Results

Table 1 presents the comparative results between our method and other baseline methods. It can be observed that although our method initially exhibits slightly inferior performance, it begins to demonstrate strong competitiveness on both benchmark datasets after step four. This result highlights the superiority of our generative replay method over other replay-based approaches. Specifically, our method exhibits a lower decline in average accuracy compared to the baseline method EMAR by 2.3% on the benchmark datasets, respectively. For the baseline method CEAR, the reductions are 1.7%. Additionally, our method performs relatively weakly in the initial stages of continual learning but improves progressively over time. We hypothesize that this trend arises due to the following factors: in the early stages, the pseudo-samples generated by our method provide a relatively weaker replay effect compared to real historical samples. However, as training progresses, baseline methods that rely on real stored samples are prone to overfitting, leading to a sharp decline in performance. In contrast, our generative replay strategy mitigates this issue by dynamically synthesizing training data, thereby preserving knowledge from previous tasks more effectively. Furthermore, unlike conventional experience replay methods, our approach eliminates the need to store real historical samples, ensuring enhanced data security and privacy protection. These findings collectively highlight the effectiveness and generalizability of our method, demonstrating its potential for practical continual learning scenarios where data storage constraints and privacy concerns are critical considerations.

Table 1. Accuracy scores over learned relation types at CRE step. GR denotes generative replay strategy and * denotes our reproduced results with the Open source code repository. Other results are directly taking from Zhao et al. [7].

FewRel										
Models	Step1	Step2	Step3	Step4	Step5	Step6	Step7	Step8	Step9	Step10
EA-EMR	89.0	69.0	59.1	54.2	47.8	46.1	43.1	40.7	38.6	35.2
CML	91.2	74.8	68.2	58.2	53.7	50.4	47.8	44.4	43.1	39.7
RP-CRE	97.9	92.7	91.6	89.2	88.4	86.8	85.1	84.1	82.2	81.5
CRL	98.2	94.6	92.5	90.5	89.4	87.9	86.9	85.6	84.5	83.1
CRECL	97.8	94.9	92.7	90.9	89.4	87.5	85.7	84.6	83.6	82.7
EMAR	**98.8**	89.1	89.5	85.7	83.6	84.8	79.3	80.0	77.1	73.8
+GR	97.1	88.4	88.0	85.2	84.0	85.1	79.7	80.3	77.5	74.4
CEAR	98.1	**95.8**	**93.6**	91.9	**91.1**	89.4	88.1	86.9	85.6	84.2
+GR	96.8	94.0	93.2	**92.1**	90.8	**89.9**	**88.6**	**87.4**	**86.1**	**84.7**

4.2 Ablation Study

In this study, we design a series of ablation experiments to comprehensively evaluate the effectiveness of our proposed method:

ST (Single Task): The generative model is trained exclusively on the first task, without exposure to subsequent tasks.

WP (Without Prototype): The relation prototypes are omitted, meaning that quality control mechanisms are not applied during the generation process of relational sentence.

FS (Fixed Old Samples): The memory buffer contains a fixed set of old samples for replay, which remain unchanged for each relation.

The experimental results, as summarized in the Table 2, reveal several key findings. Notably, the WP setting exhibits the poorest performance, underscoring the critical role of relation prototypes in maintaining sample quality during generative replay. The absence of quality control leads to the generation of low-fidelity pseudo-samples, which in turn adversely impacts overall model performance.

The FS setting achieves relatively better results; however, the performance degradation highlights the risks of overfitting when replaying a static set of old samples. This suggests that a more dynamic memory update mechanism could be beneficial in mitigating overfitting effects.

Furthermore, the ST setting demonstrates inferior performance compared to our full approach, reinforcing the advantages of multi-task training. The observed improvements suggest that learning from multiple tasks enhances the generative model's ability to produce high-quality samples, thereby boosting overall effectiveness.

Collectively, these findings substantiate the efficacy of our proposed method and emphasize the necessity of incorporating quality control and dynamic replay strategies to optimize generative replay in continual learning scenarios.

Table 2. The ablation study at CRE step. ST denotes training using only Task 1, WP indicates not using relation prototypes to control the quality of the generated relational sentences, and FS represents keeping the replay samples for each relation fixed during the continual learning process.

FewRel										
Models	Step1	Step2	Step3	Step4	Step5	Step6	Step7	Step8	Step9	Step10
ST	97.1	88.3	87.7	84.9	83.6	84.5	79.4	79.9	77.1	73.8
WP	97.1	88.1	87.5	84.7	83.3	84.1	78.9	79.7	76.8	73.2
FS*	97.1	88.4	87.8	84.8	83.4	84.3	79.2	79.6	76.9	73.5
Ours	97.1	88.4	88.0	85.2	84.0	85.1	79.7	80.3	77.5	74.4

5 Conclusion

In this work, we propose a deep generative replay strategy to address the critical challenges of continual relation extraction, including overfitting and data privacy constraints. Our approach employs a generative model to synthesize high-quality samples of previously learned relations, ensuring knowledge retention without relying on direct access to past data. To enhance the fidelity and representativeness of the generated samples, we incorporate prototype vectors that serve as anchors, guiding the generation process toward maintaining semantic consistency. Extensive experiments on two widely used benchmark datasets demonstrate that our method outperforms existing state-of-the-art approaches, significantly alleviating forgetting while preserving data security and privacy. The findings of this study underscore the potential of generative replay as a scalable and effective paradigm for CRE, enabling robust knowledge accumulation in dynamic, real-world environments.

Acknowledgments. This work was partly supported by National Natural Science Foundation of China (62306137).

References

1. Wang, H., Xiong, W., Yu, M., et al.: Sentence embedding alignment for lifelong relation extraction. arXiv preprint arXiv:1903.02588 (2019)
2. Han, X., Dai, Y., Gao, T., et al.: Continual relation learning via episodic memory activation and reconsolidation. In: Proceedings of the 58th Annual Meeting of the Association for Computational Linguistics, pp. 6429–6440 (2020)
3. Wu, T., Li, X., Li, Y.F., et al.: Curriculum-meta learning for order-robust continual relation extraction. In: Proceedings of the AAAI Conference on Artificial Intelligence, vol. 35, no. 12, pp. 10363–10369 (2021)
4. Cui, L., Yang, D., Yu, J., et al.: Refining sample embeddings with relation prototypes to enhance continual relation extraction. In: Proceedings of the 59th Annual Meeting of the Association for Computational Linguistics and the 11th International Joint Conference on Natural Language Processing (Volume 1: Long Papers), pp. 232–243 (2021)
5. Zhao, K., Xu, H., Yang, J., et al.: Consistent representation learning for continual relation extraction. arXiv preprint arXiv:2203.02721 (2022)
6. Hu, C., Yang, D., Jin, H., et al.: Improving continual relation extraction through prototypical contrastive learning. arXiv preprint arXiv:2210.04513 (2022)
7. Zhao, W., Cui, Y., Hu, W.: Improving continual relation extraction by distinguishing analogous semantics. arXiv preprint arXiv:2305.06620 (2023)
8. Han, X., Zhu, H., Yu, P., et al.: FewRel: a large-scale supervised few-shot relation classification dataset with state-of-the-art evaluation. arXiv preprint arXiv:1810.10147 (2018)

Efficient Subgraph Matching with Adaptive Filtering and Core-Set Prioritization

Heng Xiao[1(✉)] and Zi Chen[2]

[1] Nanjing University of Science and Technology, Nanjing, China
hengxiao@njust.edu.cn
[2] Nanjing University of Aeronautics and Astronautics, Nanjing, China
zichen@nuaa.edu.cn

Abstract. Subgraph matching is a fundamental problem in graph analysis. Recently, many algorithms have been developed, often using backtracking search. However, existing subgraph matching algorithms suffer from significant limitations, including prohibitive computational costs for large-scale graphs, inefficient pruning strategies that fail to scale with real-world network complexities, and suboptimal search orders that amplify computational overhead. To address these challenges, we propose an efficient framework with three key innovations: (1) multi-round candidate filtering, iteratively pruning the search space until reduction is less than 1% of the previous set; (2) core-driven dynamic programming, prioritizing core vertices to optimize the matching order and enhance efficiency; and (3) early pruning techniques to eliminate nonpromising branches. Extensive experiments on real-world datasets demonstrate that our approach outperforms state-of-the-art algorithms by one to two orders of magnitude in most cases.

Keywords: Subgraph matching · Core-Driven Dynamic Programming · Candidate filtering

1 Introduction

Subgraph matching, the process of finding isomorphic instances of a query graph within a larger data graph, serves as a fundamental primitive in domains ranging from social network analysis, biochemical structure identification [4] and knowledge graphs [14]. For example, in Fig. 1, $\{(u_1, v_1), (u_2, v_4), (u_3, v_5), (u_4, v_{10})\}$ is a match from query graph q to data graph G. Subgraph matching is a typical NP-hard problem, which means that the computational complexity is factorial in the worst case.

The algorithms (e.g., CECI [2], CFL [3], DP-iso [7], TurboIso [8], GraphQL [10], VEQ [13], BⓈX [15]) adopt the preprocessing-enumeration framework, which is widely used in recent algorithms in the database community. As highlighted in recent studies [22,26], the selection of an effective matching order is

J. Li et al. (Eds.): APWeb-WAIM 2025, CCIS 2660, pp. 44–55, 2026.
https://doi.org/10.1007/978-981-95-7996-9_5

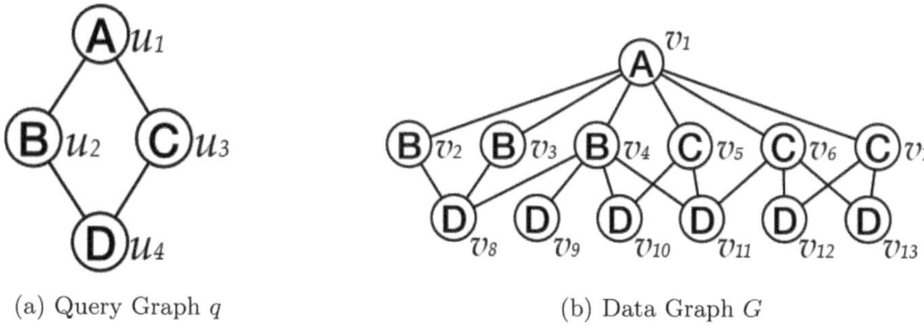

Fig. 1. Example query graph and data graph.

crucial for backtracking-based algorithms, given its significant impact on search efficiency. A poorly chosen matching order can result in excessive backtracking and unnecessary exploration of irrelevant branches, significantly increasing runtime. Many existing methods rely on heuristic-based approaches to determine the matching order, but these heuristics often overlook the structural importance of certain vertices, such as core vertices, which are pivotal in the query graph. Consequently, the matching process becomes sub-optimal, leading to inefficiencies and degraded performance.

This paper addresses the computational challenges of subgraph matching by proposing a novel and efficient framework that integrates three innovative techniques. The contributions of this paper are itemized as follows:

- **Multi-Round Adaptive Candidate Filtering:** We introduce a multi-round filtering mechanism that iteratively reduces the candidate set until the reduction rate falls below 1% of the previous round. This approach significantly minimizes unnecessary computations, thereby improving the efficiency of the subgraph matching process.
- **Core-Set Prioritization via Dynamic Programming:** We propose a dynamic programming-based sorting strategy that prioritizes core vertices in the query graph. By leveraging the structural importance of these core vertices, our method optimizes the matching order, reducing the search space and improving overall performance.
- **Efficient Pruning Strategies:** We design advanced pruning techniques to eliminate non-promising branches early in the search process. These strategies further accelerate computation while maintaining high accuracy, making our framework suitable for large-scale graphs.

By integrating adaptive filtering, core-set prioritization, and pruning optimization, our work provides a practical solution for subgraph matching and sets new benchmarks for future research in graph query processing.

Table 1. Notations with internal vertical lines.

Notations	Descriptions
g, q and G	graph, query graph and data graph
V, E and Σ	vertex set, edge set and label set
$d(u)$, $L(u)$ and $N(u)$	degree, label and neighbors of u
$e(u, v)$	an edge between u and v
$g[V]$	vertex-induced subgraph of g on V
C	candidate vertex set
f and φ	subgraph isomorphism and matching order
$N_{+}^{\varphi}(u)$ ($N_{-}^{\varphi}(u)$)	backward (forward) neighbors of u given φ
$\mathcal{A}$	auxiliary data structure
$\mathcal{A}_{u}^{u'}(v)$	neighbors of v in $C(u')$ where $v \in C(u)$

2 Related Work

The subgraph matching problem has been extensively studied in the literature, with two primary algorithmic paradigms dominating the field: **join-based methods** and **backtracking-based methods**. Join-based approaches [20,23] decompose the query graph into subcomponents and iteratively join candidate matches to construct valid embeddings. As surveys [22] on subgraph matching indicate, backtracking-based algorithms generally follow a *filter-order-enumerate framework* [2,3,5,7,12,13,15], where three key stages are sequentially executed:

- **Filtering**: Candidate vertices/edges in the data graph are pre-filtered based on structural and semantic constraints (e.g., label compatibility, degree constraints) to prune the search space [7,13].
- **Ordering**: A matching order for query vertices is determined to minimize backtracking. Traditional heuristics include selecting vertices with minimal candidate sets or leveraging minimum spanning trees (e.g., `QuickSI` [21] uses a minimum-degree-based ordering to reduce branching factors).
- **Enumeration**: A backtracking search is performed following the chosen order, with pruning strategies applied to eliminate non-promising branches [7,13,24].

Recent advancements have focused on improving the efficiency of backtracking frameworks. For instance, `GuP` [1] introduces guard-based pruning, which tracks failures encountered during backtracking and stores them as "guards". This allows the system to avoid redundant exploration when the same issues reoccur. `BICE` [6] constructs a candidate bipartite graph and utilizes bipartite matching to reduce backtracking steps. `Circinus` [11] employs similar bipartite matching techniques to optimize the search process.

In parallel with subgraph matching, related graph algorithm research has seen significant progress. Meng et al. [16,17] provide comprehensive surveys on distributed graph algorithms and benchmarking methodologies, highlighting the scalability challenges in large-scale graph processing. Yuan et al. [25] and Hao et al. [9] address hop-constrained path query processing, demonstrating efficient techniques for constrained path enumeration in large graphs. Additionally, recent work on triangle counting in hypergraph streams [19] and butterfly counting in bipartite graphs [18] showcases advanced techniques for specialized graph pattern mining tasks.

These developments in related graph algorithm domains provide valuable insights and techniques that can be adapted to improve subgraph matching efficiency, particularly in handling large-scale graphs and complex structural constraints.

3 Preliminaries

In this section, we define the notations and terminologies used throughout this paper.

We concentrate on the undirected labeled graph denoted as $g = (V, E)$, where V represents the collection of vertices and E signifies the set of edges connecting these vertices.For any vertex $u \in V$, let $N(u)$ represent the set of neighbors of u, defined as $\{u' \in V \mid e(u, u') \in E\}$. When considering a subset of vertices $V' \subseteq V$, the notation $g[V']$ denotes the subgraph induced by V' within graph g. It is important to note that graph g is labeled at its vertices; specifically, there exists a labeling function L that maps each vertex $u \in V$ to a label l from a predefined label set Σ. Additionally, in this context, the query graph and the data graph are referred to as q and G, respectively. In Table 1, we compile a summary of the notations that are frequently utilized throughout this paper. Following this, we provide formal definitions for subgraph matching along with its related foundational concepts.

Subgraph Isomorphism. Given a query graph $q = (V_q, E_q)$ and a data graph $G = (V_G, E_G)$, a subgraph isomorphism is defined as an injective mapping $f : V_q \rightarrow V_G$ satisfying the following conditions:

- For every vertex $u \in V_q$, the label of u matches the label of its image under f, i.e., $L(u) = L(f(u))$.
- For every edge $e(u, u') \in E_q$, there exists a corresponding edge in G between the images of u and u' under f, i.e., $e(f(u), f(u')) \in E_G$.

Problem Definition. The task of subgraph matching involves identifying all such subgraph isomorphisms from q to G. To simplify terminology, we refer to each subgraph isomorphism as a match. Additionally, we assume that q is connected and contains at least three vertices ($|V(q)| \geqslant 3$), since finding matches for single vertices or edges is considered trivial.

4 Methodology

Our subgraph matching framework addresses computational challenges through three core innovations: multi-round candidate filtering, core-driven dynamic programming, and efficient pruning techniques.

4.1 Multi-round Candidate Filtering

The first phase of our framework employs **multi-round iterative filtering** to aggressively prune the search space while adaptively adjusting filtering criteria.

Basic Methods. The Label and Degree Filtering (LDF) generates candidate nodes $C(u)$ for a query node u based on its label $L(u)$ and degree $d(u)$:

$$C(u) = \{v \in V(G) \mid L(v) = L(u) \wedge d(v) \geqslant d(u)\} \tag{1}$$

This ensures candidates v in the data graph G match the label of u and have a degree at least as large as u's degree. All existing algorithms incorporate LDF as a baseline filtering step. The Neighbor Label Frequency Filtering (NLF) refines $C(u)$ by analyzing neighbor label distributions. For $v \in C(u)$, if there exists a label $l \in L(N(u))$ such that:

$$|N(u,l)| > |N(v,l)| \tag{2}$$

where:

- $L(N(u)) = \{L(u') \mid u' \in N(u)\}$ (labels of u's neighbors),
- $N(u,l) = \{u' \in N(u) \mid L(u') = l\}$ (neighbors of u with label l),

then v is pruned from $C(u)$. NLF is adopted by CFL [3], CECI [2], and DP-iso [7] to enhance pruning efficiency. Unlike existing methods such as DP-iso [7] and VEQ [13], which typically perform a fixed number of filtering rounds (e.g., 3 rounds), our approach dynamically iterates until convergence, ensuring thorough candidate reduction without premature termination.

Iterative Filtering Process. Given a query graph q and data graph G, the candidate set $\mathcal{C}_t$ at iteration t is defined as:

$$\mathcal{C}_t = \{v \in V(G) \mid \text{satisfies structural/semantic constraints w.r.t. } q\} \tag{3}$$

where constraints include node attributes, edge connectivity, and core decomposition properties. Each round refines $\mathcal{C}_t$ by applying stricter constraints derived from the current candidate set. The iterative process terminates when the reduction ratio between consecutive rounds drops below a threshold:

$$\frac{|\mathcal{C}_t|}{|\mathcal{C}_{t-1}|} < \tau \quad (\tau = 0.01 \text{ by default}), \tag{4}$$

ensuring aggressive pruning while avoiding termination before meaningful reduction occurs. In contrast, DP-iso and VEQ use a fixed three-round pipeline, which may either over-prune (losing valid candidates) or under-prune (leaving excessive candidates).

4.2 Core-Driven Dynamic Programming

Subgraph matching is a computationally intensive task, often suffering from combinatorial explosion in large-scale graphs. To address this, we propose a core-driven dynamic programming (CDDP) technique that integrates structural analysis with heuristic search. The core idea is to prioritize vertices with high structural centrality (core numbers) while mitigating the branching factor caused by high-degree nodes. This dual objective ensures both search efficiency and solution quality.

Core Number and Structural Centrality. The core number $\text{core}(u)$ of a vertex u quantifies its structural importance by identifying the largest k-core it belongs to. A k-core is a maximal subgraph where every vertex has at least degree k. Vertices in higher k-cores are critical for maintaining graph connectivity, making them ideal candidates for early selection in subgraph matching.

Dynamic Programming Formulation. The CDDP algorithm employs a state transition framework to minimize computational cost. The dynamic programming state is defined as:

$$\text{DP}(S, \phi) = \min_{u \in S} \{\text{cost}(\phi \cup \{u\}) + \text{DP}(S \setminus \{u\}, \phi \cup \{u\})\} \tag{5}$$

where:

- S: Set of remaining vertices to match in the query graph Q.
- ϕ: Current partial isomorphism mapping from Q to the target graph G.
- $\text{cost}(\phi \cup \{u\})$: Estimated cost of extending ϕ with vertex u, defined as:

$$\text{cost}(\phi \cup \{u\}) = \alpha \cdot \text{pruning\_efficiency}(u) + \beta \cdot \text{branching\_factor}(u) \tag{6}$$

  where α and β are hyperparameters balancing pruning efficiency and branching reduction.

Core-Driven Vertex Selection. At each step, the next vertex u^* is selected via:

$$u^* = \arg\max_{u \in S} \left\{ \frac{\text{core}(u)}{\text{degree}(u)} \right\} \tag{7}$$

This heuristic balances two critical factors:

- **Core Centrality**: Prioritizes vertices in dense k-cores (high $\text{core}(u)$), ensuring structural relevance.
- **Degree Normalization**: Avoids overloading the search with high-degree vertices (low $\text{degree}(u)$), which can exponentially increase branching.

Discussion. The CDDP technique demonstrates that structural analysis (core numbers) can synergize with heuristic search strategies to achieve scalable subgraph matching. Future work includes extending this approach to attributed graphs and distributed computing environments.

4.3 Efficient Pruning Techniques

To further accelerate subgraph matching, we employ two **context-aware pruning strategies** that eliminate non-promising branches early in the search process. These techniques are tightly integrated with multi-round filtering and the core-driven dynamic programming technique to maximize efficiency.

Constraint Violation Check (CVC). The Constraint Violation Check (CVC) ensures that partial mappings ϕ adhere to both structural and semantic constraints. A mapping ϕ is valid if:

$$\forall(u,v) \in E(Q), \quad \phi(u) \sim_G \phi(v) \quad \wedge \quad \text{label}_Q(u) = \text{label}_G(\phi(u)) \tag{8}$$

where:

- $\phi(u) \sim_G \phi(v)$ denotes adjacency in graph G (i.e., $\{\phi(u), \phi(v)\} \in E(G)$).
- $\text{label}_Q(u)$ and $\text{label}_G(v)$ represent the labels of vertices u in Q and v in G, respectively.

If any constraint in Eq. 8 is violated, the branch is pruned immediately. Specific violation cases include:

- **Structural Violation**: If $\phi(u)$ and $\phi(v)$ are not adjacent in G, the edge $(u,v) \in E(Q)$ cannot be satisfied.
- **Semantic Violation**: If $\text{label}_{Q(u)} \neq \text{label}_G(\phi(u))$, the vertex u is mapped to an incompatible node in G.

Core-Aware Pruning (CAP). To complement the above, we introduce **core-aware pruning** that leverages structural properties:

$$\text{CAP}(u) = \begin{cases} \text{prune} & \text{if } \forall v \in \mathcal{C}(u), \text{core}_G(v) < \text{core}_Q(u) \\ \text{continue} & \text{otherwise} \end{cases} \tag{9}$$

This ensures that no candidate v can satisfy the core number requirement of u, preemptively terminating paths that violate structural constraints.

5 Experiments

In this section we evaluate the performance of the competing algorithms for subgraph matching.

5.1 Experimental Setup

We compare our MRCD framework(combining multi-round filtering, core-driven dynamic programming, and efficient pruning) with the state-of-the-art methods:

VEQ [13], CaLiG [24]. All source codes were obtained from the authors of previous papers and implemented in C++. Experiments are conducted on a machine running CentOS with two Intel Xeon E5-2680 v3 2.5 GHz CPUs and 256 GB memory.

Datasets. We conduct experiments on six real-world datasets: Yeast, Human, HPRD, WordNet, EU2005, and DBLP. These datasets are commonly used in previous studies on subgraph matching [3,7,13,22,23,26]. For DBLP, which lack node labels, we randomly assigned a label from 20 distinct classes to each vertex. Dataset characteristics (e.g., graph size, density, and label distribution) are summarized in Table 2.

Table 2. Dataset graphs.

Dataset	$\|V\|$	$\|E\|$	$\|L\|$	d	Type
Yeast	3,112	12,519	71	8.0	Protein
Human	4,674	86,282	44	36.9	Protein
HPRD	9,460	37,081	307	7.8	Protein
DBLP	317,080	1,049,866	20	6.6	Colla.
WordNet	76,853	120,399	5	3.1	Lexical
EU2005	862,664	16,138,468	4	37.4	Web

Query Graphs. For each data graph, we generate two categories of query sets: $\mathcal{Q}_{iS}$ (sparse) and $\mathcal{Q}_{iD}$ (dense), where $i \in \{4, 8, 12, 16, 20\}$ for Human dataset, and $i \in \{4, 8, 16, 24, 32\}$ for the remaining datasets (Yeast, HPRD, WordNet, EU2005, and DBLP). Each query set contains 100 connected query graphs with i vertices. All query graphs of $\mathcal{Q}_{iS}$ satisfy avg-deg(q) $\leq$ 3, where avg-deg(q) denotes the average degree of vertices in query graph q. All query graphs of $\mathcal{Q}_{iD}$ have avg-deg(q) $>$ 3. To ensure valid embeddings exist for all queries, each query graph is constructed as a connected subgraph of the data graph.

Metrics. We employ two metrics to evaluate the effectiveness and efficiency of all methods.

- **Query Processing Time (QPT).** *Query Processing Time (QPT)* measures the time required to return a specified number of subgraph embeddings, a widely adopted evaluation criterion in traditional methods [2,3,7,8,13,22]. Following this standard, we evaluate algorithm performance by measuring the QPT needed to return 10^5 embeddings. Due to the NP-hard complexity of subgraph matching, certain queries may fail to produce 10^5 results within a practical timeframe. To ensure comparability with prior work, we impose a 300 s time limit for QPT measurements.

- **Embeddings per Second (EPS).** *Embeddings per Second (EPS)* [15,26] quantifies computational efficiency by averaging the number of valid embeddings returned per second. It serves as a standardized benchmarking criterion for evaluating subgraph matching algorithms, as it simultaneously reflects temporal performance and result generation capacity. Notably, EPS emphasizes an algorithm's ability to generate embeddings rapidly.

5.2 Overall Performance

We employ two metrics to evaluate the effectiveness between MFCD and other methods.

Under the QPT Metric. MFCD demonstrates a significant performance advantage over competing algorithms on most datasets where it outperforms VEQ and CaLiG by more than one order of magnitude. However, on the HPRD dataset, VEQ achieves a superior speedup. This discrepancy arises from the unique characteristics of HPRD and the complementary strengths of the two algorithms. HPRD's sparse networks contain fewer repetitive substructures, weakening the effectiveness of MFCD's pruning mechanisms. Meanwhile, VEQ's equivalence set enables it to rapidly handle precise subgraph queries common in HPRD, where exact matches dominate.

Under the EPS Metric. To evaluate the scalability of MFCD, we conducted experiments on two real-world datasets (HPRD and Yeast). Figure 3 summarize EPS of MFCD, VEQ, and CaLiG across different query sizes. On the HPRD dataset, MFCD consistently outperforms both VEQ and CaLiG across all query sizes. Notably, in the $\mathcal{Q}_{16D}$, MFCD achieves an EPS speedup of over one order of magnitude compared to VEQ. On the Yeast dataset, MFCD demonstrates remarkable stability in EPS performance across all query sizes. Even when query complexity increases, MFCD's EPS remains nearly constant (within $\pm 5\%$ variation). This stability stems from Yeast's highly repetitive subgraph patterns. In contrast, VEQ and CaLiG exhibit significant performance degradation as query size grows, highlighting their sensitivity to structural complexity (Fig. 2).

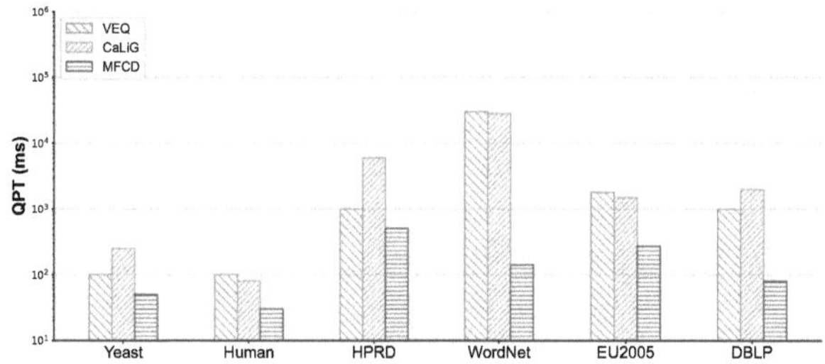

(a) Vary G on dense query sets.

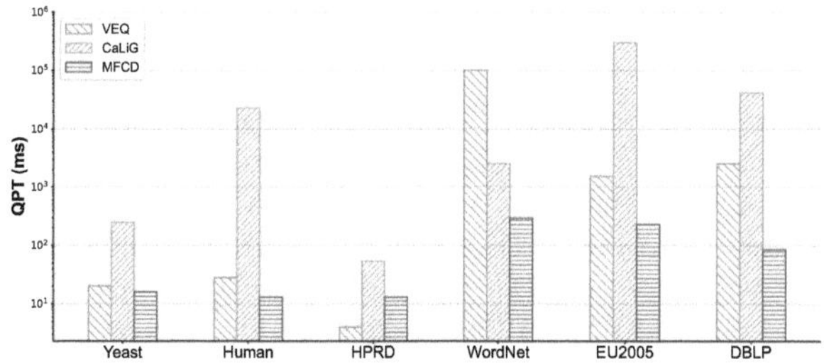

(b) Vary G on sparse query sets.

Fig. 2. Overall performance.

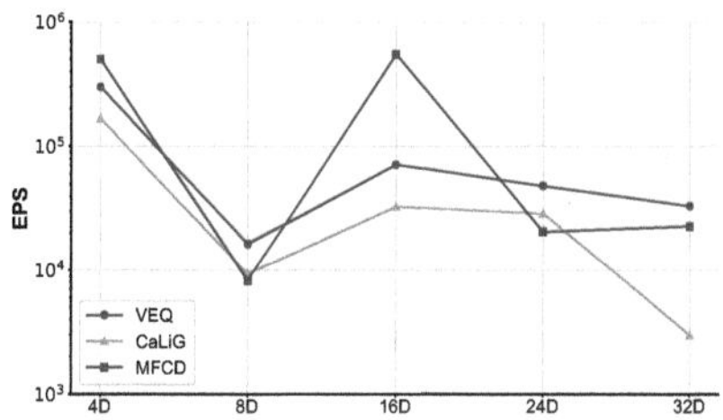

(a) Vary query sizes on HPRD.

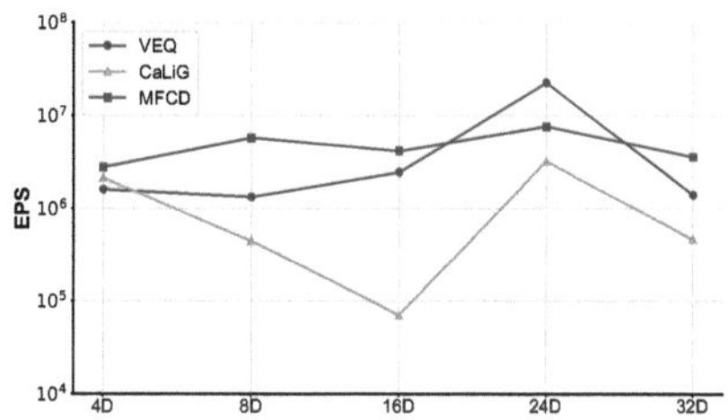

(b) Vary query sizes on Yeast.

Fig. 3. Overall performance with different query size

6 Conclusion

In this paper, we propose an efficient subgraph matching framework that integrates three core innovations to address the computational challenges of large-scale subgraph matching. First, multi-round candidate filtering reduces the search space iteratively, halting when the reduction ratio between consecutive iterations drops below 1%, ensuring minimal candidate sets without premature termination. Second, core-driven dynamic programming prioritizes structurally central vertices (core vertices) based on their core numbers, optimizing the matching order to balance computational efficiency and structural importance. Third, early pruning techniques eliminate non-promising branches using two criteria: lower-bound estimation to compare current paths with the best-known solution and constraint violation checks to prune structurally inconsistent paths. These optimizations synergistically reduce redundant computations and enhance search efficiency. Extensive experiments on real-world datasets demonstrate the framework's superiority over state-of-the-art methods (e.g., VEQ and CaLiG), achieving a 1–2 orders of magnitude speedup. By bridging theoretical efficiency and practical scalability, our approach provides a robust solution for large-scale subgraph matching in diverse applications.

References

1. Arai, J., Fujiwara, Y., Onizuka, M.: Gup: fast subgraph matching by guard-based pruning. Proc. ACM Manag. Data **1**(2) (2023). https://doi.org/10.1145/3589312
2. Bhattarai, B., Liu, H., Huang, H.H.: Ceci: compact embedding cluster index for scalable subgraph matching. In: Proceedings of the 2019 International Conference on Management of Data, pp. 1447–1462. SIGMOD '19, Association for Computing Machinery, New York, NY, USA (2019). https://doi.org/10.1145/3299869.3300086
3. Bi, F., Chang, L., Lin, X., Qin, L., Zhang, W.: Efficient subgraph matching by postponing cartesian products. In: Proceedings of the 2016 International Conference on Management of Data, pp. 1199–1214. SIGMOD '16, Association for Computing Machinery, New York, NY, USA (2016). https://doi.org/10.1145/2882903.2915236
4. Bonnici, V., Giugno, R., Pulvirenti, A., Shasha, D., Ferro, A.: A subgraph isomorphism algorithm and its application to biochemical data. BMC Bioinformatics **14**, S13 – S13 (2013). https://api.semanticscholar.org/CorpusID:1338930
5. Carletti, V., Foggia, P., Saggese, A., Vento, M.: Challenging the time complexity of exact subgraph isomorphism for huge and dense graphs with vf3. IEEE Trans. Pattern Anal. Mach. Intell. **40**(4), 804–818 (2018). https://doi.org/10.1109/TPAMI.2017.2696940
6. Choi, Y., Park, K., Kim, H.: Bice: exploring compact search space by using bipartite matching and cell-wide verification. Proc. VLDB Endow. **16**(9), 2186–2198 (2023). https://doi.org/10.14778/3598581.3598591
7. Han, M., Kim, H., Gu, G., Park, K., Han, W.S.: Efficient subgraph matching: Harmonizing dynamic programming, adaptive matching order, and failing set together. In: Proceedings of the 2019 International Conference on Management of Data, pp. 1429–1446. SIGMOD '19, Association for Computing Machinery, New York, NY, USA (2019). https://doi.org/10.1145/3299869.3319880
8. Han, W.S., Lee, J., Lee, J.H.: Turboiso: towards ultrafast and robust subgraph isomorphism search in large graph databases. In: Proceedings of the 2013 ACM SIGMOD International Conference on Management of Data, pp. 337–348. SIGMOD '13, Association for Computing Machinery, New York, NY, USA (2013).https://doi.org/10.1145/2463676.2465300
9. Hao, K., Yuan, L., Zhang, W.: Distributed hop-constrained st simple path enumeration at billion scale. Proc. VLDB Endow. **15**(2), 169–182 (2021)
10. He, H., Singh, A.K.: Graphs-at-a-time: query language and access methods for graph databases. In: Proceedings of the 2008 ACM SIGMOD International Conference on Management of Data, pp. 405–418. SIGMOD '08, Association for Computing Machinery, New York, NY, USA (2008). https://doi.org/10.1145/1376616.1376660

11. Jin, T., et al.: Circinus: fast redundancy-reduced subgraph matching. Proc. ACM Manag. Data **1**(1) (2023). https://doi.org/10.1145/3588692
12. Jüttner, A., Madarasi, P.: Vf2++—an improved subgraph isomorphism algorithm. Discret. Appl. Math. **242** (2018). https://doi.org/10.1016/j.dam.2018.02.018
13. Kim, H., Choi, Y., Park, K., Lin, X., Hong, S.H., Han, W.S.: Versatile equivalences: Speeding up subgraph query processing and subgraph matching. In: Proceedings of the 2021 International Conference on Management of Data, pp. 925–937. SIGMOD '21, Association for Computing Machinery, New York, NY, USA (2021). https://doi.org/10.1145/3448016.3457265
14. Kim, J., Shin, H., Han, W.S., Hong, S., Chafi, H.: Taming subgraph isomorphism for rdf query processing. Proc. VLDB Endow. **8**(11), 1238–1249 (2015). https://doi.org/10.14778/2809974.2809985
15. Lu, Y., Zhang, Z., Zheng, W.: Bsx: subgraph matching with batch backtracking search. Proc. ACM Manag. Data **3**(1) (2025). https://doi.org/10.1145/3709665
16. Meng, L., et al.: A survey of distributed graph algorithms on massive graphs. ACM Comput. Surv. **57**(2), 1–39 (2024)
17. Meng, L., et al.: Revisiting graph analytics benchmark. Proc. ACM Manag. Data **3**(3), 1–28 (2025)
18. Meng, L., Yuan, L., Lin, X., Li, C., Wang, K., Zhang, W.: Counting butterflies over streaming bipartite graphs with duplicate edges. arXiv preprint arXiv:2412.11488 (2024)
19. Meng, L., Yuan, L., Lin, X., Zhang, W., Zhang, Y.: Triangle counting in hypergraph streams: a complete and practical approach. arXiv preprint arXiv:2509.00674 (2025)
20. Mhedhbi, A., Salihoglu, S.: Optimizing subgraph queries by combining binary and worst-case optimal joins. Proc. VLDB Endow. **12**(11), 1692–1704 (2019). https://doi.org/10.14778/3342263.3342643
21. Shang, H., Zhang, Y., Lin, X., Yu, J.X.: Taming verification hardness: an efficient algorithm for testing subgraph isomorphism. Proc. VLDB Endow. **1**(1), 364–375 (2008). https://doi.org/10.14778/1453856.1453899
22. Sun, S., Luo, Q.: In-memory subgraph matching: an in-depth study. In: Proceedings of the 2020 ACM SIGMOD International Conference on Management of Data, pp. 1083–1098. SIGMOD '20, Association for Computing Machinery, New York, NY, USA (2020). https://doi.org/10.1145/3318464.3380581
23. Sun, S., Sun, X., Che, Y., Luo, Q., He, B.: Rapidmatch: a holistic approach to subgraph query processing. Proc. VLDB Endow. **14**(2), 176–188 (2020). https://doi.org/10.14778/3425879.3425888
24. Yang, R., Zhang, Z., Zheng, W., Yu, J.X.: Fast continuous subgraph matching over streaming graphs via backtracking reduction. Proc. ACM Manag. Data **1**(1) (2023). https://doi.org/10.1145/3588695
25. Yuan, L., Hao, K., Lin, X., Zhang, W.: Batch hop-constrained st simple path query processing in large graphs. In: 2024 IEEE 40th International Conference on Data Engineering (ICDE), pp. 2557–2569. IEEE (2024)
26. Zhang, Z., Lu, Y., Zheng, W., Lin, X.: A comprehensive survey and experimental study of subgraph matching: trends, unbiasedness, and interaction. Proc. ACM Manag. Data **2**(1) (2024). https://doi.org/10.1145/3639315

Graph Neural Network-Driven Ant Colony Optimization for Combinatorial Problems on Graphs

Yijin Xie[1], Tianle Pu[1], Li Zeng[1], Chao Chen[1], and Pengfei Zhang[2(✉)]

[1] Laboratory for Big Data and Decision, National University of Defense Technology, Changsha, China
{xyj0326,put122,zengli24,chenc1997}@nudt.edu.cn

[2] National Key Laboratory of Information Systems Engineering, National University of Defense Technology, Changsha, China
gameboy@nudt.edu.cn

Abstract. Graph Combinatorial Optimization Problems (COPs) are pervasive in domains like logistics and network design, but their NP-hard nature presents a major hurdle for exact solvers. Metaheuristics such as Ant Colony Optimization (ACO) offer a viable alternative, yet their performance critically depends on manually designed, problem-specific heuristics, which limits their adaptability and efficiency. To address this, we propose the Machine Learning-enhanced Ant Colony Optimization (ML-ACO) framework, which integrates Graph Neural Networks (GNNs) and Deep Reinforcement Learning (DRL) to automate heuristic design. ML-ACO leverages a GNN to learn powerful, data-driven heuristics from the graph structure of the problem, which then guide a traditional ACO search procedure. Crucially, our framework incorporates modular designs, including distinct solution construction mechanisms for both sequential (e.g., path planning) and partition-based (e.g., graph partitioning) problems, alongside flexible modules for integrating complex real-world constraints such as path risks and task dependencies. We demonstrate the effectiveness of ML-ACO on the classic Max-Cut problem, where it improves solution quality by 10.9% over traditional ACO. Furthermore, in complex, simulated logistics scenarios, ML-ACO outperforms traditional ACO by 13.5–42.4% across key metrics, effectively handling intricate constraints. Our results validate that ML-ACO provides a versatile and powerful approach for solving complex, real-world COPs on graphs.

Keywords: Graph Combinatorial Optimization · Ant Colony Optimization · Machine Learning · Graph Neural Networks · Reinforcement Learning

1 Introduction

Graph Combinatorial Optimization Problems (COPs), which involve finding an optimal set of nodes or edges from a vast discrete search space defined on a graph,

J. Li et al. (Eds.): APWeb-WAIM 2025, CCIS 2660, pp. 56–66, 2026.
https://doi.org/10.1007/978-981-95-7996-9_6

are fundamental to many real-world applications, including logistics, network design, and resource allocation [7]. The NP-hard nature of most graph COPs renders exact algorithms impractical for large-scale instances. Consequently, metaheuristics like Ant Colony Optimization (ACO) [3] have become indispensable tools for providing high-quality approximate solutions.

However, the efficacy of ACO is intrinsically tied to its heuristic information, which guides the probabilistic decisions of the artificial ants as they traverse the problem graph. This heuristic is traditionally designed manually, a process that demands significant domain expertise and empirical tuning, creating a critical bottleneck. This reliance on handcrafted heuristics makes it exceedingly difficult to adapt to diverse problem structures—such as the difference between a sequential path-finding problem and a partition-based graph problem—or complex, real-world constraints, thereby limiting the algorithm's performance and generalizability.

To overcome this, Neural Combinatorial Optimization (NCO) has emerged, leveraging machine learning to automatically learn heuristics from data [2]. While promising, many existing models are designed as end-to-end solutions that struggle to incorporate complex constraints or generalize to problems beyond their training distribution [1]. More recent hybrid approaches combine learning with traditional solvers, but still primarily target well-structured, sequential problems like the Traveling Salesperson Problem (TSP) [6]. This leaves a significant gap in addressing COPs with more complex graph characteristics, such as partition-based problems (e.g., Max-Cut) or those with intricate constraints found in industrial logistics [11].

To bridge this gap, we propose the Machine Learning-enhanced Ant Colony Optimization (ML-ACO) framework. Our approach enhances the classic ACO algorithm by replacing the manual heuristic component with a neural module. We use a Graph Neural Network (GNN) to learn rich feature representations from problem instances represented as graphs, which a decoder then translates into powerful, data-driven heuristic values [4]. This entire neural module is trained end-to-end using reinforcement learning, guided by the quality of the final solutions produced by the ACO search [10]. By integrating learning into the proven and flexible ACO framework, we retain its strong search capabilities while automating the most challenging part of its design. This hybrid approach of integrating a learnable GNN-based heuristic into the ACO search process is inspired by recent advances in the field, particularly the DeepACO framework [11].

Contributions. Our main contributions are: (1) We propose and implement a modular ML-ACO framework that systematically enhances ACO with GNNs, featuring distinct construction mechanisms for both partition-based and sequential problems. (2) We validate its effectiveness on diverse problems, from the classic Max-Cut to complex simulated scenarios involving vehicle routing with risks (CVRP) and resource allocation with dependencies (MKP). (3) We provide a detailed methodology for adapting the framework, offering practical guidance for researchers.

2 Related Work

Our work builds upon advancements in several fields, primarily Ant Colony Optimization and the emerging area of Neural Combinatorial Optimization, with a strong focus on graph-based problems.

2.1 Ant Colony Optimization

ACO is a population-based metaheuristic inspired by the foraging behavior of real ants [3]. Ants deposit pheromones on their paths, and subsequent ants are probabilistically attracted to paths with higher pheromone concentrations. This mechanism allows the colony to collectively find good solutions to COPs. The success of ACO depends on two main components: pheromone trails (τ) that encode shared experience and heuristic information (η) that provides problem-specific guidance. While powerful, designing an effective heuristic η is often the most challenging part, requiring deep domain knowledge and extensive tuning. Variants like the MAX-MIN Ant System (MMAS) have improved convergence by setting bounds on pheromone values to prevent premature stagnation [8]. Our work replaces the handcrafted η with a learned heuristic, automating this critical design step.

2.2 Neural Combinatorial Optimization

The idea of using machine learning to solve COPs, termed Neural Combinatorial Optimization (NCO), has gained significant traction [2]. Early works focused on end-to-end models using sequence-to-sequence architectures like Pointer Networks [9] to construct solutions autoregressively, which showed success on problems like the TSP [6] but can be difficult to adapt to non-sequential problems or complex constraints. More recently, GNNs have been used to leverage the inherent graph structure of COPs [4,5], as they can learn rich representations to guide search algorithms. Our ML-ACO framework belongs to this hybrid category, integrating GNN-based learning into the robust ACO search procedure.

3 Methodology

Our ML-ACO framework enhances traditional ACO by integrating a learnable neural heuristic module.

3.1 Problem Formulation

A graph COP can be generally defined on a graph $G = (V, E, X_V, X_E)$, where V is the set of nodes, E is the set of edges, and X_V and X_E are node and edge features, respectively. A solution S is a subset or permutation of V or E that satisfies a set of problem-specific constraints. The goal is to find a feasible solution S^* that minimizes or maximizes an objective function $f(S)$.

3.2 ML-ACO Framework Architecture

Our ML-ACO framework adopts the core design of recent neural-enhanced ACO methods [11], which replace the handcrafted heuristic η with a neural heuristic η_θ generated by a GNN. The workflow, depicted in Fig. 1, consists of three key modules: a GNN Encoder, a Heuristic Decoder, and the core ACO search procedure.

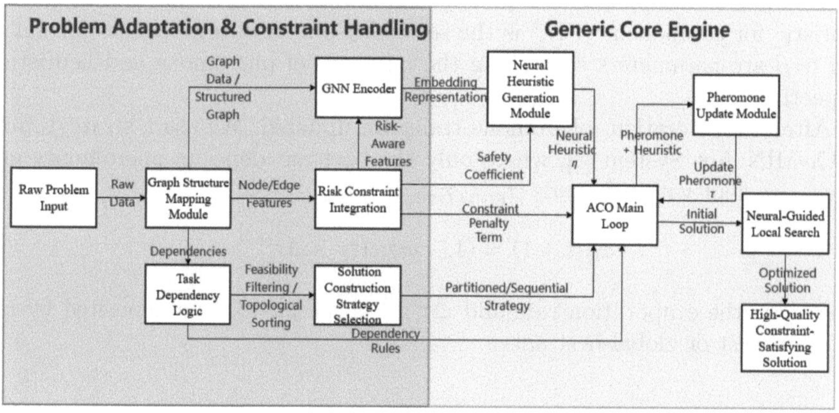

Fig. 1. The ML-ACO framework architecture.

GNN Encoder. We use a GNN to learn expressive feature representations that capture the structural properties of the input problem graph. The GNN iteratively updates the embedding h_v for each node $v \in V$ by aggregating messages from its neighbors. For a GNN with L layers, the update rule for node v at layer $k \in \{1, \ldots, L\}$ is:

$$h_v^{(k)} = \text{COMBINE}^{(k)}\left(h_v^{(k-1)}, \text{AGGREGATE}^{(k)}\left(\{(h_v^{(k-1)}, h_u^{(k-1)}, e_{vu}) | u \in \mathcal{N}(v)\}\right)\right) \quad (1)$$

where $\mathcal{N}(v)$ is the set of neighbors of v, $h_v^{(0)}$ is the initial node feature, and AGGREGATE and COMBINE are learnable functions (e.g., neural networks). After L layers, the final node embeddings $h_v^{(L)}$ encode rich contextual information about each node's role in the overall problem structure.

Heuristic Decoder. The decoder is a small neural network (e.g., a Multi-Layer Perceptron) that maps the final node embeddings $h_v^{(L)}$ to a scalar heuristic value $\eta_\theta(v)$. This value represents the learned desirability of selecting node v during the solution construction process.

$$\eta_\theta(v) = \text{Decoder}(h_v^{(L)}) \quad (2)$$

The entire GNN-Decoder module, parameterized by θ, is trained end-to-end.

ACO with Neural Heuristics. The core of the ACO algorithm remains intact. A colony of m ants concurrently constructs solutions. The transition probability for an ant k at decision point i to select component j is given by:

$$p_{ij}^k(t) = \frac{[\tau_{ij}(t)]^\alpha [\eta_\theta(j)]^\beta}{\sum_{l \in N_i^k} [\tau_{il}(t)]^\alpha [\eta_\theta(l)]^\beta} \tag{3}$$

where $\tau_{ij}(t)$ is the pheromone value at iteration t, $\eta_\theta(j)$ is our learned neural heuristic for component j, N_i^k is the set of feasible next components for ant k, and α, β are parameters controlling the influence of pheromone and heuristics, respectively.

After each iteration, pheromone trails are updated. We use a strategy from MAX-MIN Ant System [8], where only the best ant deposits pheromones and values are kept within bounds $[\tau_{min}, \tau_{max}]$:

$$\tau_{ij}(t+1) = (1-\rho)\tau_{ij}(t) + \Delta\tau_{ij}^{best} \tag{4}$$

where ρ is the evaporation rate and $\Delta\tau_{ij}^{best}$ is the pheromone deposited by the iteration-best or global-best ant.

Training with Reinforcement Learning. Following the training paradigm for similar hybrid models [11], the neural module (GNN encoder and decoder) with parameters θ is trained using policy gradient reinforcement learning. The ACO construction process is viewed as a sequential decision-making process, where the policy is defined by the transition probabilities in Eq. (3). The objective is to maximize the expected quality of the solutions found. Using the REINFORCE algorithm [10], the gradient of the expected objective $J(\theta)$ is:

$$\nabla_\theta J(\theta) = \mathbb{E}_{S \sim p_\theta(S)}[R(S)\nabla_\theta \log p_\theta(S)] \tag{5}$$

where S is a solution constructed by an ant, $p_\theta(S)$ is the probability of constructing that solution, and $R(S)$ is the reward (e.g., the negative of the solution cost). This allows the model to be trained end-to-end by rewarding it for producing high-quality solutions.

3.3 Adaptability for Diverse Problem Structures

A key feature of ML-ACO is its modular implementation, which allows it to adapt to different problem structures. Building on the general applicability shown by frameworks like DeepACO [11], we specifically designed and implemented two distinct solution construction modules.

Sequential Construction. For problems like CVRP or TSP, solutions are constructed as a sequence of nodes. Ants build a path incrementally. The neural heuristic $\eta_\theta(v)$ informs the choice of the next node to add to the path, considering constraints such as vehicle capacity.

Partition-Based Construction. For problems like Max-Cut, the goal is to partition the node set V into two disjoint subsets, S_1 and S_2. Here, ants do not build a sequential path. Instead, for each node $v \in V$, they make a probabilistic decision to assign it to either S_1 or S_2. The heuristic $\eta_\theta(v)$ in this case represents the learned preference for assigning node v to a particular partition.

3.4 Complex Constraint Handling

ML-ACO integrates complex constraints at multiple levels. First, constraint-related information (e.g., node demands, time windows, path risks) is encoded as features for the GNN. Second, hard constraints (e.g., vehicle capacity, task dependencies) are enforced by masking out infeasible actions during solution construction. Finally, soft constraints or objectives (e.g., minimizing risk) are incorporated into the reinforcement learning reward function through reward shaping, guiding the model to produce desirable solutions. For example, the reward for a solution s can be defined as $R(s) = R_{\text{objective}}(s) - \lambda \cdot \text{ConstraintViolation}(s)$.

4 Experiments

We conducted a series of experiments to evaluate the performance and versatility of our ML-ACO framework on various graph COPs.

4.1 Experimental Setup

We generated synthetic datasets for three problems: Max-Cut, a Capacitated Vehicle Routing Problem (CVRP) with path risks, and a Multidimensional Knapsack Problem (MKP) with task dependencies. We compared ML-ACO against traditional ACO and, where applicable, other strong heuristic algorithms. The primary metric was solution quality (e.g., cut value, total cost, total value), supplemented by computation time and solution stability.

For implementation, all models were trained with the Adam optimizer and a REINFORCE loss function. For the CVRP and MKP experiments, we used a 12-layer GNN encoder (32 hidden units, `SiLU`) with a 3-layer MLP decoder, setting the learning rate to 3e-4. For Max-Cut, a 4-layer GCN encoder (64 hidden units) was trained via curriculum learning over 60 epochs (LR = 1e-4, batch size 32). A population of 20 ants was used across all experiments.

4.2 Results on Max-Cut Problem

We first evaluated ML-ACO on the partition-based Max-Cut problem against a suite of well-established baselines (Table 1). Our framework consistently outperformed most baseline algorithms on 50-node instances, achieving an average cut value of **3038.6**, which was **10.9%** higher than traditional ACO. Notably, ML-ACO's performance is highly competitive with the best-performing algorithm,

Tabu Search (TS), while having a comparable runtime to traditional ACO. Its low standard deviation also indicates high solution stability.

Table 1. Max-Cut Algorithm Performance Comparison (N = 50).

Algorithm	Avg. Cut Value	Std. Dev.	Max Cut Value	Avg. Time (s)	% of Best Known
Traditional ACO	2739.9	69.72	2807.0	0.70	88.38%
Local Search (LS)	3012.0	92.54	3105.0	0.35	97.16%
Simulated Annealing (SA)	3011.5	52.52	3077.0	0.01	97.15%
Genetic Algorithm (GA)	2843.9	79.33	2928.0	0.82	91.74%
Greedy	2661.2	90.06	2805.0	<0.01	85.85%
G-W Simplified	2112.6	101.17	2249.0	<0.01	68.15%
Spectral Method	1903.0	203.41	2261.0	<0.01	61.39%
Tabu Search (TS)	3100.0	68.72	3160.0	0.86	100.00%
ML-ACO (Ours)	**3038.6**	**56.84**	**3144.0**	**0.78**	**98.02%**

The learned heuristics successfully guided the search toward better graph partitions, validating the effectiveness of our approach for non-sequential problems. The convergence plot in Fig. 2 shows that ML-ACO not only finds better solutions but also converges much faster than traditional ACO, reaching the baseline's final quality in just a fraction of the iterations. The GNN's ability to capture global graph properties allows it to generate heuristics that avoid getting trapped in local optima, a common issue for purely local or greedy algorithms.

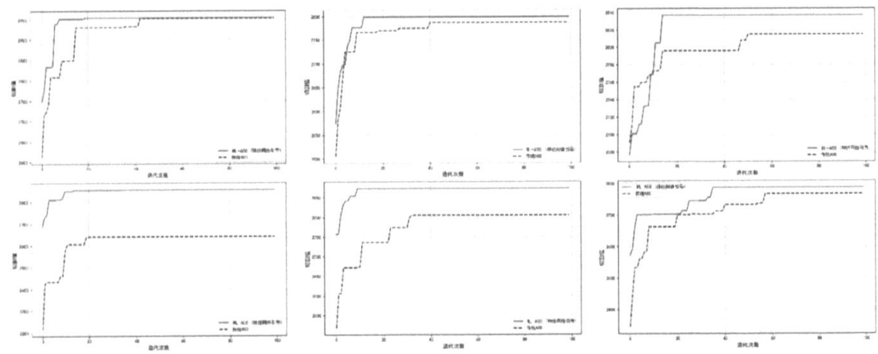

Fig. 2. Convergence of ML-ACO vs. Traditional ACO on Max-Cut.

4.3 Application in Simulated Logistics Scenarios

To test ML-ACO in more complex settings, we applied it to two simulated logistics problems.

Material Distribution (CVRP-Based). This scenario involves optimizing vehicle routes for material distribution while considering both travel distance and path risks (e.g., traversing unsafe zones). As shown in Table 2, ML-ACO significantly outperformed traditional ACO across all tested scales. On 100-node instances, it reduced total transportation costs by **42.4%** and the risk factor by **40.4%**. Unlike traditional heuristics that rely on a fixed weighting of factors, our GNN-based approach learns this trade-off dynamically from the data, leading to more robust solutions as visualized in Fig. 3.

Table 2. Results for Material Distribution (CVRP-based).

Nodes	Algorithm	Avg. Cost	Avg. Risk Score	Avg. Time (s)
20	Traditional ACO	11.27	278.56	8.87
	ML-ACO	**8.12**	**186.64**	**7.35**
50	Traditional ACO	27.01	1402.26	19.46
	ML-ACO	**16.10**	**1002.66**	**16.18**
100	Traditional ACO	58.51	4520.18	35.65
	ML-ACO	**33.71**	**2692.92**	**29.82**

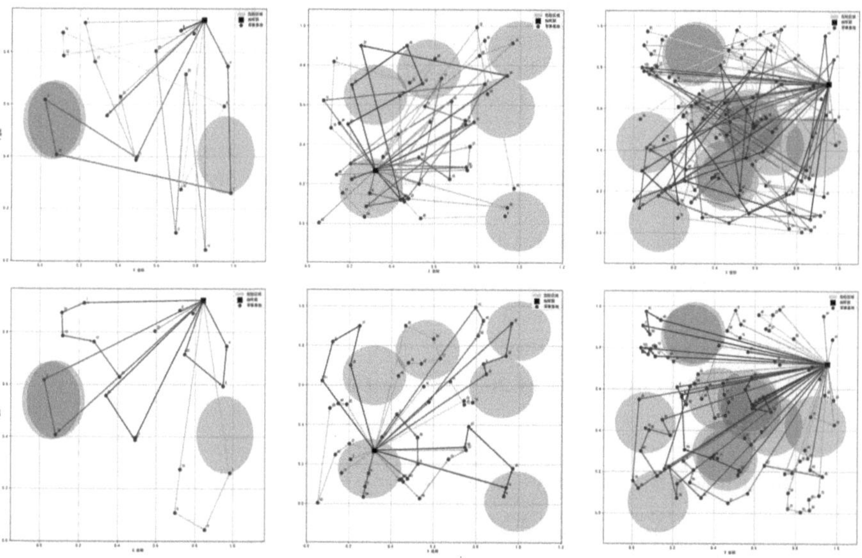

Fig. 3. Visualization of results for Traditional ACO (top) and ML-ACO (bottom) (N = 20, 50, 100).

Resource Allocation (MKP-Based). This scenario involves allocating limited, multi-dimensional resources to a set of tasks with varying values and interdependencies. ML-ACO had to maximize the total value of completed tasks while respecting resource limits and dependencies. As shown in Table 3, on instances with 100 tasks, ML-ACO increased the total expected task value by **13.5%** compared to traditional ACO. The expected value metric, defined as $\sum V_i \times (1 - R_i)$, balances the raw task value (V_i) with its associated risk (R_i). ML-ACO achieved a better balance, leading to a higher quality solution.

Table 3. Results for Resource Allocation (MKP-based) with 100 items.

Algorithm	Total Value	Avg Risk	Expected Value	Util. R1	Util. R2	Util. R3	Util. R4	Util. R5
Trad. ACO	217.51	0.20	43.94	**95.96%**	92.53%	92.52%	87.73%	90.86%
ML-ACO	**225.13**	0.22	**49.88**	92.05%	**94.73%**	**96.12%**	**95.14%**	**96.08%**

4.4 Ablation Study

To investigate the contributions of our framework's core components, we conducted an ablation study on 50-node Max-Cut instances, comparing the full ML-ACO model against versions with key modules deactivated.

The results in Table 4 clearly show the contribution of each module. The neural heuristic alone (ML-ACO (No NLS)) improved the solution quality over Vanilla ACO by 6.5%. The NLS module alone provided a larger boost of 12.9%. However, the full ML-ACO model, which synergistically combines the GNN-guided search with NLS, achieved the best performance, improving the cut value by **14.8%** over the baseline. This demonstrates that the learned heuristics provide a better starting point for the local search to exploit.

Table 4. Ablation Study on Max-Cut (50-node instances).

Configuration	Avg. Cut Value	Improvement over Vanilla ACO
Vanilla ACO	1828.80	-
ACO (NLS-Only)	2065.60	+12.9%
ML-ACO (No NLS)	1947.60	+6.5%
ML-ACO (Full)	**2100.20**	**+14.8%**

4.5 Parameter Sensitivity Analysis

We conducted a sensitivity analysis to evaluate the impact of key hyperparameters on ML-ACO's performance on 50-node Max-Cut instances (see Table 5). The analysis reveals that performance is relatively robust to changes in the pheromone factor (α) and the heuristic factor (β), though optimal values (around 1.0 for both) clearly enhance solution quality. The number of iterations shows a clear point of diminishing returns around 50 iterations, which suggests that the GNN-guided search converges quickly. Overall, the analysis confirms that while hyperparameter tuning can provide gains, the framework is not overly sensitive to their exact values, demonstrating good robustness.

Table 5. Parameter Sensitivity Analysis on Max-Cut (50-node instances).

Parameter	Value	Avg. Cut Value
Ant Quantity	5	2097.20
	10	2102.00
	20	2095.20
	30	**2116.40**
Pheromone Factor (α)	0.5	2074.60
	1.0	**2111.00**
	1.5	2098.80
	2.0	2104.20
Heuristic Factor (β)	0.1	2094.00
	0.5	2102.20
	1.0	**2136.20**
	2.0	2089.80
	5.0	2129.80
Pheromone Retention Rate ($1-\rho$)	0.80	2091.00
	0.85	2057.20
	0.90	**2100.80**
	0.95	2056.20
	0.99	2048.60
Iterations	20	2048.60
	50	**2092.60**
	100	2082.60
	150	2049.00

5 Conclusion

In this paper, we introduced ML-ACO, a versatile framework that enhances ant colony optimization with GNN-based heuristics trained via reinforcement learning. Its key contribution is a modular design with dual construction mechanisms, enabling it to effectively solve both sequential and partition-based graph COPs. Experiments on Max-Cut and complex logistics scenarios show that ML-ACO significantly outperforms traditional ACO by effectively handling diverse problem structures and intricate constraints. This validates ML-ACO as a powerful and adaptable tool for the operations research community. Future work could explore multi-objective optimization and advanced graph representation learning.

Acknowledgments. This work was supported by the National Natural Science Foundation of China under Grant No. 62302513.

Disclosure of Interests. The authors have no competing interests to declare that are relevant to the content of this article.

References

1. Bello, I., Pham, H., Le, Q.V.: Neural combinatorial optimization with reinforcement learning. In: International Conference on Learning Representations (2017)
2. Bengio, Y., Lodi, A., Prouvost, A.: Machine learning for combinatorial optimization: a methodological tour d'horizon. Eur. J. Oper. Res. **290**(2), 405–421 (2021)
3. Dorigo, M., Maniezzo, V., Colorni, A.: Ant system: optimization by a colony of cooperating agents. IEEE Trans. Syst. Man Cybern. B (Cybern.) **26**(1), 29–41 (1996)
4. Khalil, E., Dai, H., Zhang, Y.: Learning combinatorial optimization algorithms over graphs. In: Advances in Neural Information Processing Systems, vol. 30 (2017)
5. Kipf, T.N., Welling, M.: Semi-supervised classification with graph convolutional networks. In: International Conference on Learning Representations (2017)
6. Kool, W., van Hoof, H., Welling, M.: Attention, learn to solve routing problems! In: International Conference on Learning Representations (2019)
7. Korte, B., Vygen, J.: Combinatorial Optimization: Theory and Algorithms. Springer, Cham (2012)
8. Stützle, T., Hoos, H.H.: Max-min ant system. Futur. Gener. Comput. Syst. **16**(8), 889–914 (2000)
9. Vinyals, O., Fortunato, M., Jaitly, N.: Pointer networks. In: Advances in Neural Information Processing Systems (2015)
10. Williams, R.J.: Simple statistical gradient-following algorithms for connectionist reinforcement learning. Mach. Learn. **8**(3), 229–256 (1992)
11. Ye, H., Wang, J., Cao, Z., Liang, H., Li, Y.: Deepaco: neural-enhanced ant systems for combinatorial optimization. CoRR abs/2309.14032 (2023)

CXL Powered Memory Intensive Graph Processing: A Comprehensive Survey

Weibo Chen[1], Yuxuan Lu[1], Dian Ouyang[1(✉)], Guojie Ma[2], and Jinlong Li[1]

[1] Guangzhou University, Guangzhou, China
{chenwb,powerlyx}@e.gzhu.edu.cn, {dian.ouyang,jinlongli}@gzhu.edu.cn
[2] Shanghai University of International Business and Economics, Shanghai, China
mgj0616@suibe.edu.cn

Abstract. Graph processing faces severe memory bottlenecks as data scales exponentially, with traditional methods struggling to handle graphs containing billions of vertices and edges. This survey investigates how Compute Express Link (CXL) technology addresses these limitations in large-scale graph analytics. We systematically analyze memory challenges across core algorithms—including k-edge/k-core decomposition, strongly connected components, and knowledge graph embedding search and demonstrate how CXL provides transformative solutions through expanded memory capacity, tiered memory management, and shared memory support. These mechanisms enable in-memory processing of previously disk-bound graphs, intelligent hot-cold data placement, and efficient parallel processing with reduced communication overhead. However, realizing CXL's potential requires addressing critical challenges in data placement optimization, remote memory access latency, and concurrent access control. This work provides the first systematic survey for understanding CXL's role in next-generation graph processing systems, bridging the gap between emerging memory technologies and graph analytics requirements. Our analysis offers valuable insights for researchers and practitioners seeking to leverage CXL for memory-intensive graph applications.

Keywords: Graph Processing · CXL · Memory Bottleneck

1 Introduction

Graph-structured data has become a fundamental component in a wide range of domains, including social networks [19,48], recommender systems [46], biological networks [10], and knowledge graphs [43]. With the explosive growth of data volume and complexity in these fields, graphs are now used to represent relationships and interactions at unprecedented scales, playing a critical role in enabling advanced analytics and intelligent applications.

Correspondingly, graph processing tasks have grown increasingly diverse and complex. Core computational problems such as subgraph decomposition,

J. Li et al. (Eds.): APWeb-WAIM 2025, CCIS 2660, pp. 67–79, 2026.
https://doi.org/10.1007/978-981-95-7996-9_7

community detection, connectivity analysis, shortest path computation, and knowledge graph embedding underpin many real-world applications. These tasks require algorithms to efficiently traverse, update, or analyze massive and often dynamic graph structures, under stringent performance and scalability constraints.

However, as graph sizes continue to expand, traditional graph processing methods face significant system-level bottlenecks. Most high-performance algorithms depend on in-memory computation, quickly exhausting available DRAM when processing large-scale graphs, which leads to memory limitations [18,40]. When datasets exceed main memory capacity, out-of-core or distributed approaches must be used, introducing additional I/O overhead and bandwidth constraints [37,38]. As a result, memory capacity, I/O throughput, and data locality have become critical barriers to the efficiency and scalability of modern graph analytics systems.

Compute Express Link [1] (CXL) is an emerging industry-standard memory interconnect technology, designed to address the ever-growing demands for flexible, high-performance memory and device connectivity in modern systems [13]. CXL enables processors to efficiently communicate with various types of devices, such as memory expanders and accelerators, offering key features like memory pooling, cache coherency, and support for both low-latency and high-bandwidth access.

Since its initial release, the CXL specification has seen rapid evolution—from version 1.1 to the more advanced 3.0 and beyond [6]. Each generation has introduced significant improvements in scalability, functionality, and system integration, making CXL increasingly attractive for addressing scalability bottlenecks in data-intensive workloads.

CXL devices are classified into three types according to their protocol support and system roles. Type 1 devices, such as network cards, use CXL.cache to maintain cache coherence with host memory and are optimized for frequent atomic operations with minimal on-device cache. Type 2 devices [17], represented by GPUs and FPGAs, combine CXL.cache and CXL.mem to provide both cache-coherent access and high-bandwidth device memory that can be directly managed by the host, reducing data transfer overhead. Type 3 devices [21] act as memory expanders, offering large pools of additional memory to the host via CXL.mem, and are mainly used to enhance memory capacity and bandwidth. This work primarily focuses on Type 3 devices, which are particularly relevant for overcoming memory limitations in large-scale graph processing.

Motivated by these trends, this paper investigates the memory bottlenecks faced by representative graph processing tasks and related workloads, such as approximate nearest neighbor (ANN) search for knowledge graph embeddings. Leveraging recent advances in CXL technology, we analyze both the opportunities for alleviating memory and I/O constraints in large-scale graph analytics and the new challenges introduced by CXL-enabled architectures. Our main contributions are: (1) a comprehensive review of memory and I/O demands in core graph and KG embedding tasks; (2) an analysis of how CXL—especially Type

3 memory expanders—can address existing bottlenecks; and (3) a discussion of open challenges and future research directions at the intersection of CXL and graph processing.

2 Related Works

A few studies have explored the application of CXL technology in graph processing. For example, the GraCXL [49] system, investigating the feasibility and optimization of CXL-like interconnects in distributed graph processing. In addition, several works have analyzed the impact of CXL on large-scale graph computation from a system architecture or prototype perspective. [26]proposed a GPU graph processing approach using CXL-based external memory, providing the first systematic evaluation of how CXL memory affects the performance of GPU-based graph traversal algorithms and demonstrating that microsecond-level latency can deliver performance close to host DRAM.

Although there has been some initial progress, studies that combine CXL and graph processing are still rare. So far, to the best of our knowledge, there is no overall survey or review that summarizes the potential connections between CXL technology and graph processing. Most existing work focuses on specific systems or hardware prototypes, without discussing the broader possibilities.

Existing surveys have analyzed related areas relevant to this work. For example, studies such as [13] have reviewed the benefits and challenges of CXL in database systems, covering topics like buffer pool expansion and elastic memory. Although databases and graph processing are both data-intensive, they differ in structure and system bottlenecks. In the area of knowledge graph (KG) embedding, surveys [36] have summarized key methods—including both fact-based approaches and those leveraging additional information such as types or text—as well as major application scenarios. Building on insights from these related fields, this paper focuses on the opportunities and challenges of applying CXL to graph processing, aiming to address a gap in existing surveys.

3 Traditional Graph Algorithms and Potential Opportunities with CXL

3.1 Existing Graph Algorithms Solution and Problems

K-Edge Connected Components (k-ECC decomposition) refers to the task of, given a large undirected graph, efficiently identifying all maximal k-edge connected components for every possible value of k. Here, a k-edge connected component is a maximal subgraph that remains connected after the removal of any (k-1) edges. k-ECC decomposition is important for applications such as network robustness analysis and community detection, but efficiently computing it on large-scale graphs remains a significant challenge.

As shown in Figure 1, we can observe how different subgraphs exhibit different levels of edge connectivity. The entire graph is a 1-edge connected component

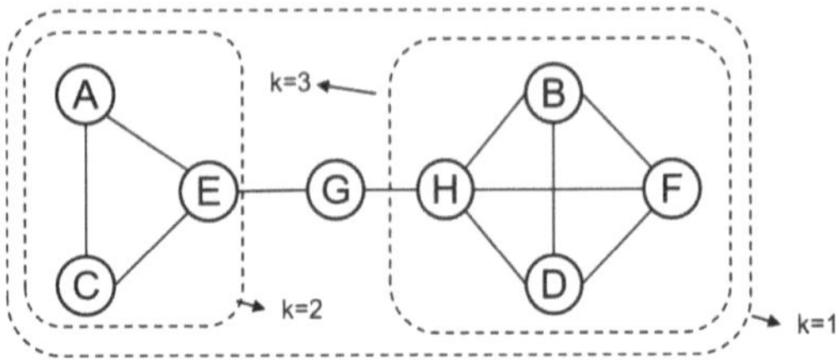

Fig. 1. An example of ECC.

(1-ECC), as it remains connected as long as at least one edge exists between any two parts. On the left, the triangle formed by nodes A, E, and C constitutes a 2-edge connected component (2-ECC), since it remains connected after the removal of any single edge within this subgraph. On the right, the subgraph formed by nodes B, D, H and F is a complete graph (k-4), which is a 3-edge connected component (3-ECC), because it stays connected even after the removal of any two edges. This example demonstrates how k-ECC decomposition reveals the hierarchical edge connectivity structure within a graph.

Most existing k-ECC decomposition algorithms [2,4,41,52] can be categorized as in-memory methods. The basic process of these methods is to run a k-ECC decomposition algorithm independently on the entire graph for each possible value of k. In each run, the algorithm recursively finds all edge cuts with size less than k, removes these edges, and then outputs the remaining maximal subgraphs as the k-ECCs. This process is repeated separately for each k.

However, in-memory methods have significant drawbacks. First, loading entire graphs and data structures into memory results in extremely high memory consumption, becoming unmanageable for large-scale graphs. Second, computations for different k values are independent and cannot share intermediate results, causing redundant computation and resource waste. Consequently, these methods are difficult to scale and impractical for graphs with millions or billions of edges.

Some algorithms focus on I/O optimization like [47]. This work introduces key improvements over traditional in-memory methods by fully leveraging hierarchical nesting structures among different k values in k-ECC decomposition to avoid redundant computations. It incorporates graph reduction and incremental maintenance to reduce processed data size, adopts efficient external memory management to decrease memory dependence, and optimizes the decomposition workflow for large-scale graphs, making it suitable for distributed processing. Although this method greatly reduces memory consumption and improves I/O performance through batch processing and efficient external memory management, I/O efficiency remains a potential bottleneck as graph sizes increase, especially in distributed or extremely large-scale scenarios.

Many graph computation tasks are memory-intensive and face similar challenges with large-scale graphs. For truss decomposition and max-truss computation, one work [18] addresses kmax-truss using semi-external memory with optimized I/O through edge support updates, while another [34] tackles k-truss decomposition via partitioning to reduce random disk accesses.

For k-core decomposition, approaches [5,40] employ semi-external models and sequential operations to minimize I/O overhead through refined scheduling and recursive processing. Strongly connected components (SCC) computation uses sequential batch processing [51] and reduced redundant edge processing [33] to transform random accesses into efficient sequential scans. Triangle listing faces significant I/O challenges, with existing frameworks [14] showing limitations highlighting the need for robust external memory solutions. Random walk algorithms encounter I/O bottlenecks on large graphs, addressed by GraphWalker [37,38] through state-aware block loading and asynchronous updates.

3.2 Potential Benefits of CXL Technology

In the following analysis, we examine three strategic approaches whereby CXL technology can mitigate memory bottlenecks in traditional graph algorithms: memory capacity expansion, tiered memory architectures, and distributed memory pooling.

For Pure In-Memory Graph Algorithms

Opportunities: CXL greatly expands server memory capacity, allowing pure in-memory ECC decomposition algorithms to handle much larger graphs than previously possible. Datasets requiring partitioned or out-of-core processing can now be loaded entirely into memory, simplifying algorithms and improving efficiency. CXL's high bandwidth and flexible expansion provide new optimization opportunities for complex graph decomposition tasks. These advantages extend beyond ECC decomposition to algorithms like k-core, k-truss (max-truss), SCC decomposition, triangle listing, and large-scale random walk/graph sampling that require maintaining massive concurrent states. Previously memory-limited algorithms suffering from frequent disk I/O can now operate entirely in memory, reducing I/O overhead and improving efficiency for large-scale network analytics.

Challenges: Although CXL increases memory capacity, remote CXL memory has inferior access latency and bandwidth compared to local DRAM. For latency- and bandwidth-sensitive in-memory algorithms, especially those requiring frequent state updates like random walk algorithms, frequent remote memory access may degrade performance. Optimizing data placement–ensuring hot data, walk states, and computation-intensive subgraphs reside in low-latency, high-bandwidth local memory–is crucial for maximizing efficiency of both ECC decomposition and random walk algorithms. Careful bandwidth resource management and concurrent access scheduling are important considerations for all pure in-memory large-scale graph algorithms in CXL environments.

For I/O-Optimized Graph Algorithms

Opportunities: I/O-optimized algorithms address memory limitations and I/O bottlenecks in large-scale graph decomposition. CXL's tiered memory architecture (near and far memory) brings further performance improvements. By intelligently utilizing CXL's near memory, frequently accessed data–such as currently processed subgraphs, active vertex/edge sets, auxiliary indexes, and k-PG structures–can be prioritized for near memory placement, while less frequently used data resides in far memory. This strategy benefits ECC, k-core, k-truss, SCC decomposition, triangle listing, and large-scale concurrent random walk algorithms. For example, in GraphWalker, walk states and active blocks can be preferentially loaded into near memory, reducing disk access and improving walk progression efficiency. CXL's tiered memory management has great potential to enhance scalability and practicality of all I/O-intensive graph algorithms in out-of-core environments.

Challenges: Fully utilizing CXL architecture requires efficient data placement and migration mechanisms for active edge sets and candidate k-ECC components in ECC decomposition, candidate node sets in k-core and k-truss decomposition, unassigned nodes in SCC, frequently accessed edge pairs in triangle listing, and walk states in random walk algorithms. Accurately identifying and scheduling hot and cold data, minimizing unnecessary migrations and bandwidth waste, and avoiding performance bottlenecks due to remote memory overload are crucial for improving performance of ECC and all I/O-optimized large-scale graph algorithms in CXL environments. Challenges such as concurrent access and memory consistency must also be addressed to ensure correctness and high availability.

For Parallel/Distributed Graph Processing

Opportunities: CXL's shared memory enables efficient data exchange for multi-CPU or multi-host coordinated processing in large-scale graph decomposition. Different subgraphs can be processed in parallel, directly exchanging intermediate results and synchronizing states through shared memory. This reduces complexity of network communication and data synchronization in traditional distributed environments, improving scalability and throughput. CXL's shared memory benefits extend to parallel algorithms for k-truss, k-core, SCC decomposition, triangle listing, and concurrent random walk analytics. In distributed systems like GraphWalker, walk states and block-level data can be efficiently exchanged via shared memory, enabling load balancing and state synchronization among computation units, significantly enhancing efficiency and scalability.

Challenges: Maintaining data consistency, efficient concurrent access, and coordination among multiple CPUs or nodes becomes challenging in shared memory environments. Efficiently managing decomposition progress, walk states, and intermediate results across multiple computing units, preventing data races and conflicts, and ensuring correctness and high availability are crucial for ECC, k-core, k-truss, SCC, and concurrent random walk algorithms. Additionally, resource isolation and security in multi-tenant scenarios must be considered to ensure stable and reliable system operation.

4 High-Dimensional ANN Search in Knowledge Graphs

4.1 Knowledge Graph Embeddings Overview

Knowledge graph embedding (KGE) methods play a crucial role in enabling high-dimensional ANN search within knowledge graphs(KGs). By mapping entities and relations to dense vectors in a continuous space, KGE methods provide the foundation for similarity-based retrieval and inference tasks.

To address diverse KG modeling requirements, various KGE methods have emerged, broadly categorized into four types:

Translation-based methods model relations as translation operations. TransE [3] and variants [9,20,39] formulate relations as vector translations from head to tail entities. RotatE [29] extends this to complex domains by representing relations as rotations, capturing wider relational patterns.
Semantic matching-based methods model entity-relation compatibility using bilinear or tensor functions. DistMult [42] employs bilinear scoring; ComplEx [31] extends DistMult to complex domains for asymmetric relations; RESCAL [25] uses full-rank matrices for richer semantic interactions.
Graph neural network-based methods leverage GNNs for structurally aware embeddings. R-GCN [27] captures multi-relational patterns via relation-specific transformations, W-GCN [28] adaptively weights neighbors, CompGCN [32] composes entities and relations through dedicated operators, and VR-GCN [45] introduces vectorized relational modeling.
Pre-trained language model-based methods exploit large-scale language models for context-aware embeddings. KG-BERT [44] reformulates triples into natural language for BERT [8] evaluation. ERNIE [50] integrates entity knowledge during pre-training, CoKE [35] uses Transformers for relational dependencies, and BERT-INT [30] models entity description interactions through deep contextualization.

The resulting embedding space possesses several important properties: it is typically of moderate to high dimensionality, supports meaningful similarity computation (e.g., via cosine or Euclidean distance), and enables the measurement of entity or relation proximity. Such properties not only facilitate traditional knowledge graph completion tasks, but also enable efficient similarity-based retrieval in large-scale KGs.

Here we present an example of knowledge graph embedding. We focus on a small school knowledge graph, as shown in Figure 2a. The graph includes entities such as Alice (a student), Dr. Smith and Dr. Lee (teachers), and courses like Python and Database. The edges in the graph represent different relationships, such as "Enrolled_in", "Teaches", and "Major_in". We are interested in understanding the possible relationships among Alice, Dr. Smith, and Dr. Lee, including some connections that might not be explicitly present in the original graph.

To make use of machine learning methods, each entity in the knowledge graph is mapped to a vector in a low-dimensional space through an embedding technique. As shown in Figure 2b, Alice, Dr. Smith, and Dr. Lee are each represented

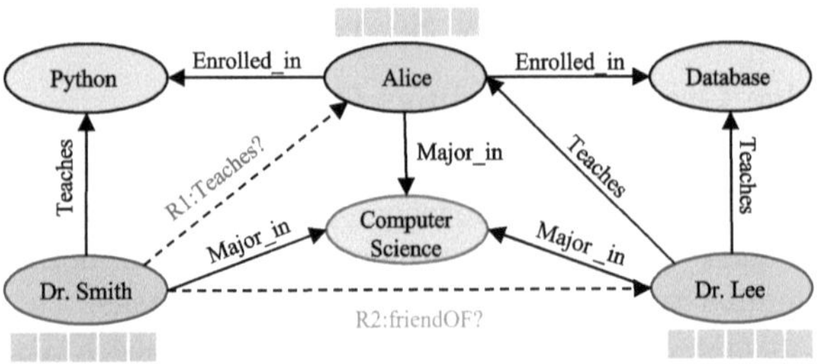

(a) School knowledge graph

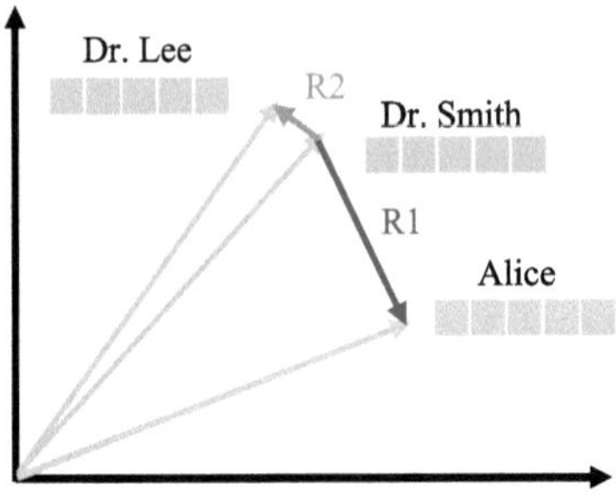

(b) Approximate nearest neighbor search

Fig. 2. An example of knowledge graph embedding.

by their own embedding vectors. The geometric distances between these vectors can reflect how closely related the entities are in the knowledge graph.

Approximate Nearest Neighbor (ANN) algorithms are used to quickly find the most relevant entities for a given query entity in the embedding space. For example, if we use Dr. Smith as the query, ANN search returns Alice and Dr. Lee as the top-k nearest neighbors. This result suggests that Dr. Smith may have strong semantic connections with both Alice and Dr. Lee, even if some relationships are missing in the original graph. For instance, Dr. Smith might be Alice's teacher, or Dr. Smith and Dr. Lee might be friends.

This example shows that knowledge graph embeddings allow us to measure the similarity between entities, and ANN algorithms help us efficiently discover potentially meaningful connections. Together, these techniques enable new knowledge discovery and relation prediction in large-scale knowledge graphs.

As knowledge graphs continue to scale, efficient vector search within the embedding space becomes increasingly crucial. Applications such as entity retrieval, relation matching, and knowledge-driven question answering require real-time retrieval of relevant entities or facts from millions of candidates. Moreover, the integration of large language models with knowledge graphs through retrieval-augmented generation (RAG) frameworks has further increased the demand for fast and accurate retrieval of relevant information from knowledge graphs. Consequently, approximate nearest neighbor (ANN) search algorithms have emerged as key enablers for scalable and responsive knowledge graph systems.

4.2 ANN Search: Existing Solutions and Limitations

Locality Sensitive Hashing (LSH) [7] maps data into multiple hash tables to increase collision probability for similar objects, but requires many tables causing high memory usage and I/O operations. C2LSH [12] uses single hash functions per table and counts collisions, reducing costs. I/O improvements include i-LSH [22] with incremental access and dynamic pruning, and SK-LSH [23] using B^+-tree indexing to minimize random I/O. Hash-based ANN methods require extensive tables for large-scale data, creating memory consumption that necessitates disk storage when insufficient, causing frequent random I/O.

Graph-based ANN algorithms have become popular for large-scale vector search. HNSW [24] constructs multi-layer sparse graphs requiring entire index and data vectors in memory. NSG [11] uses single-layer sparse graphs with strict node degree limits but still requires all data in RAM. DiskANN [16] stores graph index and vectors on SSDs, keeping only compressed vectors and frequent nodes in memory. Through I/O-friendly layout and beam search, DiskANN increases manageable data scale and reduces memory dependence.

While HNSW [24] and NSG [11] deliver excellent performance with sufficient memory, their requirements become bottlenecks for billion-point datasets, forcing distributed systems. DiskANN [16] overcomes memory limitations by minimizing SSD accesses per query but is bounded by SSD I/O bandwidth and concurrency.

4.3 Opportunities with CXL Integration

Existing research [15] shows that CXL technology significantly improves ANNS query performance, scalability, and efficiency. In the subsequent analysis, we investigate three methodological approaches through which CXL integration can address memory constraints in hash-based ANN algorithms for knowledge graph embeddings: hierarchical memory for ANN processing, scalable memory for ANN indices, and distributed memory for ANN systems.

Hierarchical Memory for ANN Processing

Opportunities: In large-scale knowledge graph applications, a small subset of entity embeddings (such as popular concepts, active users, or trending entities) are accessed with much higher frequency, while the majority remain infrequently queried. CXL-enabled tiered memory allows these high-frequency entity embeddings and their associated index structures (e.g., top-layer HNSW neighbors, critical NSG edges, or hot LSH buckets) to be resident in host DRAM, while cold data can be offloaded to CXL-attached DRAM. This approach improves the throughput and latency of popular queries while reducing overall system memory costs.

Challenges: The primary challenge is accurately and dynamically tracking the access patterns and "temperature" of entity embeddings and their related index structures, so they can be effectively migrated between memory tiers. Rapid changes in access patterns or unpredictable query workloads may lead to frequent tier transitions, causing system instability or performance jitter. The tiering strategy must also adapt to diverse workloads—such as batch processing and online serving—while supporting efficient data migration and fault handling.

Scalable Memory for ANN Indices

Opportunities: CXL memory pooling enables multiple retrieval nodes to dynamically allocate and share a large memory space for storing massive knowledge graph entity embeddings and their indexes (such as NSG graphs, HNSW layers, or LSH tables). Nodes can elastically request memory resources according to workload, supporting smooth scaling and efficient resource utilization as the volume or distribution of KG queries changes. This is especially valuable for dynamic and evolving knowledge graph workloads.

Challenges: In a pooled memory environment, simultaneous access to popular entity embeddings or hot index regions by multiple nodes can create bandwidth contention and memory hotspots. Dynamic allocation and reclamation of memory resources can also lead to fragmentation or resource contention, requiring robust scheduling and isolation strategies. The pooling infrastructure must also provide reliable monitoring and management to ensure smooth scaling and high availability during resource adjustments.

Distributed Memory for ANN Systems

Opportunities: Unified global memory access enables all collaborating nodes—whether in a distributed or multi-process setup—to directly share and access a single consistent copy of entity embeddings and index structures. This reduces data duplication and simplifies nearest neighbor retrieval, particularly for large-scale or collaborative KG applications.

Challenges: Concurrent access to shared entity embeddings and index structures introduces challenges in managing data consistency and access contention. In highly concurrent scenarios, frequent queries or updates to hot entities can result in contention, locking, or local bottlenecks. Moreover, robust fault isolation, recovery, and data protection mechanisms are essential in large-scale shared memory environments to ensure system reliability and resilience.

5 Conclusion

Traditional graph processing faces severe memory bottlenecks with large-scale data. This paper investigates how Compute Express Link (CXL) technology addresses these limitations. We analyze memory challenges in core graph algorithms and knowledge graph embeddings. CXL Type 3 memory expanders offer expanded capacity for in-memory processing, tiered memory management, and shared memory support. While CXL enables larger-scale processing and reduces I/O overhead, challenges remain in data placement and latency management. This work provides a framework for understanding CXL potential in graph analytics.

References

1. Compute express link (CXL) (2025). https://computeexpresslink.org/. Accessed 28 May 2025
2. Akiba, T., Iwata, Y., Yoshida, Y.: Linear-time enumeration of maximal k-edge-connected subgraphs in large networks by random contraction. In: Proceedings of the 22nd ACM International Conference on Information & Knowledge Management, pp. 909–918 (2013)
3. Bordes, A., Usunier, N., Garcia-Duran, A., Weston, J., Yakhnenko, O.: Translating embeddings for modeling multi-relational data. In: NIPS, vol. 26 (2013)
4. Chang, L., Yu, J.X., Qin, L., Lin, X., Liu, C., Liang, W.: Efficiently computing k-edge connected components via graph decomposition. In: Proceedings of the 2013 ACM SIGMOD International Conference on Management of Data, pp. 205–216 (2013)
5. Cheng, J., Ke, Y., Chu, S., Özsu, M.T.: Efficient core decomposition in massive networks. In: 2011 IEEE 27th International Conference on Data Engineering, pp. 51–62. IEEE (2011)
6. Das Sharma, D., Blankenship, R., Berger, D.: An introduction to the compute express link (CXL) interconnect. ACM Comput. Surv. **56**(11), 1–37 (2024)
7. Datar, M., Immorlica, N., Indyk, P., Mirrokni, V.S.: Locality-sensitive hashing scheme based on p-stable distributions. In: Proceedings of the Twentieth Annual Symposium on Computational Geometry, pp. 253–262 (2004)
8. Devlin, J., Chang, M.W., Lee, K., Toutanova, K.: BERT: pre-training of deep bidirectional transformers for language understanding. In: NAACL, pp. 4171–4186 (2019)
9. Du, Z., Du, Z., Wang, L.: Open knowledge graph representation learning based on neighbors and semantic affinity. CSCD **52**(12), 2549–2561 (2019)
10. Fratkin, E., Naughton, B.T., Brutlag, D.L., Batzoglou, S.: Motifcut: regulatory motifs finding with maximum density subgraphs. Bioinformatics **22**(14), e150–e157 (2006)
11. Fu, C., Xiang, C., Wang, C., Cai, D.: Fast approximate nearest neighbor search with the navigating spreading-out graph. arXiv preprint arXiv:1707.00143 (2017)
12. Gan, J., Feng, J., Fang, Q., Ng, W.: Locality-sensitive hashing scheme based on dynamic collision counting. In: Proceedings of the 2012 ACM SIGMOD International Conference on Management of Data, pp. 541–552 (2012)
13. Guo, Y., Li, G.: A CXL- powered database system: opportunities and challenges. In: 2024 IEEE 40th International Conference on Data Engineering (ICDE), pp. 5593–5604 (2024). https://doi.org/10.1109/ICDE60146.2024.00447
14. Hu, X., Tao, Y., Chung, C.W.: Massive graph triangulation. In: Proceedings of the 2013 ACM SIGMOD International Conference on Management of Data, pp. 325–336 (2013)
15. Jang, J., Choi, H., Bae, H., Lee, S., Kwon, M., Jung, M.: {CXL-ANNS}:{Software-Hardware} collaborative memory disaggregation and computation for {Billion-Scale} approximate nearest neighbor search. In: 2023 USENIX Annual Technical Conference (USENIX ATC 2023), pp. 585–600 (2023)

16. Jayaram Subramanya, S., Devvrit, F., Simhadri, H.V., Krishnawamy, R., Kadekodi, R.: Diskann: fast accurate billion-point nearest neighbor search on a single node. In: Advances in Neural Information Processing Systems, vol. 32 (2019)
17. Ji, H., et al.: Demystifying a CXL type-2 device: a heterogeneous cooperative computing perspective. In: 2024 57th IEEE/ACM International Symposium on Microarchitecture (MICRO), pp. 1504–1517. IEEE (2024)
18. Jiang, J., Zhang, Q., Li, R.H., Dai, Q., Wang, G.: I/O efficient max-truss computation in large static and dynamic graphs. In: 2024 IEEE 40th International Conference on Data Engineering (ICDE), pp. 3217–3229. IEEE (2024)
19. Li, Y., Fan, J., Wang, Y., Tan, K.L.: Influence maximization on social graphs: a survey. IEEE Trans. Knowl. Data Eng. **30**(10), 1852–1872 (2018)
20. Lin, Y., Liu, Z., Sun, M., Liu, Y., Zhu, X.: Learning entity and relation embeddings for knowledge graph completion. In: AAAI, vol. 29 (2015)
21. Liu, J., et al.: Systematic CXL memory characterization and performance analysis at scale. In: Proceedings of the 30th ACM International Conference on Architectural Support for Programming Languages and Operating Systems, vol. 2, pp. 1203–1217 (2025)
22. Liu, W., Wang, H., Zhang, Y., Wang, W., Qin, L.: I-lsh: I/O efficient c-approximate nearest neighbor search in high-dimensional space. In: 2019 IEEE 35th International Conference on Data Engineering (ICDE), pp. 1670–1673. IEEE (2019)
23. Liu, Y., Cui, J., Huang, Z., Li, H., Shen, H.T.: SK-LSH: an efficient index structure for approximate nearest neighbor search. Proc. VLDB Endow. **7**(9), 745–756 (2014)
24. Malkov, Y.A., Yashunin, D.A.: Efficient and robust approximate nearest neighbor search using hierarchical navigable small world graphs. IEEE Trans. Pattern Anal. Mach. Intell. **42**(4), 824–836 (2018)
25. Nickel, M., Tresp, V., Kriegel, H.P., et al.: A three-way model for collective learning on multi-relational data. In: ICML, vol. 11, pp. 3104482–3104584 (2011)
26. Sano, S., et al.: GPU graph processing on cxl-based microsecond-latency external memory. In: Proceedings of the SC 2023 Workshops of the International Conference on High Performance Computing, Network, Storage, and Analysis, pp. 962–972 (2023)
27. Schlichtkrull, M., Kipf, T.N., Bloem, P., Van Den Berg, R., Titov, I., Welling, M.: Modeling relational data with graph convolutional networks. In: ESWC, pp. 593–607. Springer, Cham (2018)
28. Shang, C., Tang, Y., Huang, J., Bi, J., He, X., Zhou, B.: End-to-end structure-aware convolutional networks for knowledge base completion. In: AAAI, vol. 33, pp. 3060–3067 (2019)
29. Sun, Z., Deng, Z.H., Nie, J.Y., Tang, J.: Rotate: knowledge graph embedding by relational rotation in complex space. arXiv preprint arXiv:1902.10197 (2019)
30. Tang, X., Zhang, J., Chen, B., Yang, Y., Chen, H., Li, C.: BERT-int: a BERT-based interaction model for knowledge graph alignment. In: IJCAI, pp. 3174–3180 (2020)
31. Trouillon, T., Welbl, J., Riedel, S., Gaussier, É., Bouchard, G.: Complex embeddings for simple link prediction. In: ICML, pp. 2071–2080. PMLR (2016)
32. Vashishth, S., Sanyal, S., Nitin, V., Talukdar, P.: Composition-based multi-relational graph convolutional networks. arXiv preprint arXiv:1911.03082 (2019)
33. Wan, X., Wang, H.: Efficient semi-external SCC computation. IEEE Trans. Knowl. Data Eng. **35**(4), 3794–3807 (2021)
34. Wang, J., Cheng, J.: Truss decomposition in massive networks. arXiv preprint arXiv:1205.6693 (2012)

35. Wang, Q., et al.: Coke: contextualized knowledge graph embedding. arXiv preprint arXiv:1911.02168 (2019)
36. Wang, Q., Mao, Z., Wang, B., Guo, L.: Knowledge graph embedding: a survey of approaches and applications. IEEE Trans. Knowl. Data Eng. **29**(12), 2724–2743 (2017). https://doi.org/10.1109/TKDE.2017.2754499
37. Wang, R., Li, Y., Xie, H., Xu, Y., Lui, J.C.: {GraphWalker}: an {I/O-Efficient} and {Resource-Friendly} graph analytic system for fast and scalable random walks. In: 2020 USENIX Annual Technical Conference (USENIX ATC 2020), pp. 559–571 (2020)
38. Wang, R., Li, Y., Xu, Y., Xie, H., Lui, J.C., He, S.: Toward fast and scalable random walks over disk-resident graphs via efficient I/O management. ACM Trans. Storage **18**(4), 1–33 (2022)
39. Wang, Z., Zhang, J., Feng, J., Chen, Z.: Knowledge graph embedding by translating on hyperplanes. In: AAAI, vol. 28 (2014)
40. Wen, D., Qin, L., Zhang, Y., Lin, X., Yu, J.X.: I/O efficient core graph decomposition: application to degeneracy ordering. IEEE Trans. Knowl. Data Eng. **31**(1), 75–90 (2018)
41. Yan, X., Zhou, X.J., Han, J.: Mining closed relational graphs with connectivity constraints. In: Proceedings of the Eleventh ACM SIGKDD International Conference on Knowledge Discovery in Data Mining, pp. 324–333 (2005)
42. Yang, B., Yih, W., He, X., Gao, J., Deng, L.: Embedding entities and relations for learning and inference in knowledge bases. arXiv preprint arXiv:1412.6575 (2014)
43. Yang, S., Han, F., Wu, Y., Yan, X.: Fast top-k search in knowledge graphs. In: 2016 IEEE 32nd International Conference on Data Engineering (ICDE), pp. 990–1001. IEEE (2016)
44. Yao, L., Mao, C., Luo, Y.: KG-BERT: BERT for knowledge graph completion. arXiv preprint arXiv:1909.03193 (2019)
45. Ye, R., Li, X., Fang, Y., Zang, H., Wang, M.: A vectorized relational graph convolutional network for multi-relational network alignment. In: IJCAI, pp. 4135–4141 (2019)
46. Yu, J., Yin, H., Xia, X., Chen, T., Li, J., Huang, Z.: Self-supervised learning for recommender systems: a survey. IEEE Trans. Knowl. Data Eng. **36**(1), 335–355 (2023)
47. Yuan, L., Qin, L., Lin, X., Chang, L., Zhang, W.: I/O efficient ECC graph decomposition via graph reduction. Proc. VLDB Endow. (2016)
48. Zhang, N., Ye, Y., Lian, X., Chen, M.: Top-l most influential community detection over social networks. In: 2024 IEEE 40th International Conference on Data Engineering (ICDE), pp. 5767–5779. IEEE (2024)
49. Zhang, X., Chang, Y., Lu, T., Zhang, K., Chen, M.: Rethinking design paradigm of graph processing system with a cxl-like memory semantic fabric. In: 2023 IEEE/ACM 23rd International Symposium on Cluster, Cloud and Internet Computing (CCGrid), pp. 25–35. IEEE (2023)
50. Zhang, Z., Han, X., Liu, Z., Jiang, X., Sun, M., Liu, Q.: ERNIE: enhanced language representation with informative entities. In: ACL (2019)
51. Zhang, Z., Yu, J.X., Qin, L., Chang, L., Lin, X.: I/O efficient: computing SCCs in massive graphs. In: Proceedings of the 2013 ACM SIGMOD International Conference on Management of Data, pp. 181–192 (2013)
52. Zhou, R., Liu, C., Yu, J.X., Liang, W., Chen, B., Li, J.: Finding maximal k-edge-connected subgraphs from a large graph. In: Proceedings of the 15th International Conference on Extending Database Technology, pp. 480–491 (2012)

Design and Implementation of a Graph Neural Network Framework for Crop Variety Classification Based on Complex Network Analysis

Zihan Shang[1], Bochang Jiang[2,3], and Yingkui Wang[2,3](✉)

[1] Nanjing Agriculture University, Nanjing 210000, China
shangzihan@stu.njau.edu.cn
[2] Tianjin Renai College, Tianjin 301636, China
[3] Tianjin University, Tianjin 301636, China
ykwang@tju.edu.cn

Abstract. With the rapid advancement of smart agriculture and precision breeding, accurate classification of crop varieties has become a fundamental prerequisite for enhancing agricultural productivity and ensuring food security. This study develops a multidimensional complex network model that integrates multi-modal information, including crop genotypes, phenotypic traits, and environmental adaptability. By employing a combined approach of node classification and graph classification through graph neural networks, this research achieves precise classification of crop varieties. The innovation lies in the deep fusion of heterogeneous data within the complex network structure and the incorporation of agricultural knowledge graphs to improve model interpretability and generalization. This methodology not only strengthens the characterization and understanding of intricate inter-varietal relationships but also provides scientific support for breeding decision-making, field management, and pest control. Consequently, it advances smart agriculture towards greater intelligence and sustainability, demonstrating broad application potential and practical value.

Keywords: Smart agriculture · Crop variety classification · Complex network analysis

1 Introduction

The rapid advancements in smart agriculture and precision breeding have made accurate classification of crop varieties a fundamental prerequisite for enhancing agricultural productivity and ensuring global food security [1]. Traditional crop breeding methods, however, are often constrained by lengthy field cycles and the delayed selection of later generations, which significantly limits the speed of new variety development [2]. With the continuous growth of the global population, projected to reach between 8.5 to 12 billion by 2100, and increasing

J. Li et al. (Eds.): APWeb-WAIM 2025, CCIS 2660, pp. 80–92, 2026.
https://doi.org/10.1007/978-981-95-7996-9_8

demands on food supply, particularly in developing countries, there is an urgent need for faster and more efficient breeding strategies to meet these demands without expanding arable land [3]. Furthermore, climate change poses adverse environmental conditions that exacerbate these challenges, making resilient and high-yielding varieties crucial for global food security [4]. In this context, precision breeding, which integrates genomics and traditional breeding techniques to introduce desirable traits, is gaining prominence as a key practice for future agricultural advancements [5]. Precise and accurate crop classification is thus vital, serving as the cornerstone of modern precision agriculture by enabling farmers to monitor crop health, predict yields, and optimize resource allocation [5]. This foundational capability also underpins effective agricultural policy-making and promotes sustainable agricultural development [6]. However, achieving high-precision crop classification remains a complex endeavor, fraught with challenges such as the inherent variability in plant growth patterns, spatial and temporal complexities in data, high data acquisition costs, and class imbalance issues [7].

1.1 The Criticality of Crop Classification and Agricultural Productivity

Accurate and timely crop classification is paramount for improving crop management and fostering sustainable agricultural development [8]. Precise crop classification provides foundational data for effective agricultural production management and food security strategies [9]. For instance, the creation of detailed crop maps through classification directly supports yield prediction and informed land-use planning [10]. Furthermore, selecting the appropriate crop variety is critical for effective disease detection and management, as disease-resistant varieties can significantly reduce the reliance on chemical interventions, thereby contributing to sustainable practices. Crop classification also aids in understanding the specific needs of different crops, including optimal soil and moisture conditions, as well as their adaptability to varying environments. Comprehensive and routinely updated information on crop acreage and spatial distribution is a fundamental requirement for agricultural monitoring. The ability to accurately identify and classify crops helps address global food challenges by improving production efficiency, ensuring food supply, and supporting the livelihoods of millions of farmers, particularly given that rice, for example, feeds over half of the world's population [11].

1.2 Limitations and Challenges of Existing Methodologies

Despite the undeniable importance of crop classification, traditional methods are often labor-intensive, time-consuming, and imprecise, leading to suboptimal agricultural outcomes [12]. Existing research indicates that precise crop classification is a challenging task that goes beyond the capabilities of simple land use/land cover techniques [13]. Significant variability in crop features, along with the presence of mixed pixels where signals stem from several land uses, makes accurate differentiation of crop varieties from remote sensing data particularly difficult. Moreover, the accuracy of early crop yield predictions is impacted by unpredictable weather, soil variability, and pest outbreaks. Even advanced deep learning models can exhibit limitations when confronted with real-world complexities, such as intricate backgrounds, multi-pest infestations, inter-class similarity, and high intra-class variance [14]. Challenges also include the dependency on costly and sparse ground-truth data, issues with cloud cover and seasonal variability in satellite imagery, and the need for fresh ground data and feature engineering for unexplored geographies or time periods [15]. These limitations underscore the pressing need for innovative methodologies that can overcome the shortcomings of traditional techniques and enhance the accuracy and robustness of crop classification.

1.3 Innovations and Contributions of This Research

This study introduces a novel Graph Neural Network (GNN)-based framework for crop variety classification, designed to address the aforementioned limitations and provide advanced solutions for smart agriculture and precision breeding. Our innovation primarily lies in two key areas. First, we achieve a deep fusion of heterogeneous data within a complex network structure, effectively integrating multi-modal information such as crop genotypes, phenotypic traits, and environmental adaptability [16]. This multi-modal data fusion, leveraging diverse data sources like synthetic aperture radar (SAR), optical remote sensing, and meteorological measurements, has been empirically shown to significantly enhance crop yield prediction accuracy, with improvements observed over previous state-of-the-art models. Second, this research incorporates agricultural knowledge graphs (KGs) into the model to enhance interpretability and generalization capabilities [17]. Knowledge graphs provide a structured representation of agricultural data, which effectively supports agricultural management and improves the precision of ontological query retrieval. By employing a combined approach of node classification and graph classification through GNNs, this research aims to achieve precise classification of crop varieties and deepen the understanding of intricate inter-varietal relationships [18]. The proposed methodology contributes a flexible, accurate, and interpretable deep learning framework capable of effectively processing multi-channel and time-series data, demonstrating superior performance across various environments. This framework provides robust scientific support for breeding decision-making, field management, and pest control, including timely intervention against pests to reduce crop losses and optimize resource utilization [19]. Ultimately, this advancements will drive smart agriculture toward greater intelligence and sustainability, showcasing broad application potential and practical value.

2 Related Works

Recently, Graph Neural Networks (GNNs) [20] have gained prominence for their ability to model relationships between data points, making them particularly well-suited for applications in precision agriculture where interconnectedness is crucial. GNNs can efficiently capture both local and global information from plants, considering their inherent connections based on spatial proximity and shared genotypes. For example, Agri-GNN, a novel Genotypic-Topological Graph Neural Network Framework built on GraphSAGE [21], has achieved an R^2 of 0.876 in yield predictions for farming fields by constructing graphs where farming plots are nodes and edges are based on spatial and genotypic similarity [22]. GNNs have also been applied to critical tasks such as soil moisture prediction, where they capture the dynamic dependency of soil moisture values in nearby locations over time by learning temporal graph structures [23–25]. Furthermore, GNNs have shown promise in crop variety suitability evaluation [26–28] and maize phenotype classification [29–31], demonstrating their ability to learn complex relationships among phenotypes and classify them to predict better yields.

The integration of knowledge graphs (KGs) with GNNs is another significant development, enhancing model interpretability and generalization [32]. KGs effectively structure agricultural data and support agricultural management, allowing for precise semantic analysis and improved query retrieval. Studies have proposed frameworks that combine BERT-GPT and Graph Recurrent Q Learning Network (GRQLN) to create dynamic and efficient ontology tools for agriculture, improving precision, accuracy, recall, and AUC while reducing delay in ontological query retrieval [33]. Knowledge distillation, for instance, has been shown to improve model interpretability by transferring class-similarity information from teacher to student models [34]. This fusion of GNNs with KGs addresses challenges in real-world applications by accommodating noise, automating dataset denoising, and providing interpretable modeling, as demonstrated in biomedical knowledge graphs for link prediction [35–37]. The development of specialized agricultural KGs, such as BDAKG [38], further exemplifies efforts to represent agricultural data semantically and multidimensionally, contributing to a more comprehensive understanding of agricultural processes and supporting informed decision-making.

3 Design and Implementation of the Proposed Framework

Building upon the theoretical foundations and research motivations discussed in the previous sections, this section details the *design and implementation* of the proposed graph neural network (GNN) framework for crop variety classification. The objective is to provide a systematic description of how heterogeneous agricultural data are transformed into graph representations, enriched with domain knowledge, and subsequently processed by advanced GNNs to achieve accurate and interpretable classifications (Fig. 1).

3.1 Multi-modal Data Processing

The input data consist of *genotypic*, *phenotypic*, and *environmental* modalities, each requiring specific preprocessing strategies:

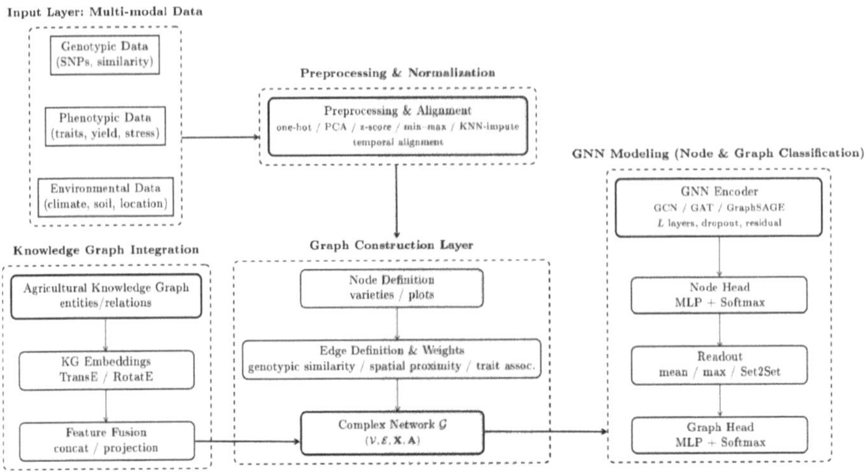

Fig. 1. End-to-end architecture of the proposed GNN framework for crop variety classification. Multi-modal inputs are preprocessed and mapped into a complex network $\mathcal{G}$; KG embeddings enrich node features. A stack of GNN layers produces node embeddings used by both a node-level head and a graph-level head (via readout). Training combines cross-entropy with knowledge consistency and regularization, optimized by Adam with learning-rate scheduling.

- **Genotypic data:** Single nucleotide polymorphism (SNP) markers are encoded via one-hot representation, followed by principal component analysis (PCA) to reduce dimensionality and alleviate multicollinearity.
- **Phenotypic data:** Measured attributes (e.g., plant height, leaf area, yield) are standardized using z-score normalization to ensure commensurable scales and balanced contributions.
- **Environmental data:** Climate and edaphic variables (e.g., temperature, rainfall, soil fertility) are min–max normalized and temporally aligned to account for seasonal effects.

Missing values are imputed via k-nearest neighbor (KNN) interpolation, and heterogeneous features are then aligned into a unified feature space suitable for downstream graph construction.

3.2 Complex Network Construction

We formalize the agricultural ecosystem as a heterogeneous complex network

$$\mathcal{G} = (\mathcal{V}, \mathcal{E}, \mathbf{X}, \mathbf{A}),$$

where $\mathcal{V}$ denotes nodes, $\mathcal{E}$ edges, $\mathbf{X}$ node feature matrix, and $\mathbf{A}$ the weighted adjacency matrix. The design choices are:

- **Nodes ($\mathcal{V}$):** crop varieties, farming plots, or environmental entities.

- **Edges ($\mathcal{E}$):** genotypic similarity, spatial adjacency, and phenotypic associations.
- **Edge weights:** strengths computed via genetic distance metrics and trait-based correlations (e.g., Pearson correlation), optionally thresholded or sparsified to enhance computational efficiency.

This construction captures multi-level interdependencies, enabling exploitation of both local neighborhoods and global structural patterns.

3.3 Knowledge Graph Integration

To enhance interpretability and generalization, we incorporate an agricultural knowledge graph (KG) that encodes relationships among varieties, traits, stress resistances, and environmental suitability.

- **Schema and triples:** domain entities and relations are represented as triples, enabling semantic consistency across data sources.
- **KG embeddings:** methods such as TransE/RotatE learn low-dimensional representations of entities and relations.
- **Fusion strategy:** KG embeddings are concatenated or projected and combined with node features $\mathbf{X}$, so the GNN can leverage structured agronomic priors alongside raw data.

This integration encourages predictions that are aligned with domain knowledge, improving trustworthiness and out-of-distribution robustness.

3.4 Graph Neural Network Architecture

We adopt a hybrid GNN design that supports both node-level and graph-level tasks.

Node-level Modeling. We stack message-passing layers such as GCN, GAT, and GraphSAGE. A GCN layer is written as

$$H^{(l+1)} = \sigma\left(\tilde{D}^{-\frac{1}{2}}\,\tilde{A}\,\tilde{D}^{-\frac{1}{2}}H^{(l)}W^{(l)}\right), \tag{1}$$

where $H^{(l)}$ are hidden features at layer l, $\tilde{A}$ is the adjacency with self-loops, $\tilde{D}$ is its degree matrix, $W^{(l)}$ trainable weights, and $\sigma(\cdot)$ a nonlinearity. GAT introduces attention-based neighbor weighting; GraphSAGE supports inductive inference on unseen nodes/varieties.

Graph-level Representation. We apply permutation-invariant readouts (mean/max pooling, Set2Set) to aggregate node embeddings into graph-level vectors, subsequently classified by a multi-layer perceptron (MLP) for variety recognition.

3.5 Model Training and Optimization

The training objective balances accuracy with robustness and interpretability:

- **Losses:** Cross-entropy for classification; a knowledge-consistency term to align GNN and KG-induced representations; optionally focal loss to mitigate class imbalance.
- **Optimization:** Adam with learning-rate scheduling and gradient clipping; early stopping on validation metrics.
- **Regularization:** Dropout, L_2 weight decay, and edge sparsification/feature dropout for graph-specific robustness.

4 Experimental Design and Results

This section presents the experimental protocol, datasets, baselines, evaluation metrics, and results of the proposed framework. We emphasize reproducibility and conduct extensive ablations and robustness analyses to characterize generalization across data modalities and environments.

4.1 Datasets

To evaluate the proposed framework, we employ two complementary multi-source agricultural datasets that jointly cover *genotypic, phenotypic, and environmental signals* (Dataset A) as well as *remote sensing imagery with field records* (Dataset B). These datasets support both node-level and graph-level classification tasks and enable assessment under spatial and temporal shifts.

Dataset A: Maize Genotype–Phenotype Graph. This dataset is derived from the our previous project. It contains extensive genotypic, phenotypic, and environmental records collected across multiple growing seasons and trial sites. *Modalities:* (i) **Genotypic**—dense SNP markers (encoded and reduced via PCA); (ii) **Phenotypic**—plant height, grain yield, leaf area, and stress tolerance traits; (iii) **Environmental**—site-level soil, rainfall, and temperature. *Graph construction:* nodes correspond to maize varieties/plots; edges encode genotypic similarity, trait-based correlation (Pearson), and spatial adjacency across trial sites. This dataset supports variety-level classification and probes how the model leverages genotype–phenotype–environment interactions.

Dataset B: Multi-modal CropType. This dataset aggregates multi-temporal Sentinel-2 optical and Sentinel-1 SAR imagery with ground-truth crop-type labels from field surveys across multiple regions. *Modalities:* (i) **Imagery**—multi-season optical and radar time series (10 m spatial resolution); (ii) **Labels**—field-level crop-type classes. *Graph construction:* nodes are field polygons; edges capture spatial contiguity and/or spectral similarity between neighboring fields. We use temporal holdout (train/test across different years) and spatial holdout (leave-region-out) to evaluate robustness and generalization (Table 1).

Table 1. Dataset statistics. Counts reflect the preprocessed splits used in our experiments.

Dataset	#Nodes	#Edges	Avg. degree	#Classes	Time span
A: Maize	~5,000 (RILs/varieties)	~45,000 (geno+pheno+spatial)	18.2	25 families (varieties)	Multi-season, multi-site
B: Multi-modal CropType	~40,000 (fields)	~120,000 (spatial+spectral)	6.0	8–10 crop types	2017–2020 (multi-country)

4.2 Experimental Setup

Tasks. We report (i) *node classification* (variety labels at node level) and (ii) *graph classification* (plot-level or region-level aggregation).

Evaluation Protocol. We use stratified k-fold CV for in-domain evaluation and leave-one-region/year-out for out-of-domain testing. Hyperparameters are tuned on validation folds only.

Implementation Details

- **Backbone:** GCN/GAT/GraphSAGE (depth L, hidden size d), residual + dropout.
- **Fusion:** KG embeddings concatenated/projection with node features.
- **Training:** Adam (lr, weight decay), batch size, max epochs, early stopping.
- **Hardware/Software:** NVIDIA H20, PyTorch 2.0, PyG, CUDA 11.8.

Metrics. For multi-class classification we report Accuracy, Macro-F1, and AUC (one-vs-rest).

$$Accuracy = \frac{1}{N}\sum_{i=1}^{N} \mathbb{1}\{\hat{y}_i = y_i\}, \quad MacroF1 = \frac{1}{C}\sum_{c=1}^{C} \frac{2\,Prec_c\,Rec_c}{Prec_c + Rec_c}.$$

Statistical Testing. We perform paired t-tests over k folds between our method and the strongest baseline; significance levels $^{\dagger}p < 0.05$, $^{\ddagger}p < 0.01$.

4.3 Baselines and Comparative Methods

We compare against classical ML, CNN-based, and graph-based approaches.

Traditional ML. SVM (RBF), Random Forest, XGBoost with engineered features (genotype PCs, trait summaries, environmental descriptors).

Deep CNNs. ResNet50/DenseNet121 on imagery streams with temporal pooling; optional multimodal late fusion.

Graph Baselines. GCN, GAT, GraphSAGE without KG; *Agri-GNN*-style topology; *KG-only* (no graph) ablation via MLP on KG embeddings.

Table 2. Baselines and our methods.

Method	Uses Graph	Uses KG	Temporal Split
SVM/RF/XGB	No	No	Optional
ResNet50/DenseNet121	No	No	Yes
GCN/GAT/SAGE	Yes	No	Optional
Ours (w/o KG)	Yes	No	Yes
Ours (Full)	Yes	Yes	Yes

Table 3. Main results (mean ± std over k folds). Best in **bold**. $^{\dagger}/^{\ddagger}$ denote significant gains over the best non-ours baseline.

Method	Acc (%)	Macro-F1 (%)	AUC (%)	OOD Macro-F1 (%)
SVM	71.2 ± 1.8	68.5 ± 2.1	75.0 ± 1.6	60.3 ± 2.7
RF	74.6 ± 1.5	71.9 ± 1.8	78.4 ± 1.9	62.7 ± 2.3
ResNet50	81.5 ± 1.7	79.2 ± 1.5	84.8 ± 1.4	69.8 ± 2.0
GCN	83.7 ± 1.4	81.1 ± 1.6	86.3 ± 1.3	71.5 ± 2.1
GAT	85.2 ± 1.6	82.9 ± 1.4	87.6 ± 1.2	73.4 ± 1.8
GraphSAGE	86.0 ± 1.5	83.5 ± 1.3	88.2 ± 1.1	74.6 ± 2.0
Ours (w/o KG)	88.4 ± 1.2	86.1 ± 1.5	90.5 ± 1.0	78.2 ± 1.7
Ours (Full)	$\mathbf{91.7 \pm 1.0}^{\ddagger}$	$\mathbf{89.8 \pm 1.2}^{\ddagger}$	$\mathbf{93.4 \pm 0.9}^{\dagger}$	$\mathbf{82.6 \pm 1.5}^{\ddagger}$

Our Variants. **Ours (Full)**: complex network + KG fusion + hybrid heads; **Ours w/o KG**: remove KG fusion; **Ours w/o spatial**: remove spatial edges; **Ours w/o genotypic**: remove genotype edges (Table 2).

4.4 Main Results and Discussion

Overall Performance. Table 3 and Table 4 summarize node-level and graph-level results. Our framework achieves state-of-the-art performance across datasets, with the largest gains on out-of-domain splits, indicating improved generalization.

4.5 Ablation Studies

We isolate the contribution of each component.

Edge-type Contribution. Removing spatial or genotypic edges degrades Macro-F1, confirming complementary neighborhood signals.

KG Fusion. Omitting KG embeddings reduces OOD performance, evidencing semantic prior benefits.

Readout and Heads. Comparing mean/max/Set2Set and single-head vs. hybrid heads demonstrates the advantage of graph-level supervision (Table 5).

Table 4. Graph-level results (mean ± std over k folds). Best in **bold**. $^\dagger/^\ddagger$ denote significant gains over the best non-ours baseline.

Method	Acc (%)	Macro-F1 (%)	AUC (%)	OOD Macro-F1 (%)
GCN	84.5 ± 1.4	82.3 ± 1.6	87.2 ± 1.3	72.9 ± 1.9
GAT	85.6 ± 1.5	83.7 ± 1.4	88.1 ± 1.2	74.2 ± 1.8
GraphSAGE	86.6 ± 1.4	84.6 ± 1.3	89.0 ± 1.1	75.6 ± 1.9
Ours (w/o KG)	89.5 ± 1.2	87.3 ± 1.4	91.2 ± 1.0	79.1 ± 1.6
Ours (Full)	$\mathbf{92.4 \pm 1.0}^\ddagger$	$\mathbf{90.6 \pm 1.2}^\ddagger$	$\mathbf{94.3 \pm 0.9}^\dagger$	$\mathbf{83.5 \pm 1.5}^\ddagger$

Table 5. Ablations on Dataset A. Removing KG or specific edge types degrades performance, confirming the complementary contributions.

Variant	Acc (%)	Macro-F1 (%)	OOD Macro-F1 (%)
Ours (Full)	**91.7**	**89.8**	**82.6**
w/o KG	88.4	86.1	78.2
w/o spatial edges	89.0	86.8	79.5
w/o genotypic edges	87.6	85.3	77.0
mean pooling	89.2	87.1	79.3
max pooling	90.1	88.0	80.4
Set2Set pooling	91.0	89.2	81.8

5 Conclusion and Future Work

This paper introduces a novel graph neural network (GNN) framework for crop variety classification that integrates multi-modal data and knowledge graphs. The framework's main contributions include a preprocessing pipeline for harmonizing genotypic, phenotypic, and environmental data, the construction of complex heterogeneous networks to represent agricultural ecosystems, and the embedding of agricultural knowledge graphs to incorporate semantic priors. A hybrid GNN architecture was designed by combining GCN, GAT, and GraphSAGE layers for both node- and graph-level tasks. To improve robustness, we employed multi-objective optimization techniques, advanced regularization, and focal loss. Extensive experiments across agricultural datasets demonstrated that this approach outperforms conventional machine learning, CNN-based methods, and baseline GNNs in terms of accuracy and generalization. Future research should explore dynamic graph modeling, self-supervised pretraining, interpretable mechanisms, and uncertainty quantification.

Disclosure of Interests. The authors have no competing interests to declare that are relevant to the content of this article.

References

1. Micheni, E.M., Machii, J., Murumba, J.: Internet of things, big data analytics, and deep learning for sustainable precision agriculture. In: 2022 IST-Africa Conference (IST-Africa), pp. 1–12, 2022
2. Sandhu, K.S., Merrick, L.F., Sankaran, S., et al.: Prospectus of genomic selection and phenomics in cereal, legume and oilseed breeding programs. Front. Genet. **12**, 829131 (2022)
3. Roy, P., Orikasa, T., Nakamura, N., et al.: Agriculture, population and their impact on the global environment, 2010
4. Akshay, D., Yewle, L., Mirzayeva, O.K.: Multi-modal Data Fusion and Deep Ensemble Learning for Accurate (2025). https://arxiv.org/html/2502.06062v1
5. Naresh, R., Singh, N.K., Sachan, P., et al.: Enhancing sustainable crop production through innovations in precision agriculture technologies. J. Sci. Res. Rep. (2024)
6. Branzova, P.: Precision agriculture: technological innovations for sustainable agriculture. Econ. J. (2024)
7. Löw, F.: Agricultural crop mapping from multi-scale remote sensing data: concepts and applications in heterogeneous Middle Asian agricultural landscapes, 2013
8. Wu, B., Zhang, M., Zeng, H., et al.: Challenges and opportunities in remote sensing-based crop monitoring: a review. Natl. Sci. Rev. **10**(4), nwac290 (2023)
9. Brown, D., Van den Bergh, I., de Bruin, S., et al.: Data synthesis for crop variety evaluation: a review. Agron. Sustain. Dev. **40**, 25 (2020)
10. Randhawa, S., Padmanaban, M., Devi, U.: Scope, extent, and challenges of an automated global crop classification model. In: IGARSS 2020 – IEEE International Geoscience and Remote Sensing Symposium, pp. 4657–4660 (2020)
11. Swarnkar, S.K., Dewangan, L., Dewangan, O., et al.: AI-enabled crop health monitoring and nutrient management in smart agriculture. In: 2023 6th International Conference on Contemporary Computing and Informatics (IC3I), vol. 6, pp. 2679–2683 (2023)
12. Elhammamy, Y.Y.S., Chung, G.C., Gan, M.T., et al.: Classification and detection of rice crop using deep learning for smart agriculture. In: Multimedia University Engineering Conference (MECON), vol. 2024, pp. 1–6 (2024)
13. Tufail, R., Tassinari, P., Torreggiani, D.: Assessing feature extraction, selection, and classification combinations for crop mapping using Sentinel-2 time series: a case study in northern Italy. Remote Sens. Appl. Soc. Environ. **38**, 101525 (2025)
14. Zheng, Y.-Y., Kong, J., Jin, X.-B., et al.: CropDeep: the crop vision dataset for deep-learning-based classification and detection in precision agriculture. Sensors (Basel, Switzerland) 19 (2019)
15. Mahesh, B.L., Shenoy, S.U.: Advances in crop classification using satellite imagery: a comprehensive review. In: IEEE International Conference on Computer Vision and Machine Intelligence (CVMI), vol. 2024, pp. 1–7 (2024)
16. Jahin, M.A., Shahriar, S., Mridha, M.F., Hossen, M.J.: Soybean Disease Detection via Interpretable Hybrid CNN-GNN: Integrating MobileNetV2 and GraphSAGE with Cross-Modal Attention (2025). https://arxiv.org/abs/2503.01284
17. Bhattacharya, S., Pandey, M.: Enhancing agricultural knowledge management using an efficient and novel ontology-based approach leveraging BERT-GPT and graph recurrent Q learning network. J. Electr. Syst. (2024)
18. Xu, K., Hu, W., Leskovec, J., Jegelka, S.: How powerful are graph neural networks? In arXiv (2018). https://arxiv.org/abs/1810.00826

19. Soultane, O.B., Salih-Alj, Y., Et-taibi, B.: Smart agriculture optimization: integrating edge computing and AI for enhanced crop management. In: 2024 10th International Conference on Applied System Innovation (ICASI), pp. 1–3, 2024. Teixeira I., Morais R., Sousa J. J., Cunha A. Deep learning models for the classification of crops in aerial imagery: a review [J]. Agriculture, 2023, 13:965
20. Scarselli, F., Gori, M., Tsoi, A.C., et al.: The graph neural network model. IEEE Trans. Neural Netw. **20**(1), 61–80 (2009)
21. Hamilton, W.L., Ying, Z., Leskovec, J.: Inductive Representation Learning on Large Graphs, 2017. ArXiv, abs/1706.02216
22. Gupta, A., Singh, A.: Agri-GNN: A Novel Genotypic-Topological Graph Neural Network Framework Built on GraphSAGE for Optimized Yield Prediction (2023). In ArXiv. https://arxiv.org/abs/2310.13037
23. Pan, Z., Xu, L., Chen, N.: Combining graph neural network and convolutional LSTM network for multistep soil moisture spatiotemporal prediction. J. Hydrol. **651**, 132572 (2025)
24. Wang, W., Wei, Y., Hao, L., Wei, Z., Zhao, T.: Soil moisture forecasting in wireless sensor networks via spatiotemporal graph convolutional networks. Vadose Zone J. (2025). https://acsess.onlinelibrary.wiley.com/doi/abs/10.1002/vzj2.70000
25. Sudhakara, B., Bhattacharjee, S.: Soil moisture prediction using multi-source datasets and graph neural network. In: 2024 IEEE India Geoscience and Remote Sensing Symposium (InGARSS). Goa, India, pp. 1–4 (2024). https://doi.org/10.1109/InGARSS61818.2024.10984286.
26. Zhang, Q., Li, B., Zhang, Y., Wang, S.: Suitability evaluation of crop variety via graph neural network. Comput. Intell. Neurosci. 5614974, 10 (2022)
27. Barvin, P.A., Sampradeepraj, T.: Crop recommendation systems based on soil and environmental factors using graph convolution neural network: a systematic literature review. Eng. Proc. **58**, 97 (2023)
28. Yang, F., et al.: Prediction of corn variety yield with attribute-missing data via graph neural network. Comput. Electron. Agric. **211**, 108046 (2023)
29. Zhang, Q., Li, B., Zhang, Y., Wang, S.: Suitability evaluation of crop variety via graph neural network. Comput. Intell. Neurosci. 5614974, 10 (2022)
30. Vatsa, A., Zhang, K., Quinche-Dutan, K.G.: Maize phenotype classification using GNN: research perspective. In: Proceedings of the 2024 IEEE Integrated STEM Education Conference (ISEC), Princeton, NJ, USA, pp. 1–4, 2024
31. Ruiz-Martinez, J.I., Mendez-Vazquez, A.: Graph-Based Deep Learning for Component Segmentation of Maize Plants (2025). https://papers.ssrn.com/sol3/papers.cfm?abstract_id=5381615
32. Raj, K.: A neuro-symbolic approach to enhance interpretability of graph neural network through the integration of external knowledge. In: Proceedings of the 32nd ACM International Conference on Information and Knowledge Management (CIKM '23), Birmingham, United Kingdom, pp. 5177–5180, 2023
33. Bhattacharya, S., Pandey, M.: Enhancing agricultural knowledge management using an efficient and novel ontology-based approach leveraging BERT-GPT and graph recurrent Q learning network. J. Electr. Syst. (2024)
34. Han, H., Kim, S., Choi, H. S., Yoon, S.: On the Impact of Knowledge Distillation for Model Interpretability (2023). In ArXiv. https://arxiv.org/abs/2305.15734
35. Neil, D., Briody, J., Lacoste, A.M.B., Sim, A., Creed, P., Saffari, A.: Interpretable Graph Convolutional Neural Networks for Inference on Noisy Knowledge Graphs, 2018. ArXiv, abs/1812.00279

36. Chen, J., et al.: Knowledge graphs for the life sciences: recent developments, challenges and opportunities (2023). https://arxiv.org/abs/2309.17255
37. Ma, T., et al.: Learning to denoise biomedical knowledge graph for robust molecular interaction prediction. IEEE Trans. Knowl. Data Eng. **36**(12), 8682–8694 (2024)
38. Nath, R.P.D., Das, T.R., Das, T.C., Raihan, S.M.S.: Knowledge graph generation and enabling multidimensional analytics on Bangladesh agricultural data. IEEE Access **12**, 87512–87531 (2024)

SemiBDMA Workshop

MSPD: Large Language Model Parallel Speculative Decoding Method Based on Multi-level Early Exiting

Chenglin Han[1], Deji Zhao[1(✉)], Xingqi Li[1], Jinyu Li[2], Haoyu Yang[1], and Haowen Xiang[1]

[1] School of Computer Science and Engineering, Northeastern University, Shenyang 110819, China
zhaodeji@stumail.neu.edu.cn,
{2301871,2371382,2301979,2301972}@stu.neu.edu.cn
[2] School of Information Science and Technology, Dalian Maritime University, Dalian 116026, China
ljy@dlmu.edu.cn

Abstract. Speculative decoding (SD) is a method that utilizes an additional draft mode to predict multiple future tokens, which are subsequently verified by the target large language model (target LLM). Despite the significant acceleration achieved by SD, several challenges remain, including the training cost of the draft mode, drafting latency, and alignment and asynchrony issues with the target LLM. To address these challenges, we propose a novel SD model called MSPD (Multi-level Speculative Parallel Decoding). MSPD is composed of two main modules: first, the multi-level early-exit model, which treats the single-level early-exit model as a new target LLM and applies the early-exit model again for drafting. This approach addresses the training cost and drafting latency issues of the draft model and enhances its alignment with the target LLM by reusing the parameters of the target LLM. Second, the parallel inference model, namely Hypothesis Verification - Continuous Draft. This model enables parallel inference between the draft mode and the target LLM, avoiding the mutual waiting that causes synchronization bottlenecks and thus reducing the impact of asynchrony. Experimental results demonstrate that MSPD significantly accelerates the inference process while maintaining text generation consistency. Compared with the SOTA model, the number of parameters is reduced by **77.4**%, and the inference speed on Spec-Bench is increased by **1.73**×.

Keywords: Large Language Models · Parallel Decoding · Early Exiting · Speculative Decoding

1 Introduction

Large Language Models (LLMs) have significantly impacted society, but their reliance on autoregressive decoding for text generation results in slow speeds [16].

J. Li et al. (Eds.): APWeb-WAIM 2025, CCIS 2660, pp. 95–106, 2026.
https://doi.org/10.1007/978-981-95-7996-9_9

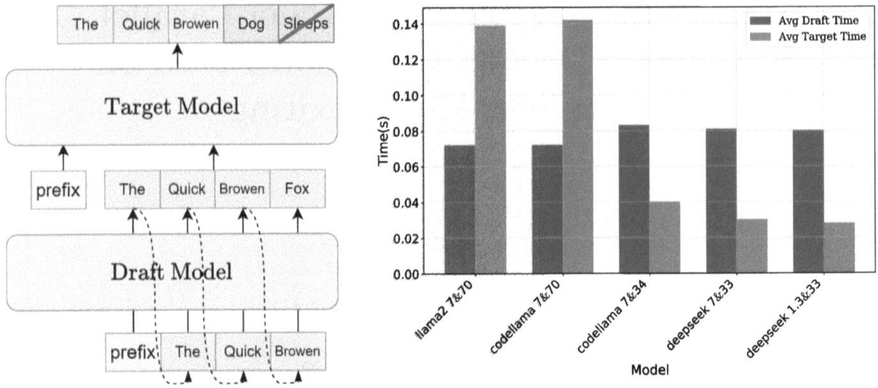

(a) Speculative Decoding (b) Average Draft and Target Times

Fig. 1. (a) In the Speculative decoding, the drafting and verification phases are waiting for each other. (b) The time of both the drafting and verification phases cannot be ignored.

This method generates tokens serially, with each token depending on the previous one, leading to increased decoding time. Additionally, it fails to fully utilize modern hardware accelerators like GPUs. Speculative Decoding (SD) has emerged as a solution to accelerate LLM inference without compromising text quality [15].

SD consists of two main components(Fig. 1a): a draft model that quickly generates initial tokens and a target LLM that verifies these tokens in parallel. However, the draft model's effectiveness is limited by its alignment with the target LLM and the latency introduced during drafting. Furthermore, the serial nature of the SD process, where drafting and verification phases wait for each other, introduces significant synchronization latency.

To address these challenges, we propose MSPD, a novel SD framework with two key modules. First, we enhance the draft model by applying a multi-level early-exit strategy, treating the single-level early-exit model as a new target LLM and further optimizing it. This reduces drafting latency and improves alignment. Second, we introduce the Hypoth-Verif&Cont-Draft framework, which allows parallel drafting and verification, overcoming the synchronization bottleneck.

The main contributions of this paper can be summarized as follows:

1. We propose a multi-level early-exit model. By regarding the single-level early-exit model as a new target LLM and applying the early-exit model again for drafting. it reduces the drafting latency of the draft mode and enhances its alignment with the target LLM by reusing the parameters of the target LLM.
2. We propose a new parallel inference model called Hypothesis Verification-Continuous Draft. This model allows the draft mode and the target LLM to perform parallel inference, thereby overcoming the synchronization bottleneck of mutual waiting.

3. The experimental results show that MSPD significantly accelerates the inference process. Compared with the Medusa model [1], the number of parameters is reduced by 77.4% (134M compared to 591M), and the inference speed on Spec-Bench [12] is increased by 1.73×.

2 Related Work

Speculative Decoding (SD) accelerates inference by predicting future tokens via a draft model and verifying them in parallel with the target LLM. The drafting phase primarily uses two methods: independent drafting and self-drafting.

2.1 Independent Drafting

Independent drafting employs a separate smaller model for speculative generation. For example, SpecDec uses a smaller model aligned with the target LLM, achieving a balance between accuracy and efficiency through knowledge distillation [7]. Leviathan [4] also uses an independent lightweight model for efficient drafting. However, these methods face challenges in aligning the draft model with the target LLM and have high hardware and training costs.

2.2 Self-drafting

Self-drafting directly uses part of the target LLM's parameters for drafting, reducing the need for an independent model and alignment challenges. For instance, Blockwise Decoding [11] and Medusa [1] add a feed-forward network (FFN) head to the target LLM for self-drafting. PPD [13] retrains the target LLM for early-exit models. Lookahead Decoding [17] introduces the Mask-Predict mechanism for self-drafting. Early-exit models like Draft&Verify [14] and Kangaroo [6] use partial layers of the target LLM for drafting. While these methods have high alignment with the target LLM, they suffer from significant drafting latency.

Current research focuses on improving speculative accuracy in self-drafting while minimizing computational overhead. MSPD employs multi-level early-exit models and parallel drafting to reduce the high latency of single-level early-exit models, enhancing alignment and efficiency.

3 Methodology

3.1 Motivated Observation

To better understand the costs associated with the drafting phase, we conducted experiments. For example, when using Llama 7B as the draft model and Llama 70B as the target LLM, the average inference time for the draft model is 0.07 s per inference, while the target LLM's average inference time is 0.14 s (Fig. 1b). This shows that the draft model's latency accounts for nearly 50% of the target

LLM's inference time, indicating that the draft model's latency is significant in the overall process.

Existing SD models have achieved notable acceleration. However, the alignment between the draft model and the target LLM and the token acceptance rate still need improvement. Moreover, as the number of drafting tokens increases, the acceptance rate of the target LLM decreases significantly. Therefore, it is crucial to enhance alignment and employ a Dynamic Adaptive Draft Length to terminate the drafting phase early.

Based on these observations, we propose MSPD, a new framework that enhances alignment between the draft model and the target LLM, reduces drafting latency, and lowers training costs through parallel inference strategies, thereby further increasing the target LLM's inference speed.

MSPD mainly consists of two modules:

1. **Multi-level Early-exit Model**: Traditional single-level early-exit models, while improving alignment with the target LLM, suffer from high drafting latency. MSPD treats the draft model as a new target LLM and applies the early-exit model to it again for drafting. This multi-level early-exit approach enhances alignment, reduces training costs, and increases token acceptance rates.
2. ***Hypoth-Verif & Cont-Draft* Model**: Inspired by Parallel Speculative Decoding [4], MSPD combines the efficient two-step strategy of the multi-level early-exit model. Specifically, the *Hypoth-Verif* phase assumes the first batch of drafting tokens is correct, allowing parallel inference between the draft model and the target LLM, thus avoiding mutual waiting delays. The *Cont-Draft* phase enables the draft model to continue drafting while the target LLM verifies tokens, instead of waiting for verification results. This strategy effectively addresses asynchronous delays and significantly improves inference parallel efficiency.

3.2 Multi-level Early-Exit Model

The multi-level early-exit model we proposed is inspired by the Kangaroo model. First, we extract the hidden state h^l from the fixed shallow network $M_b[:\, l]$ of the target LLM (Eq. 1). Then we train an Adapter Network A_2 to fit the direct mapping relationship from the $M_b[l :]$ to the final output of the LM head. The specific structure of the adapter network includes two normalization layers and a multi-head attention layer, as shown in Fig. 2. This design not only reduces the training cost but also enhances the alignment between the draft mode and the target LLM.

As the number of layers l in $M_b[:\, l]$ increases, the semantic information fusion in h^l becomes more accurate, and the acceptance rate of drafting tokens also increases (as shown in Fig. 5a). However, the increase in the fixed number of layers l also leads to an increase in drafting latency.

To address this issue, we further propose a multi-level early-exit model. We regard the fixed shallow network $M_b[:\, l]$ of the target LLM and A_2 as a new

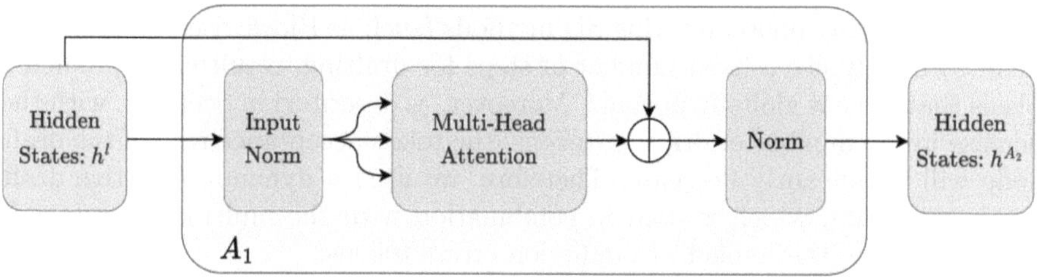

Fig. 2. The specific structure of the Adapter Network includes two normalization layers and a multi-head attention layer.

target LLM and apply the early-exit algorithm again. We extract h^n from an even shallower layer $M_b[:n]$ (where $n < l$) and train another adapter network A_1 to fit the mapping relationship from $M_b[:n]$ to $M_b[n:l] \circ A_2$, as shown in Fig. 3 and Eqs. 2. This multi-level early-exit model architecture improves the prediction accuracy of the draft mode through hierarchical extraction of hidden states while reducing the inference time overhead.

$$h^l = M_b[:l](x) \tag{1}$$

$$A_1(h^n) = M_b[n:l] \circ A_2(h^n) \tag{2}$$

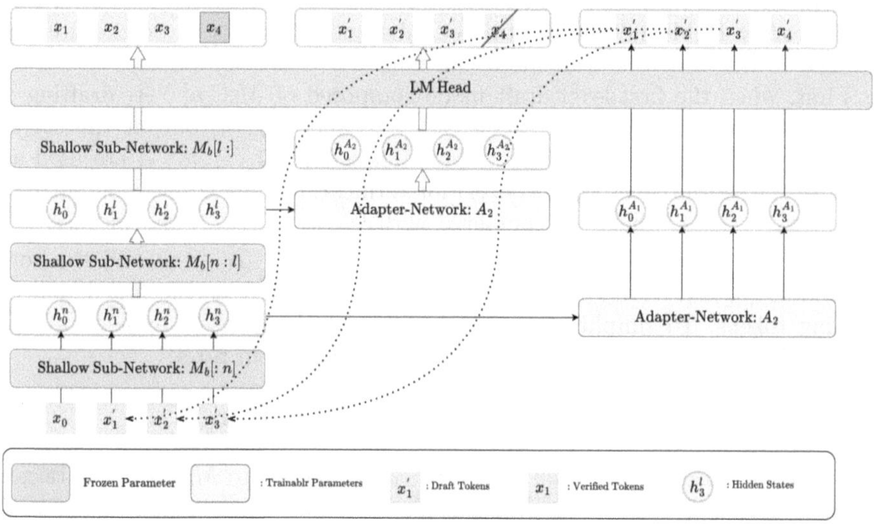

Fig. 3. The framework of Multi-level Early-Exit Model. First, A_1 drafting $x_1^{'}, x_2^{'}, x_3^{'}, x_4^{'}$ multiple times, and then passes the corresponding hidden states to A_2 for verification. Finally, the hidden states of A_2's verification process are passed to $M_b[l:]$ for final verification, and the result is that x_4 is rejected.

In the drafting phase, existing SD methods (such as Blockwise Decoding and Medusa) mostly use a fixed number of steps for drafting, resulting in predicted tokens that are not globally optimal. Moreover, as described in Sect. 3.1, with the increase in the number of drafting tokens, the token acceptance rate of the draft mode will significantly decrease. Therefore, we used a dynamic adaptive draft length algorithm, which is used in combination with the multi-level early-exit model to reduce the impact of prediction errors tokens.

Specifically, a threshold η is set in the first layer (the lowest layer) of the multi-level early-exit model. When the top-1 confidence of the token predicted by the first layer A_1 is below the predefined threshold η, the drafting of the first layer is stopped, as shown in Eq. 3:

$$\max_{n} M_n^s(x) \leq \eta \tag{3}$$

We considered using a similar threshold early-exit method in the second layer draft mode A_2But found that the number of inferences of the second layer draft mode is determined by the verification time of the target LLM. Therefore, we abandoned the use of a similar threshold early-exit method in the second layer.

3.3 Parallel Inference Model

Hypothesis Verification. The core of *Hypoth-Verif* is to eliminate the dependence of the target LLM verification phase on the drafting tokens. Specifically, to enable parallel inference of the target LLM, we assume that the first batch of drafting tokens is correct, thus allowing the verification of this batch of tokens by the target LLM and the drafting of subsequent batches to be executed in parallel. In this way, the draft mode can skip the verification phase.

First, when the first-level draft mode composed of $M_b[: n] \circ A_1$ drafting γ tokens forward, the output is $[x_1^{''}, x_2^{''}, \ldots, x_\gamma^{''}]$, and $M_b[: n]$ output the corresponding hidden states $[h_0^n, h_2^n, \ldots, h_{\gamma-1}^n]$. These $[h_0^n, h_2^n, \ldots, h_{\gamma-1}^n]$ are used as input to the second-level $M_b[n : l] \circ A_2$ for verification. The verification outputs are $[x_1^{'}, x_2^{'}, \ldots, x_\gamma^{'}]$, which are compared with $[x_1^{''}, x_2^{''}, \ldots, x_\gamma^{''}]$, and inconsistent tokens are rejected. The final result is $[x_1^{'}, x_2^{'}, \ldots, x_\alpha^{'}]$ $(\alpha \leq \gamma)$ and the corresponding $[h_0^l, h_1^l, \ldots, h_{\alpha-1}^l]$. We refer to $[x_1^{'}, x_2^{'}, \ldots, x_\alpha^{'}]$ as the first batch of drafting tokens. To simplify the model, we treat this process as a whole V_A (verification by A_2), with the specific formula as shown in Eq. 4.

$$([x_1^{'}, x_2^{'}, \ldots, x_\alpha^{'}], [h_0^l, h_1^l, \ldots, h_{\alpha-1}^l]) = V_A(x_0) \tag{4}$$

Then, the hidden states $[h_0^l, h_1^l, \ldots, h_{\alpha-1}^l]$ are passed to $M_b[l :]$ of the target LLM for verification. It is worth noting that during the verification process of $M_b[l :]$, V_A does not stop and wait for the verification results but continues to draft $[x_{\alpha+1}^{'}, \ldots, x_\delta^{'}]$ and $[h_\alpha^l, \ldots, h_{\delta-1}^l]$.

As shown in Fig. 4, if $M_b[l :]$ accepts all the drafting tokens in the first batch, the *Hypoth-Verif* is success. Similarly, $[h_\alpha^l, \ldots, h_{\delta-1}^l]$ are passed to $M_b[l :]$ for verification, and the process enters the *Cont-Draft* phase. If $M_b[l :]$ does not accept all the drafting tokens in the first batch, the *Hypoth-Verif* is failure, and the correct tokens verified by $M_b[l:]$ are used as input for re-verification. Since $M_b[l:]$ provides the correct tokes, the cost of a failed *Hypoth-Verif* is almost negligible.

Continuous Draft. *Cont-Draft* is the phase that follows a successful hypothesis verification. The specific process is as follows:

Firstly, similarly, during the verification of $M_b[l:]$, V_A does not stop but continues to draft μ tokens forward until $M_b[l:]$ completes the verification.

Secondly, if $M_b[l:]$ verifies and accepts all $[x'_{\alpha+1}, \ldots, x'_{\delta}]$, there is no need to sample new tokens from the verification results of $[x_{\alpha+1}, \ldots, x_{\delta}]$. The Continuous *Cont-Draft* continues, and $[h^l_{\delta}, \ldots, h^l_{\mu-1}]$ is passed to $M_b[l:]$ for verification. And during this process, V_A continues drafting. This entire process loops until $M_b[l:]$ does not accept all the drafting tokens, at which point the process reverts to the *Hypoth-Verif* phase and restarts the *Hypoth-Verif* stage. As shown in Fig. 4

3.4 Training of the Draft Models

In MSPD model, only two Adapter Networks A_1 and A_2 need to be trained. This structural design aims to obtain the hidden states and output the hidden states h^{A_1} and h^{A_2}, which are finally sampled through the LLM Head to generate specific tokens. as shown in Fig. 2. The ultimate goal of the draft mode is to predict tokens, this process can be regarded as a classification task. Therefore, we chose the cross-entropy loss [9] as the loss function for training the A_1 and A_2. Cross-entropy loss can effectively measure the difference between the predicted distribution and the true distribution, thereby providing a clear optimization direction for model training. The formulas are shown in Eqs. 5, 6, and 7.

During the training process, since the draft mode is relatively less sensitive to training data [5], we don't directly use the large-scale corpus identical to that of the target LLM for training. Instead, we chose a public dataset ShareGPT.

$$y_{\alpha+1} = \text{Softmax}(\text{LM Head}(h^{A_i})) \tag{5}$$

$$\hat{y}_{\alpha+1} = \text{Softmax}(\text{LM Head}(\hat{h^l})) \tag{6}$$

$$L_{\text{cls}} = \text{Cross Entropy}(y_{i+1}, \hat{y}_{i+1}) \tag{7}$$

The $\hat{y}_{\alpha+1}$ represents the true probability distribution, while the $y_{\alpha+1}$ represents the predicted distribution by the MSPD model. The h^{A_i} represents h^{A_1} and h^{A_2}. The L_{cls} represents the classification loss computed from the hidden states.

To enhance the robustness of the adapter networks, we adopt a data augmentation method similar to that used by EAGLE [5]. Specifically, we add random noise sampled from a uniform distribution $U(-0.1, 0.1)$ [3] to the fixed-layer

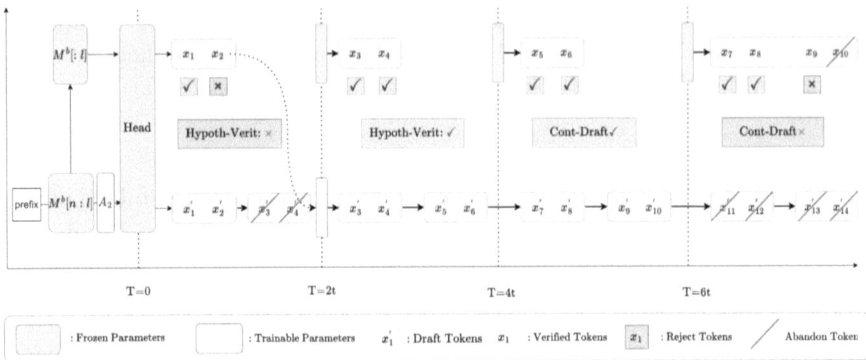

Fig. 4. The framework of Parallel Inference Model. At T=0, the draft model generates the first batch of $x_1^{'}$ and $x_2^{'}$, and the target LLM is also verified based on the hidden state, and x_2 is rejected. At T=2t, the drafting continues, and the same target LLM is verified and $x_3^{'}$ and $x_4^{'}$ are accepted. At T=4t, it is drafting because it is not rejected, until T=6t when x_{10} is re-entered into the first stage.

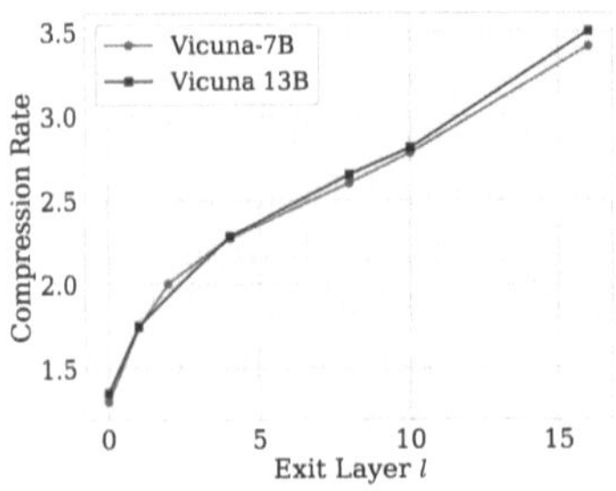

(a) CR as a Function of l

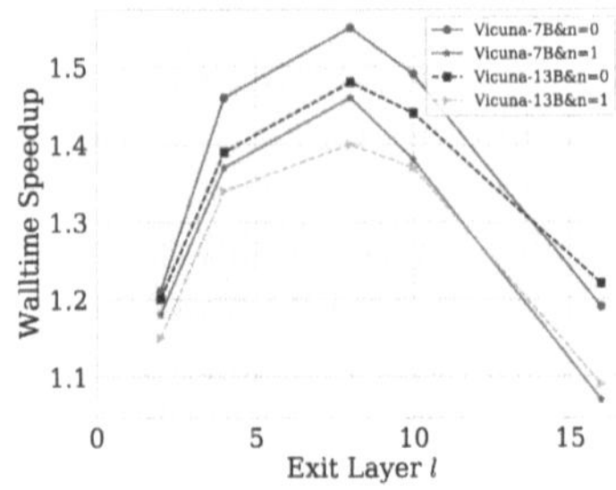

(b) WS as a Function of l and n

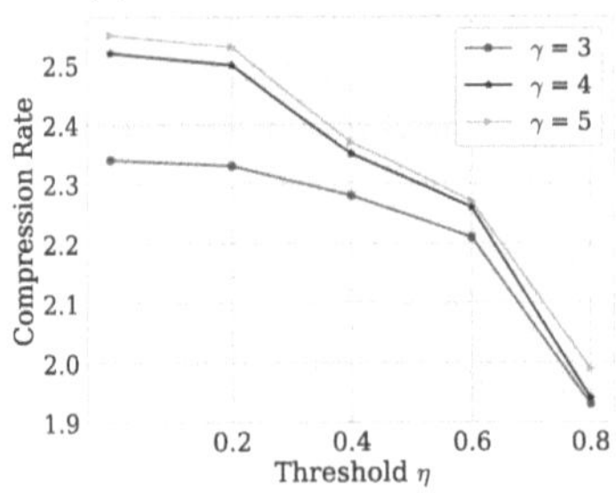

(c) CR as a Function of η and γ

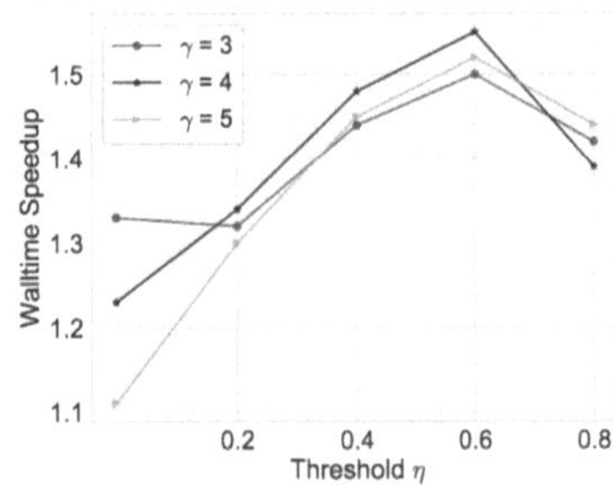

(d) WS as a Function of η and γ

Fig. 5. (a)(b) CR-WS Trade-offs in Exit Layer l (c)(d) CR-WS Trade-offs in Threshold η.

hidden states extracted by the target LLM during training. This approach enables the adapter networks to learn robustness against noise, better handling minor variations in the target LLM's hidden states during actual inference.

4 Experiments

4.1 Implementation Details

In this study, we selected two Vicuna models with different parameter scales (7B and 13B) as the target LLMs to comprehensively evaluate the performance of the proposed method. These two models are representative in terms of parameter scale and performance, providing a rich basis for comparison in our experiments.

To validate the effectiveness of our proposed method, we chose several highly-regarded and SOTA methods in the field of SD for comparison, including Lookahead [17], Medusa [1], REST [2], and Kangaroo [6].

In terms of evaluation, to ensure the objectivity and comparability of the experimental results, we conduct a comprehensive comparison of the performance of these SD methods with the recently proposed Spec-Bench benchmark testing framework. All experiments were completed on a unified hardware platform, specifically two NVIDIA A40 GPUs, to ensure the consistency of experimental conditions.

In our current algorithm, during the LLM verification phase, we employ Greedy Decoding, similar to the Kangaroo [6] model, to ensure the consistency of the final text.

Training: Regarding the training aspect of our proposed algorithm, we followed a training strategy similar to that of Medusa . First, we fixed the target LLM and then trained the Adapter Network. In MSPD, the A_1 and A_2 was trained on the public dataset ShareGPT, which contains 68,000 dialogue with a learning rate set at 3×10^{-5}. We employed the AdamW optimizer [8] to optimize the Adapter Network for 13 training epochs, ensuring that the model achieved optimal performance in the experiments. Moreover, the training of the draft mode for both the 7B and 13B Vicuna models as the target LLM could be completed within 1 to 2 days on an A40 server.

4.2 Results

Table 1 shows that MSPD significantly outperforms Lookahead, Medusa, REST, and SOTA Kangaroo on the Spec-Bench dataset, achieving an average **1.55**× speedup and **1.73**× on MT-Bench subtasks. In Table 2, MSPD's draft model requires only 134M parameters, significantly fewer than other SD methods.

4.3 Ablation Studies

To analyze the single-layer early-exit model, we fixed different numbers of layers l of the target LLM and trained the corresponding Adapter Network. The results (Fig. 5a) show that as l increases, the token acceptance rate (CR) improves. This is because more shared parameters between the draft model and the target LLM increase the likelihood of token acceptance.

Table 1. Average speedup of the MSPD model on Spec-Bench on A40

Size	Method	Translation		QA		Summarization		Math		RAG		MT Bench		Avg.
		CR	Speedup	CR	Speedup	CR	Speedup	CR	Speedup	CR	Speedup	CR	Speedup	
7B	Lookahead	1.24	1.14×	1.56	1.21×	1.53	1.25×	1.96	1.50×	1.49	1.19×	1.70	1.39×	1.28×
	Medusa	**1.58**	1.42×	1.50	1.33×	1.49	1.31×	1.73	1.54×	1.52	1.28×	1.78	1.55×	1.41×
	REST	1.54	1.25×	1.90	1.63×	1.64	1.32×	1.53	1.24×	1.92	1.46×	2.22	1.67×	1.43×
	Kangaroo	1.41	1.26×	1.87	1.45×	1.87	1.51×	2.14	1.61×	2.05	1.53×	2.22	1.68×	1.51×
	MSPD	1.53	**1.32×**	**1.96**	**1.47×**	**1.93**	**1.54×**	**2.20**	**1.65×**	**2.15**	**1.57×**	**2.26**	**1.73×**	**1.55×**
13B	Lookahead	1.24	1.01×	1.39	0.98×	1.50	0.99×	1.94	1.23×	1.52	0.92×	1.68	1.06×	1.03×
	REST	1.53	1.07×	**1.92**	1.41×	1.66	1.14×	1.55	1.06×	1.87	1.34×	1.98	1.36×	1.23×
	Medusa	**1.60**	**1.32×**	1.49	1.24×	1.52	1.25×	1.80	1.46×	1.51	1.22×	1.88	1.48×	1.33×
	Kangaroo	1.45	1.18×	1.79	1.35×	2.00	1.42×	**2.42**	**1.65×**	2.16	1.40×	2.44	1.67×	1.45×
	MSPD	1.55	1.28×	**1.92**	**1.38×**	**2.09**	**1.47×**	2.32	1.60×	**2.21**	**1.47×**	**2.48**	**1.69×**	**1.48×**

Table 2. Parameter comparison of different draft models

Architecture	Input LN	Attention	Post LN	FFN	Linear	Last LN	Head	Parameters	Speedup
Medusa	×	×	×	×	×4	×	×4	591M	1.41×
1-Layer Transformer	✓	✓	✓	✓	×	✓	×	202M	1.37×
MLP Only	✓	×	×	×	×2	✓	✓	165M	1.22×
MSPD	×2	×2	×	×	×	×2	×	134M	**1.55×**

For the multi-level early-exit model, we focused on the first-layer draft model with $n = 0$ or $n = 1$ fixed layers. The results (Fig. 5b) indicate that $n = 0$ yields higher CR regardless of the second-layer draft model's fixed layers. Thus, we chose $n = 0$ and $l = 8$ as the optimal configuration.

In our parallel decoding strategy, we determined the second-layer draft model's inference steps based on the target LLM's verification time. We analyzed the threshold η and the maximum number of steps γ for the first-layer draft model. The results (Fig. 5c and Fig. 5d) show that the optimal η remains consistent across different γ. After comprehensive analysis, we set $\gamma = 4$ and $\eta = 0.6$ as the optimal hyperparameters. The final results are shown in Table 1.

To verify the effectiveness of the parallel inference model, we keep the multi-level early-exit model unchanged and remove the parallel inference model for ablation experiments, with the results shown in Table 3. It can be found that the parallel inference model can reduce the impact of asynchronous delay.To verify the contribution of the dynamic draft length algorithm, we conducted an ablation study for Vicuna-7B on the MT-Bench subtask, comparing static draft length with dynamic draft length, where the static draft length = 4 (Table 3).

Table 3. Ablation study results of vicuna-7B on the MT-Bench subtask within Spec-Bench

Method	**CR**	**Speedup**
MSPD	2.26	1.73×
MSPD w/o Parallel	2.12 (↓ 6.19%)	1.61×(↓ 6.9%)
MSPD w/o Dynamic Length	1.92 (↓ 15.04%)	1.39×(↓ 19.65%)

5 Conclusion and Future Work

5.1 Conclusion

In this study, we introduce an innovative SD model named MSPD (Multi-level Speculative Parallel Decoding). This approach maintains text generation consistency while significantly accelerating LLM inference and reducing training costs. The core of MSPD consists of two parts: the Multi-level Early Exit Model and the Parallel Inference Model. The Multi-level Early Exit Model treats the single-level early-exit model as a new target LLM and applies the early-exit algorithm to it again for acceleration. The parallel inference Model enables the drafting phase and the verification phase to be executed in parallel, addressing the issue of mutual waiting and asynchrony. Experimental results demonstrate compare with the SOTA model, the MSPD model has an acceleration ratio of up to **1.73**×. Moreover, MSPD reduce the number of training parameters by **77.4**%, significantly lowering the training cost.

5.2 Future Work

Although the MSPD model demonstrates promising performance in current experiments, further optimization is still necessary. Future work will focus on optimizing the verification phase by introducing a tree-based verification mechanism [10] to reduce verification time through hierarchical verification and exploring more efficient LLM inference acceleration strategies.

References

1. Cai, T., et al.: Medusa: Simple llm inference acceleration framework with multiple decoding heads. In: Proceedings of the 41st International Conference on Machine Learning, ICML'24, JMLR.org (2024)
2. He, Z., Zhong, Z., Cai, T., Lee, J., He, D.: REST: Retrieval-based speculative decoding. In: Duh, K., Gomez, H., Bethard, S. (eds.) Proceedings of the 2024 Conference of the North American Chapter of the Association for Computational Linguistics: Human Language Technologies (Volume 1: Long Papers). pp. 1582–1595. Association for Computational Linguistics, Mexico City, Mexico (2024). https://doi.org/10.18653/v1/2024.naacl-long.88, https://aclanthology.org/2024.naacl-long.88/
3. Jain, N., et al.: NEFTune: Noisy embeddings improve instruction finetuning. In: The Twelfth International Conference on Learning Representations (2024). https://openreview.net/forum?id=0bMmZ3fkCk
4. Leviathan, Y., Kalman, M., Matias, Y.: Fast inference from transformers via speculative decoding. In: International Conference on Machine Learning, pp. 19274–19286. PMLR (2023)
5. Li, Y., Wei, F., Zhang, C., Zhang, H.: Eagle: speculative sampling requires rethinking feature uncertainty. In: Proceedings of the 41st International Conference on Machine Learning, ICML'24, JMLR.org (2024)

6. Liu, F., et al.: Kangaroo: Lossless self-speculative decoding for accelerating LLMs via double early exiting. In: The Thirty-eighth Annual Conference on Neural Information Processing Systems (2024). https://openreview.net/forum?id=lT3oc04mDp
7. Liu, X., et al.: Online speculative decoding. arXiv preprint arXiv:2310.07177 (2023)
8. Loshchilov, I., Hutter, F.: Decoupled weight decay regularization. In: International Conference on Learning Representations (2017). https://api.semanticscholar.org/CorpusID:53592270
9. Mao, A., Mohri, M., Zhong, Y.: Cross-entropy loss functions: Theoretical analysis and applications. In: International conference on Machine learning. pp. 23803–23828. PMLR (2023)
10. Spector, B.F., Re, C.: Accelerating LLM inference with staged speculative decoding. In: Workshop on Efficient Systems for Foundation Models @ ICML2023 (2023). https://openreview.net/forum?id=RKHF3VYjLK
11. Stern, M., Shazeer, N., Uszkoreit, J.: Blockwise parallel decoding for deep autoregressive models. Adv. Neural Inf. Process. Syst. **31** (2018)
12. Xia, H., et al.: Unlocking efficiency in large language model inference: A comprehensive survey of speculative decoding. In: Ku, L.W., Martins, A., Srikumar, V. (eds.) Findings of the Association for Computational Linguistics: ACL 2024, pp. 7655–7671. Association for Computational Linguistics, Bangkok, Thailand (2024). https://doi.org/10.18653/v1/2024.findings-acl.456, https://aclanthology.org/2024.findings-acl.456/
13. Yang, N., et al.: Inference with reference: Lossless acceleration of large language models. arXiv preprint arXiv:2304.04487 (2023)
14. Zhang, J., et al.: Draft & verify: Lossless large language model acceleration via self-speculative decoding. In: Ku, L.W., Martins, A., Srikumar, V. (eds.) Proceedings of the 62nd Annual Meeting of the Association for Computational Linguistics (Volume 1: Long Papers), pp. 11263–11282. Association for Computational Linguistics, Bangkok, Thailand (2024). https://doi.org/10.18653/v1/2024.acl-long.607, https://aclanthology.org/2024.acl-long.607/
15. Zhao, D., et al.: Enhancing math reasoning ability of large language models via computation logic graphs. Knowl.-Based Syst. 113905 (2025)
16. Zhao, D., et al.: Autograph: Enabling visual context via graph alignment in open domain multi-modal dialogue generation. In: Proceedings of the 32nd ACM International Conference on Multimedia, pp. 2079–2088 (2024)
17. Zhao, Y., Xie, Z., Liang, C., Zhuang, C., Gu, J.: Lookahead: An inference acceleration framework for large language model with lossless generation accuracy. In: Proceedings of the 30th ACM SIGKDD Conference on Knowledge Discovery and Data Mining, pp. 6344–6355 (2024)

A New Yi Font Generation Using Diffusion Model

Zedong Li[1,2], Mengdi Li[1,2](✉), Bo Lu[1], Jianxin Zhang[1], and Cunrui Wang[1,2]

[1] College of Computer Science and Engineering, Dalian Minzu University, Dalian 116650, China
lzd@dlnu.edu.cn, lmd18018901065@gmail.com
[2] Dalian Chinese Font Design Technology Innovation Center, Dalian 116650, China

Abstract. As an important carrier of the excellent culture of the Chinese nation, the Yi script has long attracted widespread attention. However, existing research mainly focuses on feature detection, recognition, and restoration of Yi characters, while issues such as preserving glyph structure and achieving stylistic transfer remain at an exploratory stage. To address these challenges, this paper proposes a Yi font generation framework based on a multi-scale conditional diffusion model. The Multi-scale Content Perception (MCP) module is designed to capture local and global spatial information in a hierarchical manner through the constructed channel-space dual-domain multi-scale attention synergistic mechanism to improve the ability to extract content information from Yi text. Meanwhile, in order to realize the cross-modal feature fusion of Yi text, the Efficient Style Insertion (ESI) module is designed, integrating a multi-head mechanism with efficient channel attention to optimize the cross-attention mechanism. Furthermore, the sliced attention strategy is adopted to alleviate the computational overhead associated with traditional cross-attention mechanisms. The experimental results on the Yi font dataset show the advantages and feasibility of the method in this paper.

Keywords: Conditional diffusion model · Yi font generation · Multi-scale attention synergy mechanisms · Cross-modal feature fusion · Cross-attention

1 Introduction

The Yi script is one of the important written symbol systems of the Chinese nation, covering many fields such as history, literature and medicine, and is an important carrier of the excellent culture of the Chinese nation [1]. The similar morphological configuration between Yi script and Chinese characters makes Yi script inherited the structural characteristics of Chinese characters [2]. Therefore, exploring the multi-style structural features of Yi fonts not only enriches the cultural connotation of Yi script but also expands its digitization pathway, posing an urgent challenge that needs to be addressed.

J. Li et al. (Eds.): APWeb-WAIM 2025, CCIS 2660, pp. 107–118, 2026.
https://doi.org/10.1007/978-981-95-7996-9_10

At present, research on Yi fonts mainly focuses on detection, recognition, and document processing. Sun et al. [3] introduced channel attention and spatial attention mechanisms in DBNet++ model, which enhanced the local feature acquisition ability of the network and increased the detection accuracy of Yi script; Jia et al. [4] proposed an automated method for segmentation and recognition of handwritten Yi text, which improved the Yi character recognition accuracy; Chen et al. [5] fused the maximum extreme value stabilization region and convolutional neural network, and proposed an improved adaptive binarization method to separate the text and non-text region separation in ancient Yi texts; Xiong et al. [6] used deep learning techniques to complete the character analysis and extraction of text features in the images of ancient Yi texts. Although the research on the generation technology of Yi fonts is still in its infancy, the research on the generation methods of Chinese fonts has made breakthroughs. Jiang et al. [7] extracted style features and migrated them to unseen characters through deep convolutional network with adversarial loss, but there are still local deformation and detail loss when dealing with complex structure or abnormal stroke; Chang et al. [8] introduced cyclic consistency loss to achieve font style migration under unpaired data conditions, which provides an effective solution for the unsupervised task, although there are still ambiguities in fine structure modeling; Wang et al. [9] introduced stroke information in generative adversarial networks and added a fusion attention mechanism, which reduces the stroke error rate of generated fonts; Xue et al. [10] introduced a skeleton generation module and combined it with a transitive-attention mechanism to achieve accurate separation and fusion of structural and stylistic features of Chinese characters under zero-sample conditions, but the method still exhibits noise and loss of fine-grained details when transferring complex font styles; Liu et al. [11] proposed an unsupervised font generation network that jointly models content and style representations. It employs a decoupled encoder architecture and a deformable convolution mechanism to enable high-quality font style transfer without relying on paired data; Wang et al. [12] introduced a multi-scale edge self-supervision module, which effectively enhances stroke edge clarity and structural fidelity. However, the method still suffers from noise artifacts and blurring of fine details, particularly in handwritten or cursive font styles. However, the advantages of extended models in terms of detail information, flexibility and stability have gradually matured in Chinese character generation techniques [13–15]. He et al. [16] introduced a diffusion model into the task of font generation and reconstructed the font image through a step-by-step denoising process, which resulted in a significant improvement in stroke accuracy and structural representation; Fu et al. [17] reformulated font generation as a multi-stage generation process by incorporating the font conversion process into a diffusion model, and used a dual network approach to process the information of different generation stages of fonts, which in turn generated font images; Yang et al. [18] fused a multi-scale content aggregation module and a style comparison refinement module, which efficiently dealt with the complex characters and font style variations to generate different styles of Chinese fonts.

Therefore, leveraging the strengths of diffusion models and the structural similarity between Yi script and Chinese characters [2], this paper adopts a diffusion-based framework. A channel-space dual-domain multi-scale attention mechanism is designed to enhance Yi font feature extraction, and a channel-optimized cross-attention module is introduced for cross-modal feature fusion, enabling the generation of multi-style Yi characters. Experimental results validate the effectiveness and advantages of the proposed method.

2 Related Work

Font generation can be viewed as a conversion process of an image from the source domain to the target domain, and its essence lies in the construction of cross-domain feature mapping functions to realize the precise migration of visual styles to the target domain. Early font generation techniques are usually based on generative adversarial networks, which provides a theoretical framework for automatic font generation tasks and is applicable to the generation of multiple font styles, but is limited by the end-to-end mapping mechanism, which can only reproduce the styles of known fonts, and lacks the ability of zero-shot generalization to unseen fonts [19–21]. And the proposal of diffusion model opens up a new way of thinking for font generation technology [17,18], which effectively integrates the conditional information into the diffusion process of the model through the introduction of the cross-attention mechanism in the UNet network, and realizes a double breakthrough in both the quality and controllability of character generation through the fine tuning in the implicit representation space and achieves good results in the field of handwritten character generation [22,23].

The diffusion model realizes the character generation process, which is mainly divided into two steps: the forward process and the reverse process. First, the forward process converts the data with a target distribution of $x_0 \sim q(x_0)$ to noise and reduces the noise $x_T \sim \mathcal{N}(0, I)$ to data by the reverse process in T time steps. Among them, the forward process is shown in Eq. (1):

$$x_t = \sqrt{\bar{a}_t} x_0 + \sqrt{1 - \bar{a}_t}\, \epsilon \tag{1}$$

where $t \sim [0, T]$, x_0 is the initial picture of the input and ϵ is the added Gaussian noise. $a_t = 1 - \beta_t$, $\bar{a}_t = \prod_{i=0}^{t}(1 - \beta_i)$, $\beta_i \sim (0, 1)$ is a fixed variance hyperparameter. In the inverse process, the model predicts the noise $\epsilon_0(x_t, t, x_s, x_r)$ by approximating the inverse mapping to obtain x_{t-1}, as shown in Eq. (2):

$$x_{t-1} = \frac{1}{\sqrt{a_t}} \left(x_t - \frac{1 - a_t}{\sqrt{1 - \bar{a}_t}} \epsilon_\theta(x_t, t, x_s, x_r) \right) + \sigma_t \epsilon \tag{2}$$

where σ_t is the hyperparameter and the noise $\epsilon \sim \mathcal{N}(0, I)$.

3 Method

Although the structure of Yi characters is similar to that of Chinese characters, the absence of authoritative standards for stroke and component segmentation

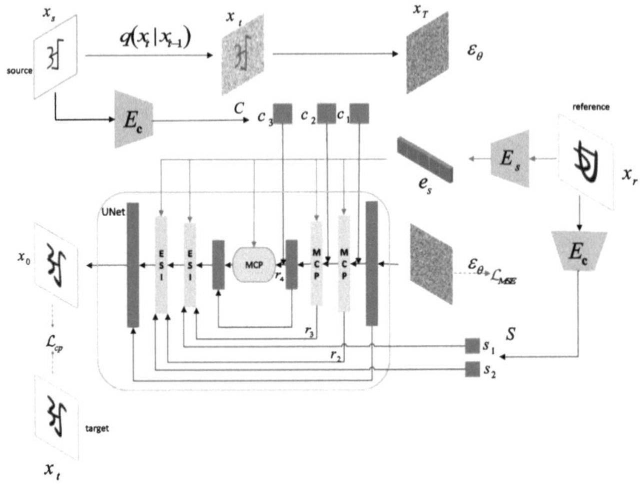

Fig. 1. Summarize the methodology of this paper.

makes it difficult to utilize character structural prior knowledge for font generation. To address this, we propose a novel Yi font generation method based on the diffusion model, named YiFont-G (Yi Font Generation). As shown in Fig. 1, the method adopts a UNet-based conditional diffusion modeling framework, where structural and style features are extracted by the content encoder and the style encoder, respectively. The model takes a source image and a reference image as inputs to generate an image that retains the textual content of the source while adopting the font style of the reference. The multi-scale content features extracted from the source image are injected into the diffusion backbone via the Multi-scale Content Perception (MCP) module. Meanwhile, the reference structural features guide spatial deformation through the Efficient Style Insertion (ESI) block, and the embedded style features are introduced into the diffusion process.

3.1 Multi-scale Content Perception

As shown in Fig. 2, the MCP module injects multi-scale content and style information into the UNet of the diffusion model. Specifically, the content encoder E_c extracts multi-scale content features $C = \{c_1, c_2, c_3\}$ from the source image x_s, while the style encoder E_c obtains the style feature e_s from the reference image x_r. These features are then injected into the UNet via MCP blocks for further processing. Each content feature c_i is embedded through a corresponding MCP block and fused with the UNet feature r_i to produce the channel feature I_c, which is subsequently refined by a channel-space dual-domain mechanism, as shown in Eq. (3–5):

$$W_c = \sigma(\text{SiLU}(S'_c \cdot W_1) \cdot W_2) \tag{3}$$

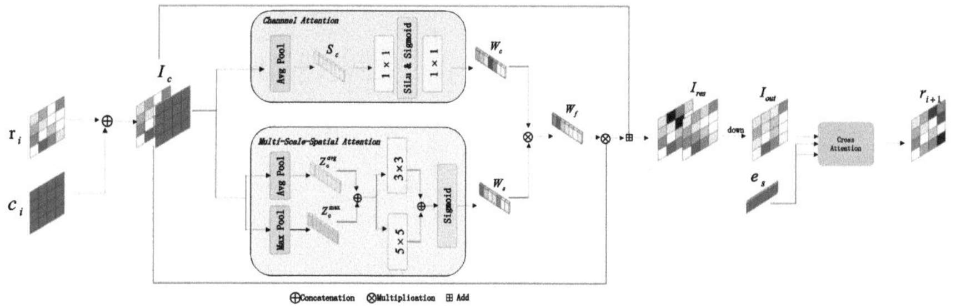

Fig. 2. Multi-scale Content Perception Block.

$$S_c' = \mathrm{Conv}_{1\times 1}(\mathrm{GN}(S_c))$$
$$S_c = \mathrm{pool}_a(I_c)$$

where, W_1, W_2 are the weight matrices of two fully connected layers, and W_c is the resulting channel-aware vector. The Swish activation is defined as $\sigma(x) = x \cdot \mathrm{sigmoid}(\beta x)$, while the SiLU function is $\mathrm{SiLU}(x) = x \cdot \mathrm{sigmoid}(x)$, and GN denotes Group Normalization. pool_a denotes average pooling, and S_c represents the global channel statistics obtained via average pooling.

$$W_s = \sigma(F_3 + F_5) \tag{4}$$

$$F_3 = \mathrm{Conv}_{3\times 3}(Z_s^{avg}) + \mathrm{Conv}_{3\times 3}(Z_s^{max})$$
$$F_5 = \mathrm{Conv}_{5\times 5}(Z_s^{avg}) + \mathrm{Conv}_{5\times 5}(Z_s^{max})$$

$$Z_s^{avg} = \mathrm{pool}_a(I_c^t)$$
$$Z_s^{max} = \mathrm{pool}_m(I_c^t)$$

where, $W_s \in \mathbb{R}^{B\times H\times W}$ is the spatial perception vector. F_3 and F_5 extract spatial information via 3×3 and 5×5 convolutions, respectively. Z_s^{avg}, $Z_s^{max} \in \mathbb{R}^{B\times H\times W}$ represent average and max pooling across channels, and pool_m denotes max pooling.

$$I_{out} = \mathrm{Conv}_{1\times 1}(\mathrm{GN}(I_{res})) \tag{5}$$

$$I_{res} = I_{attn} + I_c$$

$$I_{attn} = I_c \times W_f$$

$$W_f = W_c[:, :, \mathrm{None}, \mathrm{None}] \times W_s[:, \mathrm{None}, :, :]$$

fusing W_c and W_s yields W_f, where $[:, :, \mathrm{None}, \mathrm{None}]$ and $[:, \mathrm{None}, :, :]$ denote that W_c is applied to all feature maps and W_s acts on all channels, respectively. The weighted feature I_{attn} is then computed, followed by residual concatenation to produce I_{res}. A 1×1 convolution is used to downsample and adjust channel dimensions, yielding I_{out}. Finally, a cross-attention module fuses the reference stylized feature e_s with I_{out} to obtain I_{i+1}.

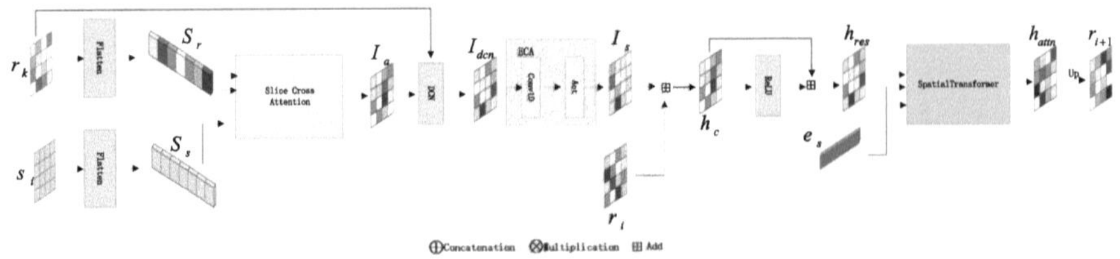

Fig. 3. Efficient Style Insertion.

3.2 Efficient Style Insertion

As shown in Fig. 3, the Efficient Style Insertion (ESI) module injects the style and structural features of the reference image into the UNet backbone. Specifically, the structural features $S = \{s_1, s_2\}$ extracted from the reference image x_r by the content encoder E_c are fed into the two skip connections to guide intermediate feature deformation and enable long-range interactions via ESI. Each s_i modulates the current skip-layer feature r_k through a cross-attention mechanism to produce I_{dcn}, which is further refined by channel attention to obtain I_s. I_s is then concatenated with the current UNet feature r_i to form h_c, which is transformed via a spatial transformer to embed the stylized feature e_s of the reference image into h_{res}, resulting in h_{attn}. Finally, r_{i+1} is produced through upsampling, as shown in Eq. (6):

$$
\begin{aligned}
r_{i+1} &= \mathrm{US}(h_{attn}) \\
h_{attn} &= \mathrm{ST}(h_{res}, e_s)
\end{aligned}
\tag{6}
$$

$$
\begin{aligned}
h_{res} &= \mathrm{ReLU}(W_1 \cdot h_c + b_1) + h_c \\
h_c &= \sigma(\mathrm{Conv1D}(I_{dcn})) + r_i
\end{aligned}
$$

$$
I_{dcn} = \mathrm{DCN}(r_k, f_{offset}(\mathrm{softmax}\left(\frac{QK^\top}{\sqrt{d_k}}\right) V))
$$

$$
Q = W_Q \cdot S_s, \quad K = W_K \cdot S_r, \quad V = W_V \cdot S_r
$$

$$
S_s = \mathrm{flatten}(s_i), \quad S_r = \mathrm{flatten}(r_k)
$$

where, US denotes upsampling, ST refers to the Spatial Transformer Networks module, and Conv1D represents one-dimensional convolution. ReLU is the activation function, and b_1 is the bias vector for input adjustment. DCN denotes Deformable Convolutional Networks, with f_{offset} as the convolutional layer for learning offsets. W_q, W_k, W_v are learnable weight matrices. flatten refers to flattening the feature map into a one-dimensional vector.

4 Experiments

The experiments employ two types of evaluation metrics: pixel-level metrics (L1 loss and SSIM) to measure pixel-wise similarity between generated images and ground truth, and perceptual metrics (FID and LPIPS) to assess visual quality based on deep feature similarity, simulating human perception. We use a publicly available Yi language dataset [24–26], containing 5,825 high-quality images of Yi characters. For evaluation, the test set is split into 360 standard font samples (ST) and 240 artistic font samples (AT). Examples rendered in different styles are shown in Fig. 4. The NuosuSil font is used as the source font in the generation task.

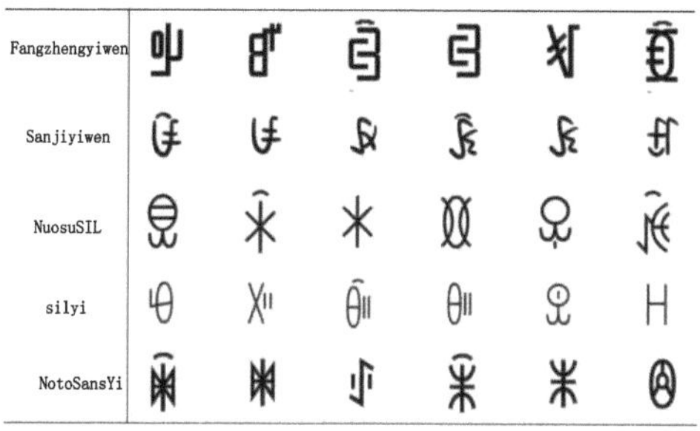

Fig. 4. Yi characters in different style fonts.

4.1 Implementation Details

In this paper, training is conducted using the AdamW optimizer ($\beta_1 = 0.9$, $\beta_2 = 0.999$) with an image resolution of 96×96, a batch size of 4, a total of 440,000 training steps, an initial learning rate of 1×10^{-4}, and a linear learning rate scheduling strategy. To enhance model robustness, the source or reference image is randomly dropped with a probability of 10% during training. All experiments were conducted in a Linux environment with CUDA 11.8, PyTorch 1.13, and Python 3.9. The hardware setup consisted of a single NVIDIA RTX 3090 GPU.

To enhance the guidance effect of the conditional inputs (content image x_s and style image x_r), a classifier-free guidance strategy is adopted during sampling. The unconditional inputs are set to pure white images (pixel value 255), denoted as $\varnothing$. The sampling process is described in Eq. (7).

$$\epsilon_\theta(x_t, t, x_s, x_r) = (1 - s)\epsilon_\theta(x_t, t, \emptyset, \emptyset) + s\epsilon_\theta(x_t, t, x_s, x_r) \quad (7)$$

where s is the guidance scale, which is set to 7.5 in the experiments. To improve the sampling efficiency, the DPM-Solver++ sampler is employed with the number of sampling steps set to 20.

4.2 Comparison with State-of-the-Art Methods

In this section, the paper proposes the YiFont-G method to compare the results with methods such as DGFont [27], CF-Font [28] and FontDiffuser [18]. The experimental results are shown in Table 1.

Table 1. Comparison of experiment result

Method	ST				AT			
	SSIM↑	L1↓	FID↓	LPIPS↓	SSIM↑	L1↓	FID↓	LPIPS↓
DG-Font	0.7663	0.1340	45.1151	0.1445	0.6848	0.1838	60.8760	0.2122
CF-Font	0.7830	0.1160	40.7634	0.1451	0.7027	0.1736	53.7360	0.2068
FontDiffuser	0.8053	0.0992	38.3557	0.1350	0.7211	0.1578	50.1168	0.1914
YiFont-G	0.8113	0.0931	37.2101	0.1352	0.7307	0.1535	49.1071	0.1842

As shown in Table 1, YiFont-G achieves overall superior performance, outperforming existing methods in both the ST and AT groups. In the experiments of font generation in the ST group, it ranks first in SSIM, L1 and FID indicators, while ranking second in LPIPS; In the experiments of font generation in the AT group, it leads across all four metrics, including SSIM, L1, FID and LPIPS. These results demonstrate the strong effectiveness of YiFont-G in both Yi font generation and multi-style font generation.

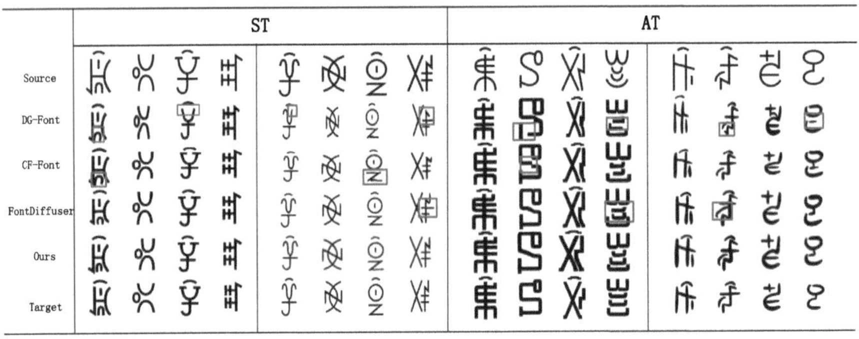

Fig. 5. Comparison of the effects of ST and AT, with red boxes indicating the shortcomings of other methods.

Traditional pixel-level metrics (e.g., SSIM and L1) often deviate from human perception, as visually superior results may not align precisely with the target image, a phenomenon that is particularly evident in the experiment of the AT group. As shown in Table 1, YiFont-G outperforms other methods in perceptual metrics while maintaining strong pixel-level scores, indicating a better balance between visual fidelity and artistic expression.

Figure 5 presents a visual comparison of fonts in the ST and AT groups. The results demonstrate that YiFont-G effectively generates high-quality Yi fonts with complete structures and consistent styles, showing clear advantages in three aspects: (1) avoiding missing strokes and preserving structural integrity; (2) maintaining correct proportions and component positions for reasonable layout; (3) transferring font styles while ensuring stylistic consistency.

4.3 Ablation Studies

This section evaluates the effectiveness of the proposed MCP and ESI modules through ablation studies on both ST and AT font groups. Each module is incrementally added to the baseline to assess its individual and combined impact on overall performance.

Table 2. Results of ablation experiments

Model		SSIM↑	L1↓	FID↓	LPIPS↓
M	E				
×	×	0.7632	0.1285	44.2363	0.1633
✓	×	0.7665	0.1264	43.9458	0.1620
×	✓	0.7670	0.1269	43.2699	0.1604
✓	✓	0.7710	0.1233	43.1586	0.1594

As shown in Table 2, M and E denote the MCP and ESI modules, respectively. The first row represents the baseline, which does not include the additional modules. After incorporating the MCP and ESI modules, the four metrics (SSIM, L1, FID and LPIPS) show significant improvement. Moreover, both the MCP and ESI modules contribute to enhanced visualization quality.

As shown in Fig. 6, red boxes mark defects in the generated fonts, while blue boxes show improvements after adding different modules. The results confirm that introducing MCP and ESI effectively addresses stroke sticking and breaking in generated Yi fonts.

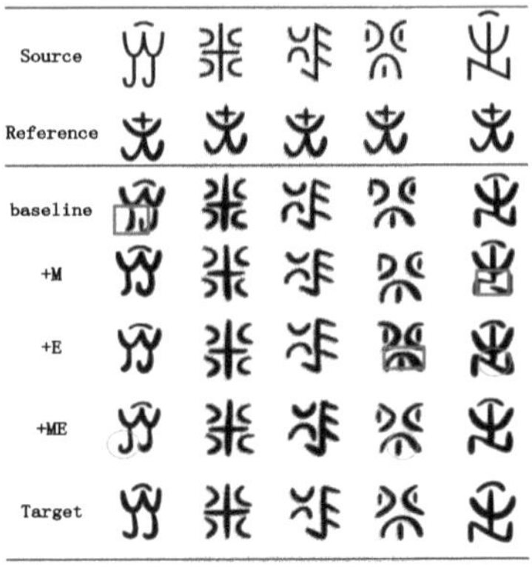

Fig. 6. Visualization results of ablation studies. (Color figure online)

5 Conclusion

This paper proposes a Yi font generation method tailored for complex morphological writing systems. The model is based on a multi-scale conditional diffusion mechanism, enabling collaborative optimization of glyph structural integrity and style transfer accuracy. By dynamically weighting key semantic features through channel attention and employing a multi-scale spatial attention network for efficient feature extraction, the proposed method improves the efficiency of cross-modal style transfer for Yi fonts and effectively alleviates feature breakage issues that often arise in stroke connections and nested structures. However, considering the structural similarity between Yi and Chinese fonts, constructing feature vectors such as radicals or strokes as prior information for font generation will be the focus of future research.

References

1. Zhou, Y.L., Liu, H., Chen, J.W., Zhang, W.: The current situation and promotion strategy of minority archives documentary heritage conservation under the perspective of ethnic memory-based on the investigation in Yunan-Guizhou Region. Arch. Sci. Study (Chin.) **176**(5), 101–109 (2020)
2. Li, G., Wang, S.: Study on the morphological intermingling relationship between Yi Scripts and Chinese characters. Des. Res. (Chin.) **14**(6), 135–140 (2024)
3. Sun, H., Ding, X., Sun, J., Yu, H., Zhang, J.: A method for detecting and recognizing Yi character based on deep learning. Comput. Mater. Contin. **78**(2) (2024)
4. Jia, X., Gong, W., Yuan, J.: Handwritten Yi character recognition with density-based clustering algorithm and convolutional neural network. In: IEEE International Conference on Computational Science and Engineering (CSE) and Embedded and Ubiquitous Computing (EUC), vol. 1, pp. 337–341. IEEE, USA (2017)
5. Chen, S.X., Han, X., Lin, X.Y., Liu, Y., Wang, M.G.: MSER and CNN-based method for character detection in ancient Yi books. J. South China Univ. Technol. (Nat. Sci. Edit.) (Chin.) **48**(6), 123–133 (2020)

6. Xiong, T.N., Qiu, J.F., Hu, J.: Collection and sorting method of ancient Yi character images based on deep learning technology. CAAI Trans. Intell. Syst. (Chin.) (2025). https://doi.org/10.11992/tis.202406036
7. Jiang, Y., Lian, Z., Tang, Y., Xiao, J.: DCFont: an end-to-end deep chinese font generation system. In: SIGGRAPH Asia 2017 Technical Briefs, pp. 1–4. ACM, New York (2017)
8. Chang, B., Zhang, Q., Pan, S., Meng, L.: Generating handwritten Chinese characters using CycleGAN. In: IEEE Winter Conference on Applications of Computer Vision (WACV), pp. 199–207. IEEE, USA (2018)
9. Wang, C., Ding, Y., Liu, Y., Zhan, G., Li, Z.: Chinese font generation from stroke semantic and attention mechanism. J. Comput.-Aided Des. Comput. Graph. **34**(8), 1229–1237 (2022)
10. Xue, M., Du, J., Zhang, Z., Zhang, X.: Skeleton and Font Generation Network for Zero-shot Chinese Character Generation. In: arXiv preprint arXiv:2501.08062 (2025)
11. Liu, Y., Ding, Y., Li, X., Zhao, Q., Xu, M.: Unsupervised font generation network integrating content and style representation. J. Comput.-Aided Des. Comput. Graph. **37**(5), 865–876 (2025)
12. Wang, Y., Xiong, K., Yuan, Y., Zhan, X., Sun, H.: EdgeFont: enhancing style and content representations in few-shot font generation with multi-scale edge self-supervision. Expert Syst. Appl. **262**, 125547 (2025)
13. Choi, J., Kim, S., Jeong, Y., Gwon, Y., Yoon, S.: ILVR: conditioning method for denoising diffusion probabilistic models. In: IEEE/CVF International Conference on Computer Vision (ICCV), pp. 14347–14356. IEEE, USA (2021)
14. Sasaki, H., Willcocks, C.G., Breckon, T.P.: Unit-DDPM: unpaired image translation with denoising diffusion probabilistic models. Comput. Vis. Pattern Recogn. (2021). https://doi.org/10.48550/arXiv.2104.05358
15. Saharia, C., et al.: Palette: image-to-image diffusion models. In: ACM SIGGRAPH 2022 Conference Proceedings, pp. 1–10. ACM, New York (2022)
16. He, H., et al.: Diff-Font: diffusion model for robust one-shot font generation. Int. J. Comput. Vis. **132**(11), 5372–5386. Springer, Heidelberg (2024)
17. Fu, B., et al.: Generate like experts: multi-stage font generation by incorporating font transfer process into diffusion models. In: Proceedings of the IEEE/CVF Conf. on Computer Vision and Pattern Recognition (CVPR), pp. 6892–6901. IEEE, USA (2024)
18. Yang, Z., Peng, D., Kong, Y., Zhang, Y., Yao, C., Jin, L.: FontDiffuser: one-shot font generation via denoising diffusion with multi-scale content aggregation and style contrastive learning. In: AAAI Conference on Artificial Intelligence, vol. 38, no. 7, pp. 6603–6611. AAAI, USA (2024)
19. Liu, M.Y., Breuel, T., Kautz, J.: Unsupervised image-to-image translation networks. Adv. Neural Inf. Process. Syst. **30** (2017)
20. Sun, D., Ren, T., Li, C., Su, H., Zhu, J.: Learning to Write Stylized Chinese Characters by Reading a Handful of Examples. arXiv preprint arXiv:1712.06424 (2017)
21. Zhang, Y., Zhang, Y., Cai, W.: Separating style and content for generalized style transfer. In: IEEE Conference on Computer Vision and Pattern Recognition (CVPR), pp. 8447–8455. IEEE, USA (2018)
22. Gui, D., Chen, K., Ding, H., Huo, Q.: Zero-shot generation of training data with denoising diffusion probabilistic model for handwritten Chinese character recognition. In: Fink, G.A., Jain, R., Kise, K., Zanibbi, R. (eds.) Document Analysis

and Recognition - ICDAR 2023. ICDAR 2023. LNCS, vol 14188. Springer, Cham (2023). https://doi.org/10.1007/978-3-031-41679-8_20
23. Nikolaidou, K., et al.: WordStylist: styled verbatim handwritten text generation with latent diffusion models. In: Fink, G.A., Jain, R., Kise, K., Zanibbi, R. (eds.) Document Analysis and Recognition - ICDAR 2023. ICDAR 2023. LNCS, vol. 14188, pp. 384–401. Springer, Cham (2023). https://doi.org/10.1007/978-3-031-41679-8_22
24. Walters, S.: The interaction of digital technology and literacy practices in nuosu script, China. J. Southeast Asian Linguist. Soc. 154–170 (2024)
25. SIL International: Nuosu SIL Font Project. https://github.com/silnrsi/font-nuosu. Accessed 25 Oct 2024
26. FounderType: FounderType Homepage. https://www.foundertype.com. Accessed 25 Oct 2024
27. Xie, Y., Chen, X., Sun, L., Lu, Y.: DG-Font: deformable generative networks for unsupervised font generation. In: IEEE/CVF Conf. on Computer Vision and Pattern Recognition (CVPR), pp. 5130–5140. IEEE, USA (2021)
28. Wang, C., Zhou, M., Ge, T., Jiang, Y., Bao, H., Xu, W.: CF-Font: content fusion for few-shot font generation. In: IEEE/CVF Conf. on Computer Vision and Pattern Recognition (CVPR), pp. 1858–1867. IEEE, USA (2023)

As AI Meets Gaokao: Fine-Tuning GPT-4o for Enhanced Performance in Objective Questions

Hu Shiye, Hu Xiaolong, Lai Shuhong, and Waqar Ali(✉)

Department of Computer Science, College of Science, Mathematics, and Technology, Wenzhou-Kean University, Wenzhou 325060, China
waali@kean.edu

Abstract. This study explores the application of artificial intelligence (AI) in education by fine-tuning a large language model, GPT-4o. We fine-tuned GPT-4o and proposed an integrated prompt engineering framework to accommodate the diverse formats and cognitive demands of Gaokao questions. Experimental results show that the fine-tuned GPT-4o achieves an accuracy of 84.04% on objective questions, demonstrating substantial gains in language-related tasks. However, performance remains limited in science subjects requiring multi-step logical reasoning and precise numerical computation. These findings highlight both the promise and limitations of LLMs in educational contexts. This work offers practical insights into optimizing AI for academic use and lays a foundation for future development of intelligent, personalized learning tools aimed at reducing educational inequality and improving global access to quality education.

1 Introduction

In modern educational systems, standardized testing plays a crucial role, especially in college admissions and graduate school selection. In China, the National College Entrance Examination (Gaokao) remains the primary pathway to higher education, deeply impacting students' academic and career trajectories. As Howlett (2018) noted, many students view the Gaokao as their only opportunity for social mobility or to "change their fate" [5]. Despite the significant time and effort devoted to preparation, data from Zhejiang's 2024 Gaokao show that nearly half of the candidates scored below 60% of the total, even in regions with relatively strong educational resources. Meanwhile, the number of Gaokao participants rose by 510,000 nationwide in 2024, intensifying the competition among examinees [9]. These trends raise pressing questions about how to improve student performance on the Gaokao to expand access to quality higher education.

Recent advances in AI, particularly the rise of large language models, have opened new possibilities for enhancing education through AI-powered content creation, personalized learning materials, and automated assessment. The introduction of transformer architectures in 2017 revolutionized machine learning,

J. Li et al. (Eds.): APWeb-WAIM 2025, CCIS 2660, pp. 119–130, 2026.
https://doi.org/10.1007/978-981-95-7996-9_11

especially in text processing [6]. Models such as OpenAI's GPT-4o have shown high accuracy in language and reasoning tasks [12], generating human-like responses. These capabilities align well with the demands of standardized tests like the Gaokao, making the fine-tuning of GPT-4o for Gaokao applications a promising research direction.

To this end, this study investigates the application of a fine-tuned GPT-4o model to address the challenges of the Gaokao. Our primary objective is to enhance the model's utility for academic use through a dual approach that combines targeted fine-tuning with comprehensive prompt engineering. We train and evaluate the model on the GAOKAO-Bench dataset, focusing on its effectiveness in handling objective questions that require complex reasoning. Through a critical analysis of the model's performance, we identify its strengths in language-based tasks and its limitations in scientific reasoning, aiming to provide practical insights for students and educators. Our experimental results demonstrate the significant potential of AI in education and point toward the development of intelligent, equitable learning technologies. Specifically, our work makes the following contributions:

1. We develop a novel fine-tuning strategy for GPT-4o on the GAOKAO-Bench dataset, achieving a state-of-the-art accuracy of 84.04% on objective Gaokao questions.
2. We propose a comprehensive prompt engineering framework that integrates zero-shot prompting, chain-of-thought reasoning, and dynamic prompt adaptation to effectively handle the diverse formats and cognitive challenges of Gaokao tasks.
3. We provide an in-depth analysis of GPT-4o's performance, offering valuable insights to guide future advancements in AI-driven educational technologies.

The remainder sections are organized as follows: Sect. 2 reviews related work on the application of large language models in education and standardized testing. Section 3 details our proposed methodology, including the fine-tuning strategy and integrated prompt engineering framework. Section 4 presents our experimental results and key findings across various Gaokao subjects. Finally, Sect. 5 discusses future research directions and concludes the paper with reflections on the broader implications of AI-enhanced learning systems.

2 Related Work

This section reviews the literature foundational to our work. We cover the use of Large Language Models (LLMs) in educational benchmarks, the specifics of the GAOKAO-Bench dataset, and the core optimization techniques of fine-tuning and prompt engineering.

2.1 Large Language Models in Education

Large Language Models have gained increasing attention for their applications in educational domains. Recent research continues to explore advanced techniques to enhance these capabilities, for instance, through mutual correlation networks for few-shot learning, which is particularly relevant in educational settings where extensive labeled data for every specific task might be scarce [1]. In parallel, Ross et al. fine-tuned GPT-4o on a curated set of studentteacher dialogues to develop GuideLM—an LLM that delivers scaffolding, Socratic guidance, and concise explanations in programming education. They demonstrated a notable improvement in pedagogical alignment metrics. [11]. Furthermore, LLMs have also been embedded in intelligent tutoring systems and educational tools such as Duolingo[1] and Khan Academy[2].

Several benchmark datasets have been proposed to evaluate these capabilities systematically. Notably, the Massive Multitask Language Understanding (MMLU) dataset [3] and BIG-Bench [13] are widely used to measure LLM performance across diverse academic disciplines such as mathematics, history, and computer science. However, such datasets often fail to capture the complexity, domain specificity, and linguistic structure of real-world, high-stakes examinations. As a result, they provide limited insights into how LLMs perform in culturally and linguistically distinct testing environments, such as Gaokao.

2.2 GAOKAO-Bench

Zhang [15] introduced the GAOKAO-Bench to address the limitations of existing benchmarks in evaluating LLM performance on real-world examinations. The dataset consists of 2,811 exam questions collected from China's Gaokao between 2012 and 2022, spanning both science and liberal arts subjects, as summarized in Table 1. The questions are categorized into objective formats and subjective formats, making GAOKAO-Bench a more comprehensive benchmark for assessing both factual knowledge and complex reasoning abilities in LLMs.

Using GAOKAO-Bench, Zhang evaluated the performance of several large language models, including GPT and E'miebot, across humanities and science domains. Among the tested models, GPT-4o demonstrated the highest overall accuracy, particularly in humanities-related tasks. These results highlight the potential of advanced LLMs to tackle multi-layered educational challenges, while revealing their current limitations in science-oriented reasoning tasks.

2.3 Domain-Specific Optimization of LLMs

Fine-tuning is a crucial method for adapting pretrained LLMs to specialized tasks. Task-specific fine-tuning [2] and domain-adaptive pretraining [4] have demonstrated significant success in improving performance on downstream tasks. For instance, recent work has increasingly focused on parameter-efficient fine-tuning (PEFT) methods like Low-Rank Adaptation (LoRA) [10], which enable models like GPT-4 to be specialized for diverse applications with minimal computational overhead, making large-scale adaptation more feasible.

[1] https://www.duolingo.com/.

[2] https://www.khanacademy.org/.

Table 1. Overview of question types in the GAOKAO-Bench dataset, highlighting subject distribution and the balance between objective and subjective formats for evaluating LLM capabilities.

	Question Type	Number	Percentage
Objective	Single Choice	1418	50.5%
	Multi-Question Choice	273	9.7%
	Multi-Choice	64	2.3%
	Five out of Seven	26	0.9%
Subjective	Open-ended Question	786	28.0%
	Cloze	218	7.8%
	Correction	26	0.9%

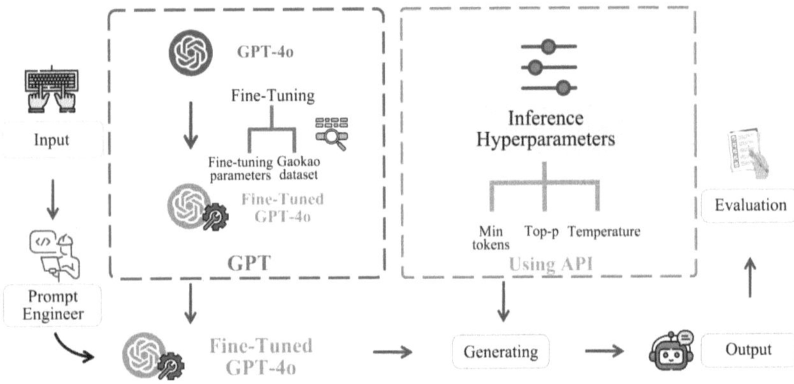

Fig. 1. Overall workflow of the proposed approach, illustrating the fine-tuning process on Gaokao-Bench, the interactive prompt engineering interface, and the end-to-end pipeline from question input to solution-oriented model output.

In addition to modifying model weights, prompt engineering has emerged as a powerful tool for enhancing an LLM's intrinsic reasoning capabilities. Chain-of-thought (CoT) prompting [14] has been particularly influential. By encouraging the model to "think step-by-step," CoT significantly improves performance on tasks requiring multi-step reasoning, such as solving mathematical problems. This ability to generate explanation-rich responses makes it highly suitable for the demands of educational assessment.

3 Methodology

In this section, we present our dual approach to aligning GPT-4o with Gaokao-style educational tasks by combining targeted fine-tuning on GAOKAO-Bench and a customized prompt engineering framework. Building on prior research, we fine-tune GPT-4o using OpenAI's platform, and systematically evaluate hyperparameter settings and prompt formats as illustrated in Fig. 1.

3.1 Problem Statement

The Gaokao, China's rigorous national college entrance exam, demands extensive preparation and often exposes gaps in student support despite their efforts. While LLMs like GPT-4o offer promise for educational tools, they face challenges with Gaokao's complexity-particularly in science subjects requiring multi-step reasoning and domain-specific precision [15]. This study explores improving LLM performance on such tasks through targeted fine-tuning and prompt engineering.

3.2 Fine-Tuning

Fine-tuning is an important intermediate stage in adapting large language models (LLMs) for Gaokao-style objective questions. We fine-tuned GPT-4o on a subset of the GAOKAO-Bench dataset, composed of multiple topics. We formatted the dataset into JSONL files that match the OpenAI fine-tuning requirements.

3.3 Hyperparameter Selection

Key hyperparameters for the fine-tuned GPT-4o were optimized empirically. We set the Temperature to 0.3 to balance factual consistency with creativity. The Top-p was maintained at the default of 1.0, and Max-tokens was set to 4096 to allow for comprehensive and detailed responses.

3.4 Prompt Engineering

To further boost performance, we designed our framework includes Zero-Shot, Chain-of-Thought (CoT), and Dynamic Prompting.

Zero-Shot Prompting: This approach provides only the question, with no examples. For example: **Question:** What is the main idea of the passage below? **Passage:** Deforestation causes soil erosion, disrupts the water cycle, and reduces carbon absorption.

Chain-of-Thought Prompting: Proposed by Wei et al. [14], CoT guides the model through intermediate steps. **Example:** Deforestation causes erosion, water cycle disruption, and climate change. These are all negative effects. Therefore, deforestation leads to serious environmental harm. **Answer:** Deforestation causes significant environmental damage.

Dynamic Prompting: Dynamic prompts are tailored to question type and complexity, offering structured guidance to improve clarity and performance, as illustrated in Table 2.

Integrated Prompt Design Strategy: Our prompting strategy integrates three techniques: zero-shot prompting, dynamic prompt selection, and chain-of-thought (CoT) reasoning. In practice, for each multiple-choice question, a subject-specific prompt is dynamically selected to instruct the model to derive a step-by-step rationale (CoT) without any preceding examples (zero-shot).

Table 2. Dynamic Prompting Examples in Liberal Arts and STEM

Liberal Arts	**Q:** Analyze the emotional tone of a poem. **Prompt:** 1. Identify imagery. 2. Describe emotional impact. 3. Infer author's intent.
STEM	**Q:** Why does a balloon expand when heated? **Prompt:** 1. State the physical law (e.g., gas law). 2. Explain heat's effect on particles. 3. Link to balloon expansion.

3.5 Implementation and Evaluation Tracking

To ensure objectivity and reproducibility, our data collection pipeline was fully automated. For each task, we logged the input prompt, the model's raw JSON response, and the associated OpenAI API request ID, all of which were subsequently parsed and stored for analysis without any manual intervention.

4 Result

This section analyzes the performance of our fine-tuned GPT-4o on GAOKAO-Bench, highlighting a key disparity between the model's strengths in language tasks and its limitations in science-related reasoning.

4.1 Experimental Setup

To explore the applicability of large language models in educational assessment, we developed a systematic setup based on fine-tuning GPT-4o with Gaokao-related data. This section describes the selected model, fine-tuning strategy, and evaluation benchmark.

Selected LLM Model. We employed GPT-4o, a cutting-edge language model capable of handling complex reasoning and multilingual understanding. The model was chosen for its balanced trade-off between performance and cost, and its compatibility with educational tasks requiring both factual knowledge and contextual interpretation.

Fine-Tuning Process. The fine-tuning process was conducted using GAOKAO-Bench. The dataset includes a wide range of subjects, with both multiple-choice and short-answer formats. The model was fine-tuned using supervised learning on this dataset to enhance its accuracy. Table 3 outlines the training configuration selected specifically for GAOKAO-Bench.

GAOKAO-Bench-2023. GAOKAO-Bench-2023 was adopted as the evaluation benchmark in this study. It consists of real 2023 Gaokao exam objective questions across multiple subjects. The test set contains a total of 183 objective questions distributed, as shown in Table 4.

Table 3. Fine-tuning aspects and training outcomes for GPT-4o.

Tuning Aspect	Specification
Training tokens	1,781,259
Batch size	2
Learning rate multiplier	2
Number of epochs	3
Final training loss	1.248
Initial training loss	2.5
Training loss curve	A concave shape, indicating effective convergence

Table 4. Objective Questions Distribution in GaokaoBench-2023 Test Set

Category	Sub-category	Question Count
STEM Subjects	Math	49
	Physics	11
	Chemistry	9
	Biology	15
	Geography	6
	Political Science	29
	History	33
English Tasks	Cloze Test	4
	Fill-in-the-Blanks	4
	Reading Comprehension	16
Chinese Tasks	Language and Usage	3
	Modern Literature	4
Total		**183**

4.2 Evaluation Results

Overall Performance. This study evaluates a fine-tuned GPT-4o model on objective questions from the GAOKAO-Bench-2023 dataset. As shown in Fig. 2, the model achieved a weighted mean accuracy of 84.04%, demonstrating strong natural language reasoning. Notably, the performance contrast between language and science subjects highlights the model's key strengths and weaknesses (Table 5).

Language Subjects: Exceptional Understanding and High Accuracy. On language-related tasks such as Chinese and English reading comprehension and English cloze tests, the fine-tuned model achieved excellent performance. The model was able to comprehend complex articles, extract key details, and answer questions accurately through logical reasoning.

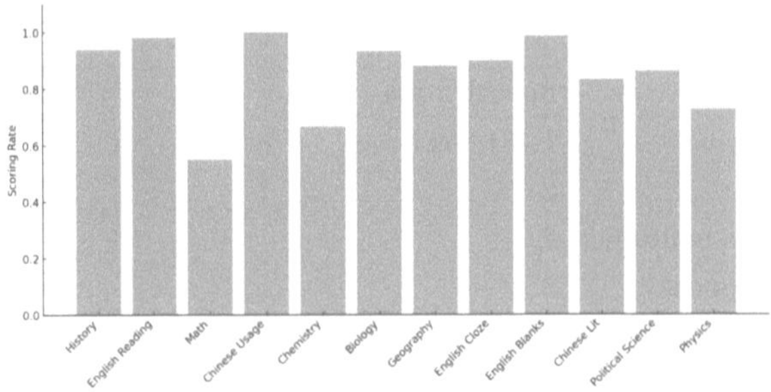

Fig. 2. Scoring rates of the fine-tuned GPT-4o across subjects on the GaokaoBench-2023 objective dataset. The model excels in language tasks but lags in STEM fields like Math and Physics, reflecting challenges in numerical and symbolic reasoning.

Table 5. Accuracy by Subject on GAOKAO-Bench-2023 Objective Questions

Subject	Correct Score	Total Score	Accuracy (%)
Chinese (Lang. Usage + Modern Lit.)	42.0	48.0	87.5
English (Cloze + Fill + Reading)	266.5	275.0	96.9
Math	145.0	245.0	59.2
Physics	48.0	66.0	72.7
Chemistry	42.0	54.0	77.8
Biology	84.0	90.0	93.3
Geography	60	68.0	88.2
History	124.0	132.0	93.9
Political Science	108.0	116.0	86.2
Overall (Weighted)	**919.5**	**1094.0**	**84.04**

Chinese: The model showed a profound knowledge of modern literature and historical texts. That explained why it scored well on nuances like tone, context, and thematic elements, displaying a deep understanding of the language.

English Reading Comprehension: The model accurately identified crucial reasoning clues and inferred from passages adequately. It showed a strong capability to extract and process essential information.

English Cloze Tests: We can see the advanced natural language processing that the model has available as it was doing so well on cloze tests, where the model was able to make accurate predictions and fill in missing context with the goal of grammatical and semantic coherence.

The model's high accuracy across both Chinese and English tasks confirms the value of targeted fine-tuning. This approach effectively adapts GPT-4o for the nuances of bilingual academic assessments, highlighting its potential for future educational tools.

Science Subjects: Challenges in Reasoning and Precision. The fine-tuned model, despite its strong language-related capabilities, has shown significant weaknesses in science-related subjects like mathematics, chemistry, and physics.

Mathematics: The model frequently errs in topics such as set theory, vectors, and trigonometric functions. While it often identifies relevant formulas, it consistently fails in their algebraic execution, accurate application, and maintaining logical continuity in multi-step problems.

Chemistry: Key weaknesses include understanding chemical properties and material compositions. The model frequently misinterprets reaction data, overlooks common interactions like hydrolysis, and incorrectly deduces elements from their properties, suggesting unreliable application of core chemical principles.

Physics: The model struggles with problems of motion under varying forces and applying energy conservation principles, particularly in thermodynamics and nuclear reactions. It often misapplies fundamental laws like Newton's, and its reasoning is frequently contradictory or incomplete.

This indicates that even after fine-tuning, the model struggles with tasks requiring high numerical precision and logical rigor. We posit this limitation likely stems from insufficient exposure to complex scientific problems in the training data. Future research could tackle these multi-step reasoning challenges by integrating external knowledge sources, such as knowledge graphs enhanced via graph-based non-sampling [7], or by employing advanced reasoning frameworks like hierarchical reinforcement learning for complex inference tasks [8].

Other Subjects: Stable and Conceptually Sound Performance. While challenges remain in science subjects, the fine-tuned model exhibited stable and accurate performance in areas such as political science, biology, geography, and history. This shows the power of this model to perform conceptual reasoning, fact-based analysis, and interpretive tasks.

Political Science: Topics concerning labor value, social roles, and institutional frameworks were all left to the model. For example, it grasped questions on labor productivity and costs throughout various environments, giving logical and brief clarifications. That implies that it can deal with theoretical concepts together with practical instances.

Biology: The model was able to handle biological processes, e.g., enzyme activity, cellular transport, energy metabolism (ATP production). It was able to explain topics such as the movement of substances through membranes and the role of biological components in metabolic pathways, and did so with a firm grasp of fundamental biological principles.

History: The model was shown to have robust reasoning in analyzing historical events and their cultural impact. For example, it was able to understand questions about rituals such as Chunqiu Justice and place them in a historical narrative. It demonstrates the capacity to weave historical facts together with ways of seeing and making political meaning.

This strong conceptual understanding is evidenced by the consistent performance of the model across these subjects.

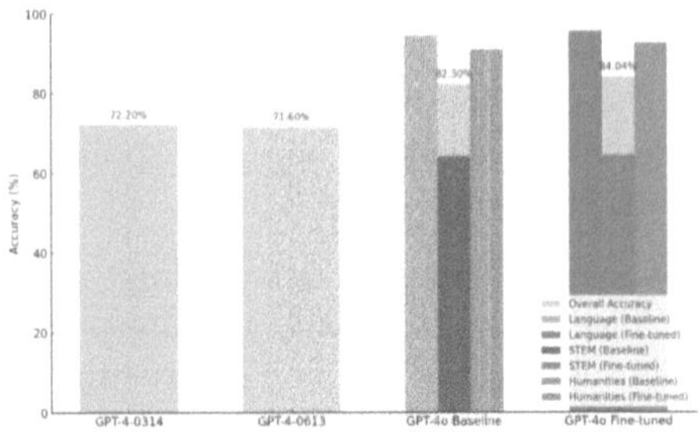

Fig. 3. Overall accuracy comparison of GPT models on GAOKAO-Bench objective questions. GPT-4-0314 and GPT-4-0613 were evaluated on the full dataset, while GPT-4o Baseline and Fine-tuned models were tested on the GAOKAO-Bench-2023 subset. For GPT-4o, results are further categorized by domain: language, STEM, and humanities.

Case Study: Chemistry Question (2023). To better understand the result, we provide a representative example from the Chemistry section of the GAOKAO-Bench-2023 dataset.

Question: Chemistry is closely related to daily life. Which of the following statements is correct?
A. Sodium benzoate can be used as a food preservative because it is acidic.
B. Soy milk produces the Tyndall effect due to the scattering of light by colloidal particles.
C. SO_2 can be used to bleach silk fabrics because it oxidizes the colored substances in them.
D. Vitamin C can be used as an antioxidant in canned fruit because it is difficult to oxidize.

Model Output (Fine-tuned GPT-4o + Prompt Engineering)

Analysis:
A. Sodium benzoate acts as a preservative due to its antibacterial properties, not because it is acidic. Incorrect.
B. Soy milk is a colloid, and the Tyndall effect results from the scattering of light by colloidal particles. Correct.
C. SO_2 bleaches by reduction, not oxidation. Incorrect.
D. Vitamin C is easily oxidized, hence useful as an antioxidant. Incorrect.
Therefore, the correct answer is B.
Answer: B

Discussion: The model demonstrates an accurate understanding of colloidal chemistry and correctly identifies misleading causal logic in distractor choices. This indicates that our fine-tuned model, equipped with step-by-step prompting, not only improves answer accuracy but also generates interpretable and pedagogically valuable explanations.

4.3 Comparative Analysis

To provide a broader perspective on model performance, we compare four versions of the GPT family: GPT-4-0314 and GPT-4-0613 (evaluated on the full GAOKAO-Bench dataset), and GPT-4o Baseline and GPT-4o Fine-tuned (evaluated on the GAOKAO-Bench-2023). As shown in Fig. 3, the overall accuracy improves significantly across versions. While GPT-4-0314 and GPT-4-0613 achieved 72.2% and 71.6%, respectively, the GPT-4o Baseline reaches 82.3%, and the Fine-tuned version further increases to 84.04%.

We further break down performance across 3 major domains on GAOKAO-Bench-2023: language, STEM subjects, and humanities. Figure 3 summarizes domain-specific accuracies. Fine-tuning shows improvement in all domains, language subjects, rising modestly from 94.42% to 95.51%, and STEM subjects, improving from 64.11% to 64.38%. Humanities performance increased about 1.72%, from 90.85% to 92.57%.

5 Conclusion

This study demonstrates that fine-tuned large language models, GPT-4o, could benefit Gaokao-style objective questions. Using fine-tuning and prompt engineering, the model was able to excel at language-based tasks, continuing the model's commitment to advanced natural language understanding and reasoning. Nevertheless, its shortcomings in science subjects spoke to its inability to address multi-step reasoning and complex calculations of a rich domain-specific dataset. This finding highlights the transformative work of AI in education, disentangling the path to enhancing learning efficiency and decreasing disparities through personalised support. In conclusion, this research provides a strong foundation for fine-tuning LLMs on standardised testing.

Future work will enhance the GAOKAO model's real-world applicability by expanding the dataset with diverse questions from various years, regions, and including subjective formats. Advanced techniques such as Retrieval-Augmented Generation and Reinforcement Learning from Human Feedback will be employed to improve reasoning and reduce bias. Efforts will also focus on enhancing interpretability, contextual understanding, and meta-learning adaptability. Finally, the model will be deployed in real educational settings for personalized learning and automated grading, allowing evaluation of its impact while ensuring responsible design.

Acknowledgments. This work was supported by the Wenzhou-Kean University Internal Start-up Research Grant under Grant No. ISRG2024005.

References

1. Chen, D., Chen, F., Ouyang, D., Shao, J.: Mutual correlation network for few-shot learning. Neural Netw. **175**, 106289 (2024)
2. Gururangan, S., et al.: Don't stop pretraining: Adapt language models to domains and tasks. In: Proceedings of the 58th Annual Meeting of the Association for Computational Linguistics, ACL 2020, Online, 5–10 July 2020, pp. 8342–8360. Association for Computational Linguistics (2020)
3. Hendrycks, D., et al.: Measuring massive multitask language understanding. arXiv preprint arXiv:2009.03300 (2020)
4. Hou, Z., Salazar, J., Polovets, G.: Meta-learning the difference: preparing large language models for efficient adaptation (2022)
5. Howlett, Z.: Fateful rite of passage: the national college entrance examination and the myth of meritocracy in post-mao China (2018). https://hcf.fas.harvard.edu/wp-content/uploads/2018/05/HOWLETT-Fateful-Rite-of-Passage-REV-2.pdf
6. II, M.J.B., Katz, D.M.: GPT takes the bar exam. CoRR **abs/2212.14402** (2022)
7. Liang, S., Shao, J., Zhang, J., Cui, B.: Graph-based non-sampling for knowledge graph enhanced recommendation. IEEE Trans. Knowl. Data Eng. **35**(9), 9462–9475 (2023)
8. Luo, X., Zhu, A., Zhang, J., Shao, J.: HierarT: multi-hop temporal knowledge graph forecasting with hierarchical reinforcement learning. Knowl.-Based Syst. **300**, 112164 (2024)
9. Online, C.E.: Statistics of the number of applicants for the college entrance examination in 2024 (2024). https://news.eol.cn/yaowen/202406/t20240606_2614724.shtml
10. Park, H., Kwak, J., Jang, D., Park, S., Choi, J.: Loramap: harnessing the power of lora connections (2024)
11. Ross, E., Kansal, Y., Renzella, J., Vassar, A., Taylor, A.: Supervised fine-tuning llms to behave as pedagogical agents in programming education. arXiv preprint arXiv:2502.20527 (2025). https://arxiv.org/abs/2502.20527, preprint
12. Shahriar, S., et al.: Putting gpt-4o to the sword: a comprehensive evaluation of language, vision, speech, and multimodal proficiency. Appl. Sci. **14**, 7782 (2024). https://doi.org/10.3390/app14177782
13. Suzgun, M., et al.: Challenging big-bench tasks and whether chain-of-thought can solve them. ArXiv.org (2022)
14. Wei, J., et al.: Chain-of-thought prompting elicits reasoning in large language models. In: Advances in Neural Information Processing Systems, vol. 35, pp. 24824–24837. Curran Associates, Inc. (2022)
15. Zhang, X., Li, C., Zong, Y., Ying, Z., He, L., Qiu, X.: Evaluating the performance of large language models on gaokao benchmark (2023)

Scene-Aware Multi-agents on Collaborative Retrieval Augmented Generation Model (CRAG)

Yuefeng Du(✉), Shurong Zhou, Kaiming Zhou, Xiaoguang Li, Tingting Liu, and Hongxin Yang

Liaoning University, Shenyang 110036, Liaoning, China
duyuefeng@lnu.edu.cn

Abstract. The RAG question answering of large model LLMs is currently a hot research direction, widely used in artificial intelligence question answering, such as ChatGPT. RAG Q&A is the process of decomposing and collaborating RAG tasks through multi-agents. However, due to the poor correlation between the retrieved knowledge sources and the poor adaptability of the generated answers to the scene, it brings a problem of semantic illusion when generating semantics in RAG, that is, answers that are not the question asked. In response to this issue, this article proposes a scene-aware multi-agents on collaborative retrieval augmented generation (CRAG) model. CRAG is a master-slave network model, where the main network and the slave network learn semantic information in the scene through the hint module. Specifically, multi-agents achieve consensus among all agents by deploying the same slave network to generate a shared perception of scene information. All agents, with the assistance of scene semantic information, think around the same question when answering RAG Q&A. Finally, comparative experiments were conducted on the TruthfulQA dataset to demonstrate the effectiveness of the proposed method.

Keywords: Multi-Agents collaboration · Second Keyword · Third Keyword

1 Introduction

Multi-agent collaboration has emerged as a pivotal paradigm in artificial intelligence, driven by its transformative potential in domains such as autonomous systems, smart grids, and distributed sensor networks. Unlike normal AI systems, collaborative multi-agent frameworks emphasize decentralized decision-making, where autonomous agents interact, negotiate, and coordinate to achieve shared objectives in dynamic, uncertain environments. This paradigm addresses challenges inherent in single-agent systems, such as limited computational capacity and narrow task-specific adaptability, by distributing intelligence across specialized agents. For instance, in smart manufacturing, agents collaboratively optimize production schedules, resource allocation, and predictive maintenance, demonstrating superior scalability and fault tolerance compared to monolithic architectures.

J. Li et al. (Eds.): APWeb-WAIM 2025, CCIS 2660, pp. 131–142, 2026.
https://doi.org/10.1007/978-981-95-7996-9_12

Currently, multi-agent collaboration [1] has received widespread concerned and wildly application. However, it has gradually gained popularity as a research field: how to control multiple agents to work together to solve the same problem simultaneously? Intelligent agents can make decisions based on their own observations when performing task segmentation, orchestration, and execution independently, demonstrating good functionality and performance. However, when multiple agents collaborate together, the lack of common guidance is particularly prominent for partially observable Markov decision processes (POMDP) [2]. Meanwhile, in the process of multi-agent collaboration, each agent may have a certain degree of perspective deviation compared to other agents due to its own parameter settings. This deviation is not only from the perspective of our observation and understanding of the problem, but also an objective requirement. At the same time, we also do not want significant differences in the understanding of this scenario-based problem, which may lead to ambiguity in the understanding of the problem. This ambiguity can cause RAG to perform contextual semantic merging, where although they are all describing the same subject, the merged context belongs to different contexts and semantics, resulting in a lack of logical coherence and creating a phantom semantics. In this case, the semantic inference given by RAG will give people an illusion that they seem to be saying the same thing, but it is not the same thing.

What the facts hope for is that when describing the same problem, although different perspectives are used, the core and essence of the described problem are consistent. In other words, although the research perspectives may be different, they are studying the same problem. Faced with this problem, during the collaboration process of multi-agent systems, a large amount of communication is required, which not only consumes a significant amount of computing resources, but also causes huge delays in computation, which is unacceptable. Our goal is to enable multi-agent systems to realize that they are only observing and understanding the same thing from different perspectives during collaboration. If multi-agent systems understand this invariance, they can infer the same state consensus from different local observations during decentralized execution and explicitly choose cooperative actions. Try to minimize communication between intelligent agents, or avoid communication altogether.

In addition, for AI interactive human-computer Q&A questions, it is worth studying how RAG can make the questions answered consistent with the user's questions in terms of content and intention, and even influence and guide users to provide more accurate and usable follow-up questions during the process of question inference. Fortunately, users' questions often contain contextual information and contextual information that complements their intentions and thoughts. Therefore, this information can be utilized to enhance collaboration between intelligent agents, enabling RAG to more accurately understand users. This article proposes a scenario driven multi-agent collaborative RAG (CRAG) model. CRAG adopts a master-slave collaborative structure to ensure that multiple agents discuss and communicate around the same problem. By capturing the contextual information of user questions, deep semantic understanding can be achieved.

Specifically, when conducting RAG Q&A through multi-agent collaboration, we face the following challenges:

- The illusion problem of multi-agent collaboration. How to ensure that all intelligent agents are collaborating around the same problem, rather than performing tasks on their own.
- Communication issues in scene collaboration. How to use lower communication costs and ensure efficient collaboration among multiple agents when solving single RAG and multi round RAG problems.

Specifically, the contributions of this article are as follows:

1. A master-slave collaborative structure has been proposed, in which intelligent agents collaborate around the same problem while ensuring their independence.
2. An improved teacher student scenario model was proposed, in which the teacher model is responsible for capturing unified problems, the student model transfers the teacher model, and expands the problems based on the scenario, effectively achieving scenario collaboration among multiple agents.
3. Proposed a cross-scenario RAG improvement method to effectively enhance contextual problem communication.
4. Experiments were conducted on multiple datasets to validate the effectiveness and efficiency of the method.

The organizational structure of the article is as follows: Sect. 2 introduces the relevant work. Section 3 provides the definition of the problem. Section 4 proposes the specific content of CRAG method and scene model for multi-agent collaborative tasks. Section 5 conducted experiments on the research content. Section 6 provides a summary.

2 Related Work

This article studies the multi-agent collaborative RAG problem under scene perception. The most closely related work to the problem studied in this article includes two aspects: collaborative consensus and knowledge distillation in large language models (LLMs).

2.1 Collaborative Consensus

Collaborative consensus is a technical framework for implementing multi-party coordinated decision-making in a distributed environment, with the core goal of local rules and global consistency. Individuals make autonomous decisions based on simple rules (such as obstacle avoidance and resource competition), and global order emerges through local interactions, ensuring that heterogeneous entities take consistent actions under dynamic and uncertain conditions. This concept integrates consensus algorithms, game theory, and social collaboration theory in computer science, and is widely applied in fields such as blockchain, robot collaboration, and medical collaborative governance. As a representation learning framework based on the similarity between samples, contrastive learning [3] has recently emerged as the most popular field in self-supervised learning. While ensuring that the learned representation can retain as much information as possible, contrastive learning promotes similar samples to be close together in the representation space. Many recent studies have focused on mining knowledge from data through self-supervised learning to improve the ability of models to solve downstream tasks. As a typical discriminative method, contrastive learning is easier to train and understand than other generative self-supervised learning methods [4, 5].

Although contrastive learning demonstrates excellent performance, the excessive state input conditions make the contrastive process difficult to control, which is computationally wasteful for multi-agent collaboration. Consensus algorithms and federated learning require separate consensus processes, and the process of method execution is too cumbersome, which is unacceptable for multi-agent collaboration. So, there is still a need for simpler deployment and execution of collaborative consensus methods.

2.2 Knowledge Distillation in LLMs

Distilling the knowledge from a larger teacher into a smaller student has been successfully used to train more efficient and stronger models across a range of applications. Particularly for LLMs, this setting has several advantages– The larger, more expensive teacher only needs to run once, and the saved representations can then be used to train a family of smaller models of various sizes. Due to the large vocabulary size of modern LLMs, naively storing all of these probabilities is infeasible (e.g., requiring 128 PetaBytes of storage for 1T tokens for Llama3 [6]). Instead, sparse knowledge distillation approaches store an efficient Top-K subset of logits from the teacher's distribution [7]). However, these methods still require a large number of logits (6400) to be stored, or even observe a fall in model performance. Multiple research efforts, in the recent past, can be connected to our proposed work [8]. Our work uses similar method of context distillation proposed in [9], wherein the teacher model receives detailed instructions including few-shot examples to generate intermediate reasoning outputs along with the target answer. The student model is then fine-tuned using a simple instruction template on the target answer.

In summary, knowledge distillation using student models can deploy multi-agent systems more quickly and help achieve task collaboration. This study focuses more on how to utilize scene information to ensure consistency in multi-agent collaboration at a deep semantic level.

3 Problem Statements

This article studies how multi-agent systems collaborate in scene RAG. This section provides the relevant definitions of the problem. For a Q&A in RAG, which includes multiple rounds of conversations. A conversation initiated by the user is denoted as Q_i.

Definition 1. Agent. An intelligent agent is a unit in RAG that handles the scheduling and execution of scene tasks, denoted as A_i. For a conversation Q_j, there are $n+1$ agents participating in task collaboration, including a top intelligent agent denoted as A_t and n sub agents denoted as $A_0 \sim A_{n-1}$.

Definition 2. Conversation Semantic. The combination of dialogue and scene information will manifest as scene semantics, denoted as a quadruple $S_{Q_j} = (u, t, l, S_{Q_{j-1}})$, where u presents the participation parties, t presents the participation time, l presents the participation environmental information. S_{Q_j} contains all the scene information related to this round of conversation, among which $S_{Q_{j-1}}$ is used to track the contextual connection between S_{Q_j} and the previous round of conversation $S_{Q_{j-1}}$. This helps to connect contextual information and prevent digression.

For multiple rounds of conversations $Q_0 \sim Q_m$, the first round of conversations is denoted as $S_{Q_0} = (u, t, l, \emptyset)$ because of no other conversation information before it.

Definition 3. Conversation Semantic Label. For scene semantics S_{Q_j}, an intelligent agent A_i will obtain its own understanding when interpreting it, and classify this result as a certain semantic label, denoted as $label_{S_{Q_j}}(A_i)$. How an intelligent agent obtains scene semantic labels will be explained in detail in Sect. 4.2.

Based on this, the definition of scene aware multi-agent collaborative consensus studied in this article is provided.

Problem Statements. For a conversation Q_j, all agents $A_t, A_0 \sim A_{n-1}$ have scene semantic labels of $label_{S_{Q_j}}(A_t)$, $label_{S_{Q_j}}(A_0) \sim label_{S_{Q_j}}(A_{n-1})$, and these agents have collaborative consensus on scene semantics. The sufficient and necessary condition is that all scene semantic labels satisfy $label_{S_{Q_j}}(A_t) = label_{S_{Q_j}}(A_0) = \cdots = label_{S_{Q_j}}(A_{n-1})$.

This question explains whether all agents maintain a consistent understanding of the content of the RAG question and answer they receive. Noted that, the collaborative consensus of multi-agent scene semantics does not mean that the scene semantics of all agents must be completely consistent. These intelligent agents only need to maintain consistency in their understanding of the problem subject. That is to say, a single intelligent agent maintains local independence and retains individual local features during observation and execution. However, all intelligent agents exhibit consistency in the global state.

4 Collaborative Retrieval Augmented Generation (CRAG)

This paper proposes a Collaborative Retrieval Augmented Generation (CRAG) model for the collaborative problem of multi-agent systems, which can enhance the consistency of collaborative task execution by perceiving scene information.

4.1 Architecture

The communication between nodes can effectively ensure the consistency of consensus states among multiple agents, but this will bring high costs and communication delays. Compared to using a unified task allocation, it can maintain the recognition and consensus of all tasks on the execution of the scenario. This paper proposes a CRAG framework based on multi-agent collaboration to address the consensus problem of scene semantic understanding in RAG, as shown in Fig. 1.

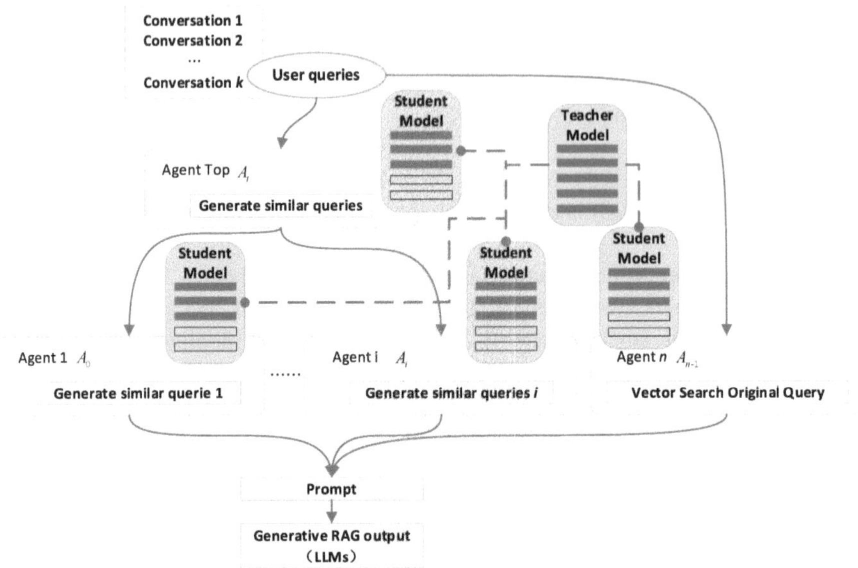

Fig. 1. The overall architecture of CRAG

The input of the framework is the user's multiple rounds of conversations in RAG. In this framework, the LlamaIndex model is used to block, vectorize, and search index the input conversations, which is the same as the general process of RAG.

Hierarchical indexing and content enhancement. For the vectorized queries obtained from the input, this framework uses two modules: the original query and the approximate query of vector search to understand the vectorized queries. The task allocation and control work of specific understanding is completed by multi-agent collaboration. For approximate queries, any top-level intelligent agent is selected, which is responsible for coordinating the generation of a topic related to the conversation and assigning the topic to other secondary intelligent agents, guiding them to generate specific similar queries. This master-slave structure allows all secondary agents to revolve around a main topic and guide them to expand their respective main topics. The original query is responsible for generating cross reference queries to prevent approximate queries from getting off topic.

In order to enhance each agent's understanding of scene tasks, CRAG includes a scene semantic collaborative perception model. This model is a scenario aware student teacher knowledge distillation model. Train a precise network for scene task perception using a teacher model, and then use knowledge transfer techniques to transfer some of the perception networks in the teacher model and deploy them to all intelligent agents. Intelligent agents use student models to perceive scene semantics. The common part of all student models comes from the perception network of the same teacher model, so that each agent uses the same model to perceive scene information when receiving scene tasks. In this way, all intelligent agents understand tasks around the same scenario, enhancing the collaboration of multiple agents.

Finally, all agents input their generated approximate queries and the content results of the original queries into prompt based on their understanding of the scenario task, and LLMs generate RAG outputs.

When generating RAG output in the CRAG framework, each agent uses a unified distilled knowledge model to understand the same scenario task, ensuring consistency in collaboration among agents.

4.2 Scene-Aware Knowledge Distillation

The general model of knowledge distillation includes teacher networks and student networks. Teacher networks typically have higher accuracy, more parameters, and higher latency. The student network is a small network obtained through knowledge distillation through the teacher network, with accuracy close to that of the teacher network, but with fewer parameters and lower deployment costs, suitable for the requirements of multi-agent consensus scenarios. In order to enhance the model's understanding of scene tasks, we have added the hint4scene module to capture the content of scene information. Enable multi-agent systems to enhance collaborative consistency based on scene information during task execution. The scene-aware knowledge distillation model is shown in Fig. 2.

From the figure, it can be seen that for the same scene feature dialogue input S, T. For conversation Q_j,S, T contains the same scene semantics S_{Q_j}. The observation values obtained by the teacher network and the student network are $\hat{T}, \hat{S}$ respectively. Dialogue input using M3E model [10] for embedding representation. Firstly, the loss functions for training the teacher network and the student network are given. As shown in formulas (1) and (2).

$$L_T = L_{CE}(T, \hat{T}) \tag{1}$$

$$L_S = \alpha L_{CE}(S, \hat{S}) + \beta L_{KD}(S, T) \tag{2}$$

where L_{CE} represents cross entropy loss, L_{KD} represents distillation loss, L_S represents student network loss, and L_S is composed of cross entropy loss L_{CE} and model distillation loss L_{KD}. Distillation loss represents the process of student networks transferring learning from teacher networks. The distillation process includes intermediate distillation of soft labels and scene semantics, as shown in Formula (3).

$$L_{KD}(S, T) = L_{KD-sl}(S, T) + L_{KD-hint}(S, T) \tag{3}$$

Knowledge Distillation of Soft Labels. The knowledge distillation of soft labels adopts the method of target distillation, which enables student networks to master the learning process of distinguishing features in teacher networks by learning the target features carried by negative labels. This can enable student networks to not only generate accurate target classification results, but also understand how the network distinguishes negative labels. The loss function of the knowledge distillation of soft labels L_{KD-sl} is shown in formulas (4) ~ (6).

$$L_{KD-sl}(S, T) = L_{CE}(P_T(z_T)^{(i)}, P_S(z_S)^{(i)}) \tag{4}$$

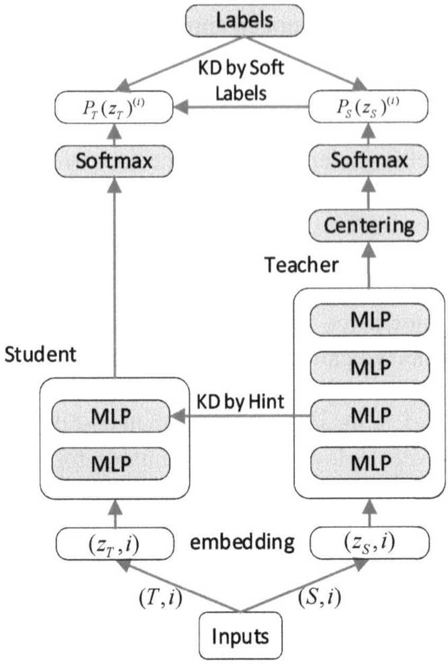

Fig. 2. The scene-aware knowledge distillation model

$$P_T(z_T)^{(i)} = \frac{\exp((z_T)^{(i)}/\tau_T)}{\sum\limits_{k=1\sim K} \exp((z_T)^{(k)}/\tau_T)}, \forall i \in \{1, 2, ..., K\} \tag{5}$$

$$P_S(z_S)^{(i)} = \frac{\exp((z_S)^{(i)}/\tau_S)}{\sum\limits_{k=1\sim K} \exp((z_S)^{(k)}/\tau_S)}, \forall i \in \{1, 2, ..., K\} \tag{6}$$

where $P_T(z_T)^{(i)}$ and $P_S(z_S)^{(i)}$ respectively represent the probabilities of category labels output by the teacher network and the student network. In the teacher network, $(z_T)^{(i)}$ represents the logit value of the i-th classification via the input dialogue feature value, where τ_T is the distillation temperature in teacher networks. K represents the total number of classifications. The output process of student networks is basically the same as that of teacher networks. For the i-th classification, student networks use $(z_S)^{(i)}$ as input and set the network's own distillation temperature τ_S.

The $P_T(z_T)^{(i)}$ in the teacher network may be very similar to the encoded representation of the real labels of the actual samples, so we recommend using a relaxed distillation temperature $\tau_T > 1$, to soften the output information in the teacher network and ensure that more distillation information is provided during training. In addition, during the distillation process, the student network used the same architecture as the teacher network, but with fewer hidden layers. Of course, the more hidden layers there are, the more accurate the classification effect for student networks will be. However, the greater the difficulty of network training and deployment, the more obvious the network latency that comes with it. The main task of multi-agent collaboration is to achieve consistency in semantic understanding of the scene. The hidden layer contains semantic information of the scene, and it is recommended to maintain a semantic depth of 3 to 4 layers.

Knowledge Distillation of Scene Semantics (KDSS). Multi-agents' collaboration requires ensuring consistency in task understanding. The semantic context of task execution helps to enhance the collaboration between intelligent agents at a deep level. When communicating scene semantic collaboration, we use the hint method to distill knowledge of scene semantics from the corresponding hidden layers. Specifically, for the hidden layers at the same level in the teacher network and student play, we use the teacher network for prompt learning and guide the student network. The loss function of KDSS $L_{KD-hint}$ is shown in formula (6).

$$L_{KD-hint}(S,T) = ||W_{V_T - V_S}||_2^2 \tag{7}$$

where V_T is the intermediate layer representation of the teacher network, V_S is the intermediate layer of the student network, corresponding to the teacher network, and V_S is guided by V_T. The result of V_T, V_S vs is in matrix W, and through matrix norm transformation, the teacher student network learns and trains scene semantics by minimizing the value of $L_{KD-hint}$.

Algorithm 1. `FDSS Training`

`Input:` S, T

`Output:` W_S^*

1: $W_T \leftarrow \{W_T^1, ..., W_T^h\}$

2: $W_S \leftarrow \{W_S^1, ..., W_S^g\}$

3: `Initialize a convolutional network` $W_r : W_T^g \rightarrow W_S$

4: $W_S \leftarrow f(W_S, W_r)$

5: $W_S^* \leftarrow \arg\min L_{KD-hint}(W_T, W_S)$

Line 1 ~ 2 initializes the teacher network W_T and the student network W_S, where W_T^h and W_S^g represent the parameters of each intermediate layer in the network, and h and g denote the number of layers for W_T and W_S, respectively. Line 3 defines W_r as the state matrix for transfer learning from W_T to W_S. Line 4 injects the scene information of W_T into W_S using W_r. Line 5 updates all parameters in the student network using Formula (7).

5 Experiments

We will evaluate the various performances of CRAG in comparison with other methods on multiple datasets, to illustrate the performance and efficiency of CRAG.

5.1 Experimental Setup

Datasets. This paper uses **TruthfulQA** [11] as the experimental dataset to evaluate the ability of LLMs to generate truthful answers. The accuracy of truthful answers can reflect the consistency issue of multi-agent collaboration, and only multi-agents with strong collaboration abilities can provide problem assignments that better meet the needs of the scenario. We will utilize three evaluation metrics, namely **GPT-Judge**, **BLEU**, and **Rougle-L**, **Accuracy** [13] to assess the authenticity of the responses generated by RAG, as well as the lexical overlap between the generated responses and the ground-truth references.

Baseline Methods. We will conduct a comparison with the RAG generation models of the following LLMs.

(1) **ChatGPT**: ChatGPT is used as the original big data model for RAG generation, and the version number of the ChatGPT model used is gpt-3.6-turbo.
(2) **Multi-agents Debate** [10]: Multi agents Debate uses multiple LM agents, each of which learns from multiple rounds of dialogue to generate their own answers to questions. Multiple intelligent agents are combined to cover the original question and achieve the effect of generating RAG answers.
(3) **ChatGPT-CRAG**: ChatGPT-CRAG uses the CRAG model proposed in this paper to capture semantic information of scenes based on the general model ChatGPT.
(4) **ChtaGPT-CRAG-sl**: ChatGPT-CRAG-sl represents a network that only learns target features. Compared to ChatGPT-CRAG, ChatGPT-CRAG-sl lacks a hint network and does not understand the semantics of the scene.

Parameters. To address the issue of semantic perturbations, this article sets the distillation temperature $\tau_T = \tau_S = 1.1$, which ensures the generation of diverse perturbation semantics and the accuracy of the output. The classification category K is set to 6 equivalent semantic classes. The intermediate layers of the basic teacher network and student network are set to $h = 5$ and $g = 3$ respectively.

5.2 Experimental Result

Overall Performance. Table 1 shows the performance of each model by comparing the overall effectiveness of different methods in TruthfulQA. Among them, regarding the authenticity judgment of question and answer, ChatGPT CRAG has a high GPT-Judge score of 57.33, which is nearly 10 points higher than the original model ChatGPT and nearly 5 points higher than ChtaGPT-CRAG-sl that only learns target features. This indicates that in the process of RAG generation, considering both the target features of the scene theme and the information features of the scene has a significant improvement effect on RAG generation. For BLEU and Rougle-L, ChatGPT scores 5 points higher than ChatGPT-CRAG-sl, indicating that ChatGPT-CRAG-sl may cause duplicate keyword searches during the target distillation process in the scene, which can also affect the quality of RAG generation. Overall, ChatGPT-CRAG has shown good performance in semantic generation, vocabulary overlap, and accuracy of scene understanding, indicating that the CRAG model can achieve collaborative consensus among multiple agents through semantic fusion of scenes during RAG generation.

Table 1. Overall performance of RAG generated responses

Models	TruthfulQA			
	GPT-Judge	BLEU	Rougle-L	Accuracy
ChatGPT	47.92	10.17	**31.31**	78.40
Multi-agents Debate	50.83	3.94	21.05	84.77
ChatGPT-CRAG	**57.33**	**11.27**	28.44	**89.22**
ChtaGPT-CRAG-sl	52.29	5.44	20.95	86.31

Scalability of Network Training. We analyze the scalability of CRAG model training from the number of intermediate layers and parameters in the network. When comparing the number of parameters, we use ChatGPT-CRAG (50) and ChatGPT-CRAG-sl (50) to represent simplified models of the original model, with ChatGPT-CRAG (50) and ChatGPT-CRAG-sl (50) having half the number of parameters of the original model. FitNets is a classic image layered teacher network model proposed by [13], which serves as a comparative model for this experiment. Figure 3 shows the scalability of network training.

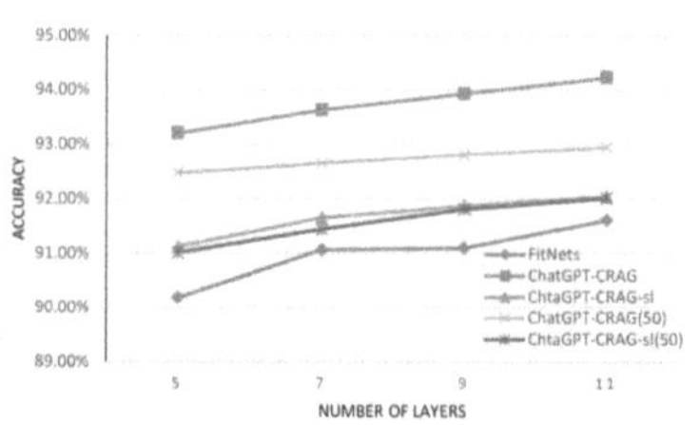

(a) Accuracy *w.r.t* number of layers

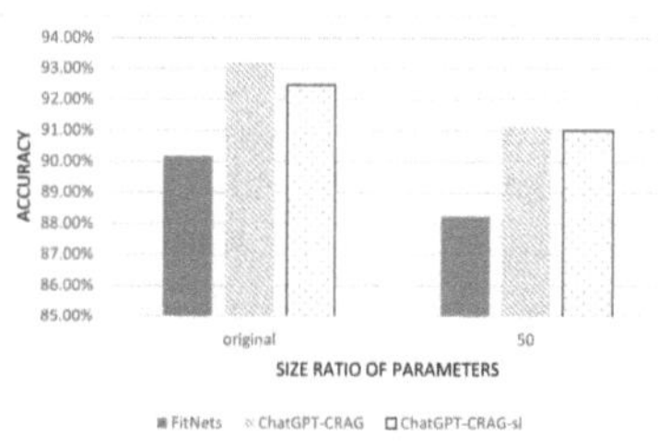

(b) Accuracy *w.r.t* size of parameters

Fig. 3. Scalability of Network Training

Figure 3(a) shows all network models achieved an accuracy of over 90% in capturing scene semantics, demonstrating good perception ability of scene information. Among them, ChatGPT-CRAG exhibits good adaptability to scene semantics due to capturing both target features of the scene and potential features of the intermediate layer simultaneously. Figure 3(b) shows that, for the same model, the number of network parameters also has a significant impact on the accuracy of the model. Overall, the depth of the model's hierarchy and the size of its parameters both have an impact on the accuracy of the model, but the effect is not significant and acceptable.

6 Conclusion

This article studies the problem of multi-agent scene collaboration in RAG and proposes a master-slave structure multi-agent collaborative control model CRAG. In actual RAG multi-agent collaboration, each agent uses the student network to perceive deep level information of the scene, which enables smaller models and easier deployment.

Acknowledgments. This work is supported by the Liaoning Collaboration Innovation Center for CSLE, the National Natural Science Foundation of China (62472204), Undergraduate Teaching Reform Project of Liaoning University (JG2024KCSZ16), and the 3rd Curriculum Ideology and Politics Model Course Project of Liaoning University under Grant No. KCSZSFKC 10, No. KCSZSFKC 13, No. KCSZSFKC 14 and No. KCSZSFKC 20.

References

1. Wang, Y., Wei, Z., Huang, Z., et al.: Dependent task offloading for air-ground integrated MEC networks: a multi-agent collaboration approach. Clust. Comput. **28**(2), 1–23 (2025)
2. Gimbert, H., Oualhadj, Y.: Deciding the value 1 problem for #-acyclic partially observable Markov decision processes. **13**(5), 1923–1933 (2012)
3. Dibeklioglu, H.: Visual transformation aided contrastive learning for video-based kinship verification. In: IEEE International Conference on Computer Vision, pp. 59–67 (2017)
4. Xu, L., Xia, L., Pan, S., et al.: Triple generative self-supervised learning method for molecular property prediction. Int. J. Mol. Sci. **25**(7), 1–12 (2024)
5. Wu, J., Zhang, T., Zha, Z.J., et al.: Prototype-augmented self-supervised generative network for generalized zero-shot learning. IEEE Trans. Image Process. **4**(11), 33–47 (2024)
6. Jayaraman, A.K., Muthappagounder, M., Ranganathan, V.K.: Llama3 with parameter efficient fine-tuning approach for 5G cellular network security text classification. Procedia Comput. Sci. **258**, 3203–3210 (2025)
7. David, W., Nikita, J., Julius, S., et al.: Sparsesurv: a Python package for fitting sparse survival models via knowledge distillation. Bioinformatics **3**(9), 1–9 (2024)
8. Hsieh, K.S., Wang, C.M.: Multi-hider reversible data hiding using a weighted color transfer and modulus operation. Appl. Sci.-Basel **13**(2), 24–31 (2023)
9. Li, X., Ding, X., Zhu, Q., et al.: An enterprise portrait tag extraction method based on context embedding and knowledge distillation. Soft. Comput. **28**(20), 11823–11834 (2024)
10. Han, C., Zheng, W., Tang, X.: Debate-to-detect: reformulating misinformation detection as a real-world debate with large language models. **4**(23), 1375–1388 (2025)
11. Lin, S., Hilton, J., Evans, O.: TruthfulQA: measuring how models mimic human falsehoods. **5**(11), 1–11 (2021)
12. Cheng, S., Guo, Z., Wu, J., et al.: EgoThink: evaluating first-person perspective thinking capability of vision-language models. **5**(12), 387–399 (2023)
13. Romero, A., Ballas, N., Kahou, S.E., et al.: FitNets: hints for thin deep nets. Comput. Sci. **1**(12), 21–27 (2015)

Cascading Dense Network for Face Image Super-Resolution

Wei Wei[1,2](✉), GuoGuo Shi[3], Zhaojie Qi[4], Yan Ma[1,2], and Fangping Cai[1,2]

[1] Weihai Ocean Vocational College, Weihai 264200, Shandong, China
shuiyouhan2002@foxmail.com
[2] Weihai Big Data Intelligent Application Engineering Technology Research Center, Weihai 264200, Shandong, China
[3] Chenghong Pharmaceutical (Weihai) Limited Liability Company, Weihai 264200, Shandong, China
[4] Rongcheng Municipal Committee of the Social Work Department, Weihai 264200, Shandong, China

Abstract. Recently, Convolutional neural network (CNN) based methods have achieved great success in the field of face image Super Resolution (SR). However, most of these methods tend to produce over-smoothed outputs, and suffer from degradation when dealing with very Low Resolution (LR) face images. To solve these problems, this paper proposes an accurate CNN architecture based on contourlet transform, which can ultra-resolve a very low resolution face image of 16×16 or smaller pixel size to its larger version. First of all, we present multi-scale fusion CNN (MSFC) to fully detect and exploit features from LR inputs. And then, we formulate the SR problem as the prediction of contourlet transform coefficients, which is able to make MSFC further capture the texture details for super-resolve face images. Extensive experiments demonstrate that the proposed method achieves superior face SR results and outperforms the state-of-the-art.

Keywords: Cascading Dense networks · multi-scale fusion CNN (MSFN) · super-resolution

1 Introduction

Face Super-Resolution (SR), a.k.a. face hallucination, refers to the technique of reconstructing a High-Resolution (HR) face image from an observed Low-Resolution (LR) input. It is a domain specific general image SR method, which focus on the human face, and can greatly facilitate face-related tasks [1, 2], since most existing techniques would degrade substantially when given LR face images.

Due to the powerful learning ability, Convolutional Neural Network (CNN) based methods have demonstrated superiority over other learning paradigms [3]. However, most of these CNN-based methods tend to produce blurry and over-smoothed outputs. Besides, they seem to only work well on small scaling factors (e.g. $2\times$or $4\times$), and degrades greatly when ultra-resolving very small image (like 16×16 or smaller pixel

J. Li et al. (Eds.): APWeb-WAIM 2025, CCIS 2660, pp. 143–151, 2026.
https://doi.org/10.1007/978-981-95-7996-9_13

size).From the first SR network SRCNN [4] to the latest RCAN [6], the number of convolutional layer increases from 3 to 400, which proves that deeper CNN can achieve better SR results. In addition, the current deep SR network is a series of identical feature extraction blocks. The ability of each block to extract features plays a crucial role in the final SR performance. Based on this consideration, this paper proposes an efficient multi-scale fusion block (MSFB) to effectively exploit features.

Due to image transform domain can reserve context and texture information of image at different levels, image SR reconstruction in the transform domain has attracted some attention. As a classic image transformation, wavelet transform has been used for SR in natural and face images. However, the direct extension of a wavelet to 2D by the tensor product of two 1D wavelets is no longer optimal for representing image that has features along smooth curves. To overcome this limitation, we introduce contourlet for SR in this paper, which is a real 2D image representation tool with multi-scale, multi-directional and anisotropic features.

This paper makes the following contributions: (1) we proposed a new feature extraction module, Multi-Scale Fusion Block (MSFB), to construct our MSFC architecture in cascaded manner. In each MSFB, the features of multiple receptive fields are efficiently exploited through convolution kernels of different sizes. By using of local residual learning (LRL) and global feature fusion (GFF), our MSFC can jointly and adaptively learn hierarchical features in a holistic manner. (2) Existing CNN-based face SR methods mostly concentrate on the spatial domain, leading to over-smooth reconstruction results; therefore, contourlet transform is applied to effectively restore global topology and local edge detail information of HR images; (3) Results showed that our MSFC with contourlet transform prediction is significantly better than other excellent methods in terms of peak signal-to-noise ratio (PSNR), structural similarity (SSIM) and augmentation of texture detail and edge structure of face images.

2 Proposed Method

2.1 Network Structure

Our MSFC structure is shown in Fig. 1, it consists of three parts, the shallow feature extraction module, the multi-scale feature extraction module, and the up-sample module. We solve the following problem:

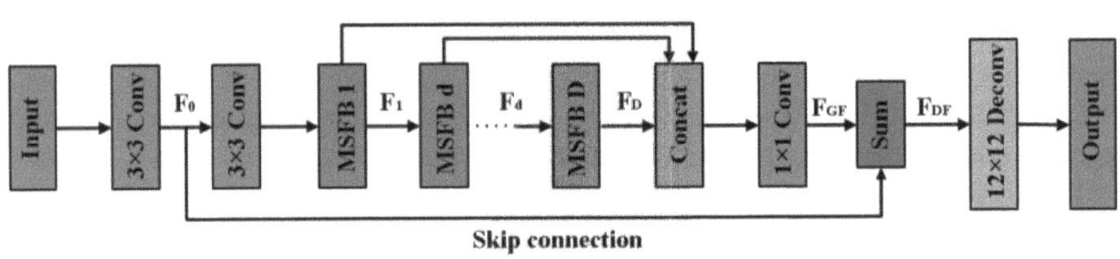

Fig. 1. The network structure of the proposed MSFC

$$\hat{\theta} = \arg\min_{\theta} \frac{1}{N} \sum_{i=1}^{N} L^{SR}(F_{\theta}(I_i^{LR}), I_i^{HR}) \quad (1)$$

where θ indicates the weights and bias convolutional layer. L^{SR} stands for the loss function for minimizing the difference between LR input I_i^{LR} and target HR I_i^{HR}. After shallow feature module, our feature extraction module contains a set of cascaded MSF blocks. Here, we use global feature G by fusing features from all the MSFBs.

$$F_{GF} = H_{GFF}([F_1, ..., F_D]) \quad (2)$$

where $[F_1, ..., F_D]$ denotes the concatenation of feature maps produced by MSFBs 1,..., D. H_{GFF} is a composite function of 1×1 and 3×3 convolutional layers. Global residual learning is then utilized to obtain the feature maps before conducting up-scaling by

$$F_{DF} = F_0 + F_{GF} \quad (3)$$

where F_0 represents the shallow feature. All the other layers before GFF are fully utilized with our MSFBs. Finally, we input to 12×12 deconvolutional layer to obtain the output of SR.

2.2 Multi-scale Fusion Blcok

The proposed MSFB is shown in Fig. 2. In each MSFB, we construct a three-bypass network and different bypass use different convolutional kernel. In this way, the information between those bypass can be shared with each other so that able to detect the image features at different scales.

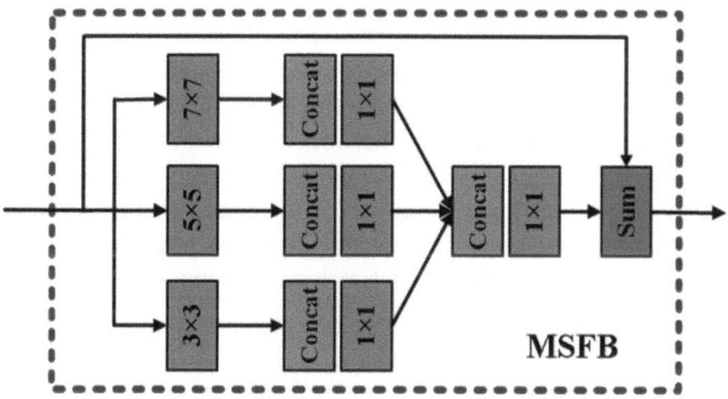

Fig. 2. The structure of the proposed MSFB

The operation can be defined as:

$$
\begin{aligned}
C_3 &= \sigma\left(w_{3\times 3}^1 * F_{d-1} + b^1\right) \\
C_5 &= \sigma\left(w_{5\times 5}^1 * F_{d-1} + b^2\right) \\
C_7 &= \sigma\left(w_{7\times 7}^1 * F_{d-1} + b^3\right) \\
H_3 &= \sigma\left(w_{1\times 1}^2 * [C_3, C_5] + b^4\right) \\
H_5 &= \sigma\left(w_{1\times 1}^2 * [C_3, C_7] + b^5\right) \\
H_7 &= \sigma\left(w_{1\times 1}^2 * [C_5, C_7] + b^6\right) \\
F_d &= w_{1\times 1}^3 * [H_3, H_5, H_7] + b^7 + F_{d-1}
\end{aligned}
\tag{4}
$$

where w and b indicate the weights and bias respectively. F_{d-1} and F_d are the input and output of the d-th MSB respectively. σ denotes the ReLU function.

2.3 Dual Branch Face Super Resolution

The input and output of the network are the contourlet transform sub-band corresponding to the low-resolution image and the reconstructed image, respectively. The main part of the network contains three branches, which are feature extraction branch, prior estimation branch and image reconstruction branch.

In the feature extraction branch, the low-resolution face image is transformed by first-order contourlet transform to obtain the corresponding low-frequency and high-frequency contourlet subbands. Then, these subbands are sent into the coding network, which contains 5 MSFB in series for feature extraction of low-resolution image. In the prior estimation branch, the contourlet transform sub-band in the feature extraction branch is taken as the input, and the face geometry prior knowledge is predicted as the output. In the image reconstruction branch, firstly, the features extracted from the feature extraction branch are fused with the prior estimated branch prediction. Here, the output results of the two branches are fused by channel splicing. Then, the high-resolution face features are learned through the decoding network containing 5 MSFB in series, and the face super-resolution task is completed by contour wave reconstruction.

The input of the prior estimation network is the output characteristic graph of the coding network. The prior estimation network consists of 3 MSFB, followed by a convolution layer with convolution kernel of 1×1, and outputs 7 channels. Each channel corresponds to different face geometric priors (Fig. 3).

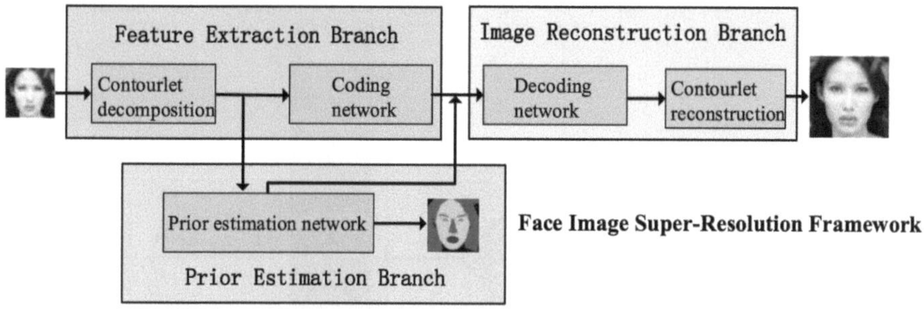

Fig. 3. The two branch structure of face image super-resolution

Figure 4 shows two face images and the geometric prior knowledge of human face used in the model. The geometric priori can see the semantic segmentation map of adult face. The left most column is the original face image. The second column is to clearly display the different segmented parts of the face with different colors. The third column from 3 to 9 shows seven kinds of geometric prior images, which are background area, hair and face Cheeks, eyebrows, glasses, nose and mouth. This kind of geometric priors can well describe the specific structure of the face and the clear edge of each part, which will play a good role in guiding the super-resolution process.

Fig. 4. The label of facial geometric prior

2.4 Contourlet Transform

Wavelet analysis cannot "optimally" represent image functions with straight lines and curves. It is well known that many signal processing tasks benefit from having a sparse representation of signals. Do and Vetterli [9] developed CT, which aimed at improving the representation sparsity of signals compared to the wavelet transform.

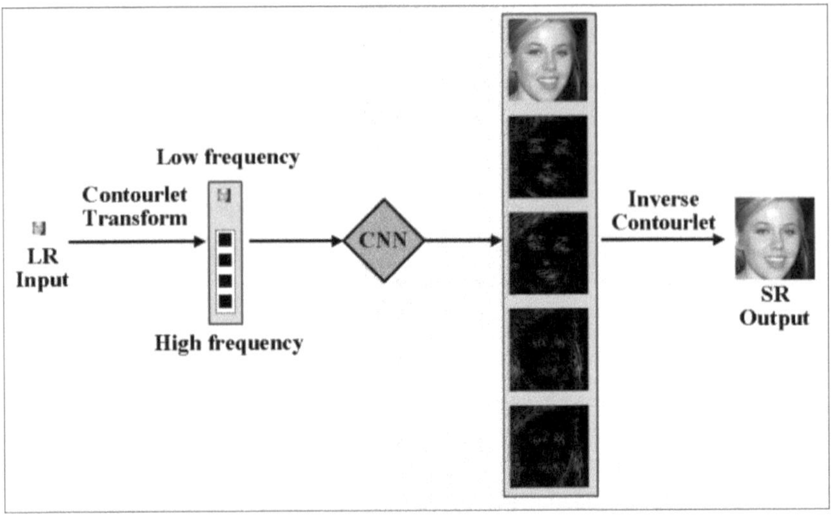

Fig. 5. The prediction of the contourlet transform subbands

As shown in Fig. 5, a low frequency subband and four high-frequency subbands of one level contourlet transform are entered in the network structure as "Input". The five subbands of the SR image are as "Output". Low-frequency subband is applied to represent global topology while high-frequency subbands capture important structural information. It is worth mentioning that contourlet transform can be used in different SR networks, which is a simple and effective way to improve the SR performance. The effectiveness of contourlet transform prediction will be evaluated in Sect. 3.2. The detailed process of contourlet transform implementation can be found in [9].

3 Experiments

The performance of the proposed algorithm has been evaluated on the large-scale Celebrity Face Attributes (CelebA) dataset [12], and we compared our method with the state-of-the-arts qualitatively and quantitatively on the dataset. We adopt the widely used Peak Signal-to-Noise Ratio (PSNR) and structural similarity (SSIM) [5] as our evaluation measurements.

3.1 Implementation

We select 10% of the CelebA datasest, which includes 20K training images and 260 testing images. And then, these images are aligned and cropped to 128 $\times$ 128 pixels as HR images. One level contourlet transform is adopted in our method. Our network contains 10 MSFBs. The learning rate initialized to 0.0001 for all layers and decreases half for every 50 epochs. Training our model takes roughly 12 h with Tesla P100GPU.

3.2 Effectiveness of Contourlet Transform

Given that in this paper, we introduce to predict the contourlet subbands in the field of face image SR, we will evaluate of this contribution. We use three methods (IDN, MNCE and our method) and integrate them with contourlet transform prediction. Figure 6 shows the average comparison result of IDN, LCGE, MNCE and our method, where we can see that all of these methods improve significantly when integrated with contourlet transform. Experimental results demonstrate that contourlet transform prediction is superior to spatial domain for SR task.

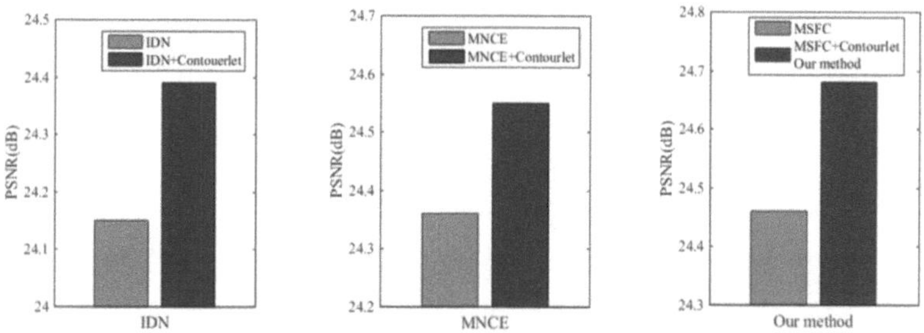

Fig. 6. The effectiveness of contourlet transform

3.3 Qualitative and Quantitative Comparisons

We compare our method with several representative methods, which include the most recently proposed CNN-based general image SR methods IDN [11], and three face specific image super-resolution methods, i.e., LCGE [7], UR-DGN [10], and MNCE [8]. The PSNR and SSIM values for comparison are shown in Table 1. We can clearly find that our method obtains higher PSNR and SSIM than other methods, which demonstrate the superior performance of the proposed method.

Table 1. Average score in terms of PSNR (dB) and SSIM of different face SR methods.

Metric	Bicubic	IDN	LCGE	UR-GAN	MNCE	Our method
PSNR	22.61	24.15	23.35	23.55	24.36	24.68
SSIM	0.6134	0.6854	0.6673	0.6696	0.6943	0.7122

Figure 7 shows the visual effects on scale 8 $\times$. We see that the face images reconstructed by the proposed method have abundant detail, which are very close to the target HR face image. We can thus conclude that the method proposed in this paper can be effectively applied to face image SR.

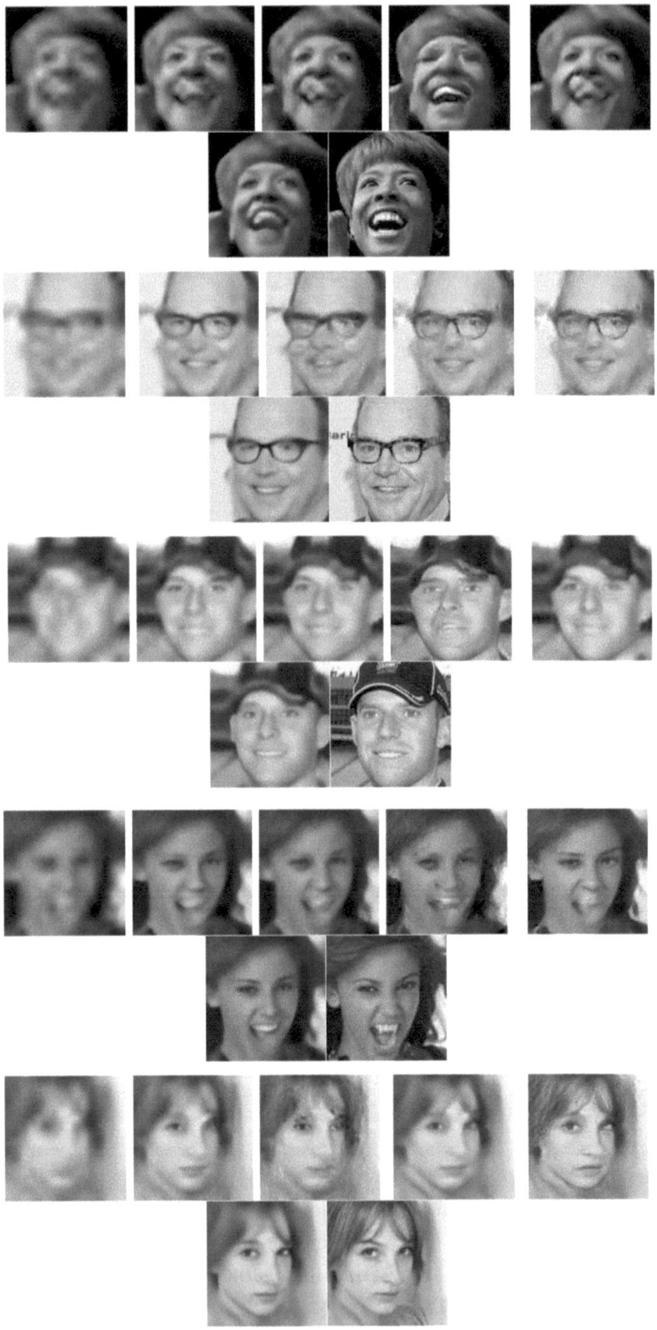

Fig. 7. Face super resolution comparisons with the state-of-the-arts methods. (a) Bicubic interpolation, (b) IDN, (c) LCGE, (d) UR-GAN, (e) MNCE, (f) Our method and (g) Ground-truth.

4 Conclusion

In this paper, a CNN-based face image SR method is proposed. Our end-to-end network contains a set of cascaded MSFBs, which effectively exploit multi-scale feature to improve the SR performance. In addition, contourlet transform is applied to the proposed network to effectively restore missing texture and edge information in the LR image. Quantitative results show that the proposed method is much better than other state-of-the-art methods, remarkably boosting restoration ability of textural structure and edge details of face images.

References

1. Chen, Y., Tai, Y., Liu, Y.M., Shen, C.H., Yang, J.: FSRNet: End-To-end learning face super-resolution with facial priors. In: Proceedings of the IEEE International Conference on Computer Vision and Pattern Recognition (CVPR), Salt Lake City, Utah, USA, 2018, pp. 2492–2501 (2018)
2. Yang, J., Luo, L., Qian, J.J., Tai, Y., Zhang, F.L., Xu, Y.: Nuclear norm based matrix regression with applications to face recognition with occlusion and illumination changes. IEEE Trans. Pattern Anal. Mach. Intell. (PAMI) **39**(1), 156–171 (2017)
3. Schulter, S., Leistner, C., Bischof, H.: Fast and accurate image upscaling with super-resolution forests. In: CVPR, Boston, USA, pp. 3791–3799 (2015)
4. Dong, C., Perez-Pellitero, E., He, K.M., Tang, X.O.: Image super-resolution using deep convolutional networks. PAMI **38**(2), 295–307 (2016)
5. Wang, Z., Bovik, A.C., Sheikh, H.R., Simoncelli, E.P.: Image quality assessment: from error visibility to structural similarity. IEEE Trans. Image Process. (TIP). **13**(4), 600–612 (2005)
6. Zhang, Y.L., Li, K.P., Li, K., Wang, L.C., Zhong, B.N., Fu, Y.: Image super-resolution using very deep residual channel attention networks. In: Proceedings of the European Conference on Computer Vision (ECCV), Munich, Germany, pp. 294–310 (2018)
7. Song, Y.B., Zhang, J.W., He, S.F., Bao, L.C., Yang, Q.X.: Learning to hallucinate face images via component generation and enhancement. In: Proceedings of the International Joint Conference on Artificial Intelligence (IJCAI), Melbourne, Australia, pp. 4537–4543 (2017)
8. Jiang, J.J., Yu, Y., Hu, J.H., Tang, S.H., Ma, J.Y.: Deep CNN denoiser and multi-layer neighbor component embedding for face hallucination. In: IJCAI, Stockholm, Sweden, pp. 1542–1548 (2018)
9. Nand, D.M., Vetterli, M.: The contourlet transform: an efficient directional multi resolution image representation. TIP **14**(12), 2091–2106 (2005)
10. Yu, X., Porikli, F.M.: Ultra-resolving face images by discriminative generative networks. In: ECCV, Amsterdam, Netherlands, pp. 318–333 (2016)
11. Hui, Z., Wang, X.M., Gao, X.B.: Fast and accurate single image super-resolution via information distillation network. In: CVPR, Salt Lake City, Utah, USA, pp. 723–731 (2018)
12. Liu, Z.W., Luo, P., Wang, X.G., Tang, X.O.: Deep learning face attributes in the wild. In: Proceedings of the IEEE International Conference on Computer Vision (ICCV), Santiago, Chile, pp. 3730–3783 (2015)

Image Super-Resolution via Dense Convolutional Neural Network in Curvelet Transform Domain

Kaibo Yang[1(✉)] and Wei Wei[2]

[1] Southwest Jiaotong University, Chengdu 611756, Sichuan, China
541673188@qq.com
[2] Weihai Big Data Intelligent Application Engineering Technology Research Center, Weihai 264200, Shandong, China

Abstract. In recent years, Convolutional Neural Network (CNN)-based image super-resolution methods have made significant progress in reconstruction performance. However, they still suffer from issues such as blurry reconstructed images and loss of texture details. This paper proposes an image super-resolution method that combines the curvelet transform with a cascaded dense convolutional neural network, transforming the traditional spatial domain pixel prediction problem into a sub-band coefficient prediction problem in the curvelet transform domain. Through multi-scale geometric analysis, the curvelet transform can represent curved edges and texture details in images more sparsely. The proposed dense network contains multiple Residual Dense Blocks (RDBs), which fully utilize features at different levels through dense connections, thereby addressing the long-term dependency problem in deep networks. Experimental results on standard datasets such as Set5, Set14, and BSD100 demonstrate that the proposed method achieves optimal objective evaluation metrics (PSNR and SSIM) for 2 × , 4 ×, and 8 × super-resolution tasks, while also producing more realistic texture details in subjective visual effects. Furthermore, evaluation on image classification tasks indicates that images reconstructed by this method are more beneficial for high-level vision tasks.

Keywords: Image super-resolution · Curvelet transform · Dense convolutional neural network · Residual dense block · Sub-band prediction

1 Introduction

Image super-resolution technology, initially conceptualized by Harris et al. in the 1960s, aims to algorithmically restore high-definition images from low-resolution inputs. Early traditional methods such as elliptic wave functions and sinusoidal templates yielded suboptimal reconstructions plagued by artifacts and distortions [1]. Huang et al. innovatively leveraged complementary information from multiple low-resolution images, significantly enhancing reconstruction quality through Fourier transform techniques, thus accelerating the technology's development and enabling breakthroughs in applications like remote sensing and medical imaging [2]. Traditional methodologies include:

J. Li et al. (Eds.): APWeb-WAIM 2025, CCIS 2660, pp. 152–163, 2026.
https://doi.org/10.1007/978-981-95-7996-9_14

interpolation-based methods (e.g., nearest-neighbor, bilinear, and bicubic interpolation) which offer computational efficiency but suffer from blocking artifacts or over-smoothing during high-magnification upscaling; reconstruction-based approaches (e.g., iterative back-projection, maximum a posteriori probability) that recover high-frequency details by integrating image priors at high computational cost; learning-based techniques where Katsuki enhanced edge realism via Gauss-Markov random fields, Wu estimated high-frequency details using contourlet transforms with Markov fields, and Yang et al. improved reconstruction through sparse representation and dictionary learning frameworks inspired by compressive sensing; and regression-based methods such as Kim's kernel ridge regression and He's Gaussian process regression that achieve resolution enhancement via mapping functions. Although bicubic interpolation remains an industry standard (e.g., Photoshop's default algorithm) due to efficiency-quality balance, refinements like Hwang's reverse gradients, Romano's sparse coding with non-local similarity integration, and Kim's curvature interpolation optimized edge preservation. Nevertheless, traditional methods still face limitations in severe detail loss and blurring during high-magnification reconstruction. The deep learning revolution, driven by end-to-end trained CNNs [3], GANs [4], and Transformers, has enabled cross-domain advancements from feature extraction to high-frequency detail generation, establishing new pathways for complex-scene image reconstruction [5].

2 Proposed Method

2.1 Principle of Curvelet Transform

The Curvelet Transform changes the way of multi-scale geometric analysis of images through a single transformation [6]. Its technical route is to decompose the image into a series of wavelet subbands and then perform ridgelet transform on each subband. Multi-scale ridgelet pyramids with different decomposition levels are used to represent different subbands of the filter bank output. At the same time, an anisotropy is strengthened between the width and length of the important frame elements of the curvelet subband [7], so that it obeys the parabolic scaling law, that is. The ridgelet transform can represent the singularity characteristics of the straight line in the image and has excellent approximation performance for the singular function of the straight line in the image [8]. However, it is not significantly stronger than the traditional wavelet transform in describing the curved edge in the image [9]. In order to find a better representation method for the curved edge in the image, Candes et al. proposed the curvelet transform based on the ridgelet transform. For the two-dimensional image target function with singular curves, the curvelet transform provides a stable and efficient near-optimal representation. Compared with wavelet transform and ridgelet transform, curvelet transform has been proved to be a more efficient sparse representation method for images [10]. In Fig. 1, a schematic diagram of the first-level curvelet transform of a color image with a resolution of 256×256 is given. It can be seen that the high-frequency sub-bands of the curvelet transform in six directions well depict the curved edges and contours of the butterfly from different directions.

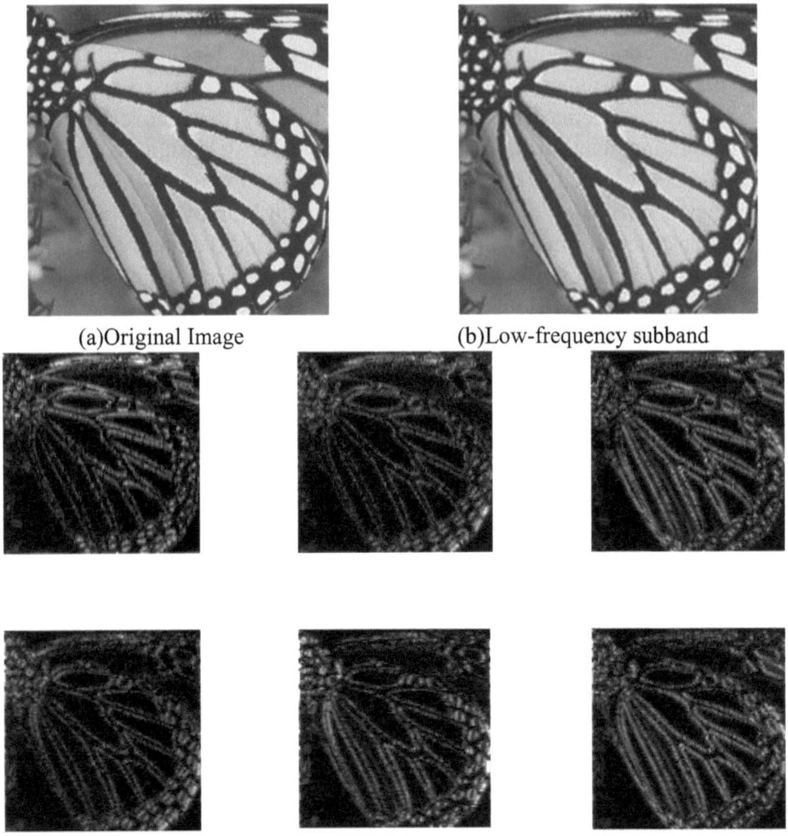
(a)Original Image (b)Low-frequency subband

(c)High frequency sub-band

Fig. 1. The diagrams of curvelet transform sub-bands

The low-frequency sub-bands describe the global topological relationship of the image, and each high-frequency sub-band can be regarded as capturing important features of the image in different directions, including the main edge contours of the butterfly and local detail features [11–14, 15]. The coefficient distribution on the high-frequency sub-band is sparse. Using it as the input and output of the convolutional neural network can not only make the network learning easier, but also more targeted. This is because the differences between high-resolution images and low-resolution images are mainly concentrated in the edge and texture areas, and the high-frequency sub-bands of the curvelet transform can capture these important areas well.

2.2 Dense Convolutional Neural Network

In order to solve the long-term dependency problem of deep networks, that is, to make full use of the shallow features mined by the network, a dense convolutional neural network with tighter connections is proposed, as shown in Fig. 2. The proposed network mainly includes a shallow feature extraction module, a deep feature extraction module and an upsampling module.

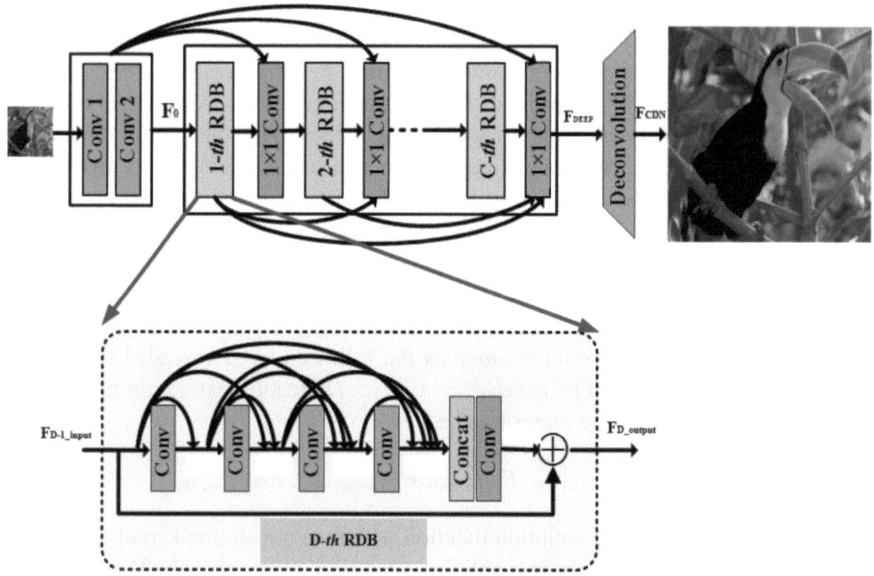

Fig. 2. The structure of the proposed Dense Convolutional Nerual Network

In the dense convolutional neural network, the shallow feature extraction module contains two concatenated convolutional layers, and each layer contains 64 convolution kernels for 3 × 3 filtering. The deep feature extraction module is composed of D residual dense blocks (RDBs) cascaded, and the structure of each dense block is shown in Fig. 2.3. Each RDB contains 4 layers of convolution. Compared with the residual block feature extraction block, the convolution layers in each RDB are more tightly connected. Through the channel splicing operation, the output of each convolution layer is used as the input of the subsequent convolution layer, realizing the maximum flow of information.

In the entire dense convolutional neural network, RDBs at different depths can fully mine image features at different levels. In addition, for each RDB, its input feature is the concatenation of the shallow feature extraction module and the output feature maps of all the RDBs in front of it. This design allows the different levels of features mined by each RDB to act on all the subsequent RDBs, making full use of all levels of features including shallow features, which not only provides more feature mining clues for the subsequent RDBs, but also solves the long-term dependency problem in deep CNN.

Assume that the degraded image and the reconstructed target image are represented as and respectively. The purpose of the model in this chapter is to solve the following problem:

$$\hat{\theta} = \arg\min_{\theta} \frac{1}{N} \sum_{i=1}^{N} L(F_{CDN}(I_i^{LR}), I_i^{HR}) \tag{2.11}$$

where, N represents the number of training samples, F_{CDN} represents the restoration network function, $F_{CDN}(I^{LR})$ represents the restored image I^{LR} reconstructed after the low-resolution image passes through the network, θ represents the set of convolution parameters, and L represents the objective function. Hui et al. experimentally proved that in the image restoration task, training with the L1 loss function first and then optimizing the parameters with the L2 loss function can achieve better reconstruction results. In order to avoid introducing additional training techniques, the training process will only use the L1 loss function.

After the degraded image I^{LR} passes through the shallow feature extraction module, the output feature map can be represented as F_0, followed by D cascaded RDBs. Let F_{d_output} represent the output of the $d(d = 1, 2, ..., D)$ feature extraction block, then the D input of the RDB can be represented as:

$$F_{D\_input} = F_{1\times1}([F_0, F_{1\_output}, F_{2\_output}, ..., F_{D-1\_output}]) \tag{2.12}$$

where, $F_{1\times1}$ represents the convolution function with the convolution kernel 1×1, and $[\cdot]$ represents the channel-by-channel splicing operation. Next, after the D h RDB, the shallow features and the output of all RDBs are fused by channel splicing, and the channel dimension reduction is performed through a convolution layer with a convolution kernel of 1×1 to obtain the mined deep features. Finally, the deep features F_{DEEP} are upsampled through a deconvolution layer to obtain the final reconstructed super-resolution image $F_{CDN}(I^{LR})$.

2.3 Coefficient Prediction Based on Curvelet Transform

Compared with wavelet transform, curvelet transform provides better sparse representation for images. The method proposed in this paper combines the first-level curvelet transform with the CNN structure. Based on the dense convolutional neural network, this chapter introduces the curvelet transform to recover more high-frequency details, and transforms the prediction problem of traditional CNN spatial domain pixels into the prediction problem of curvelet transform domain coefficients. As shown in Fig. 3.

First, the first-level curvelet transform is performed on the degraded image to obtain a low-frequency subband and four high-frequency subbands. The low-frequency subband is the approximate coefficient, which captures the global topological relationship of the input image, and the high-frequency subband is the detail coefficient, which sparsely represents the high-frequency information of the image in different directions.

Then, the low-frequency subband and the high-frequency subband are input into two dense convolutional neural networks respectively.

Finally, the output subbands of the two dense convolutional neural networks are combined, and the super-resolution reconstructed image can be obtained by inverse curvelet transform. This method is simple and effective, and is also applicable to other CNN-based image super-resolution methods.

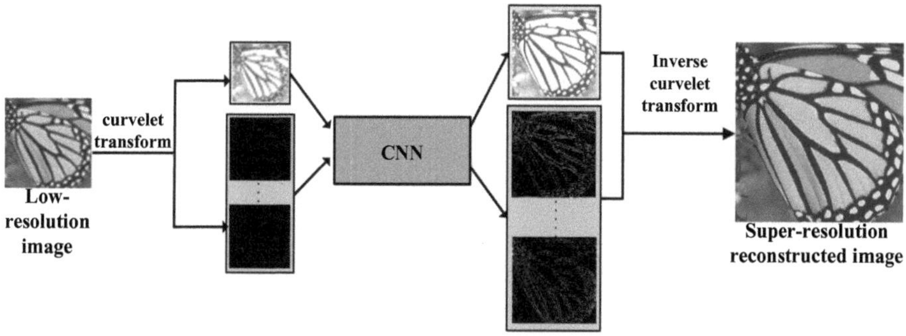

Fig. 3. The curvelet coefficient prediction based on CNN.

3 Experiments and Conclusions

In this section, the image super-resolution reconstruction capability will be measured from two aspects: objective evaluation indicators and subjective reconstruction effect. The training images are from the DIV2K dataset, which contains 800 high-resolution images without any data augmentation. The test images are from three common test sets: Set5, Set14, and BSD100. During the training process, a sliding window of size 48 $\times$ 48 and step size 36 is used to select corresponding image blocks on the low-resolution image and the high-resolution image. Then, a first-level curvelet transform is performed on each image block to obtain a matching training sample pair. The model training uses the Adam optimizer, and the initial learning rate is set to 0.0001. The learning rate is kept decreasing by half every 50 training cycles (epochs). Training on a single Tesla P100 GPU for about 100 epochs can make the model basically converge.

3.1 Model Analysis

This section will verify the effectiveness of the dense convolutional network structure design and the curvelet transform. As shown in Fig. 4, the convergence curves on the Set5 and BSD100 test sets at four times the scale are given. It can be seen that the proposed method has a rapid performance improvement in the first 40 training cycles (epochs), and basically reaches convergence when it reaches 80 epochs.

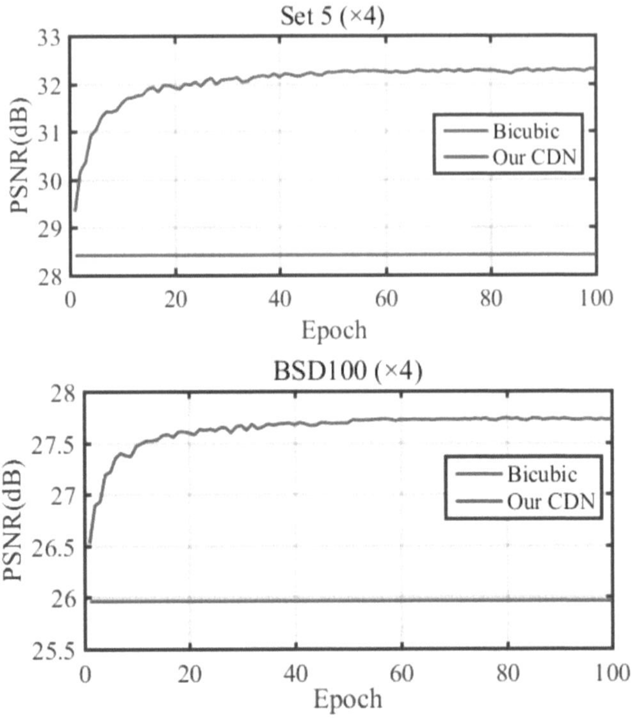

Fig. 4. The convergence curves on the dataset

Next, we will analyze the relationship between network depth and performance. Increasing network depth plays a crucial role in improving performance, but it also increases the number of parameters. Here, we will give the relationship between different depths and super-resolution performance, and select a reasonable structural design from them. Table 1 shows the PSNR results of reconstructed images of dense convolutional networks with different depths at 4 times the scale on the BSD100 dataset and the number of parameters of the corresponding model. In Table 2.1, A represents the number of dense blocks contained in the model, and B represents the number of convolutional layers in each dense block. Each convolutional layer contains 32 filters. It can be seen that when a certain depth is reached, the performance improvement begins to slow down, while the number of parameters continues to increase rapidly. In order to achieve a good balance between restoration performance and network parameter quantity, the proposed method contains 7 dense blocks, each of which contains 6 convolutional layers. In addition, the MultiAdds indicator is also used. MultiAdds counts the number of compound multiplication and accumulation operations of a single image and is widely used in the evaluation of the computational complexity of image restoration methods. When performing 4x super-resolution on a low-resolution image of size 320 × 180, the MultiAdds of the model finally selected in the experiment is 76.8G.

Table 1. Comparison results of different network depths on BSD100

PSNR(dB) /Parameter quantity(K)	A = 6	A = 7	A = 8
B = 4	27.06 / 555	27.14 / 679	27.19 / 812
B = 6	27.34 / 1063	27.43 / 1272	27.46 / 1490
B = 8	27.50 / 1793	27.52 / 2123	27.48 / 2463

Finally, the effectiveness of the curvelet transform is verified. This chapter introduces the curvelet transform based on the deep convolutional network, which transforms the prediction problem of spatial pixels into the prediction problem of curvelet subband coefficients, significantly improving the super-resolution performance of the model, especially in the ability to restore texture details. Here, the 4x super-resolution on the Set5, Set14 and BSD100 datasets will be tested. Figure 5 shows the super-resolution results in different transform domains, and a model containing 7 dense blocks is selected as the benchmark. Under different datasets, the performance of the combination of dense convolutional neural network and curvelet transform is significantly better than the two models based on wavelet transform and spatial domain, which proves the effectiveness of the curvelet transform.

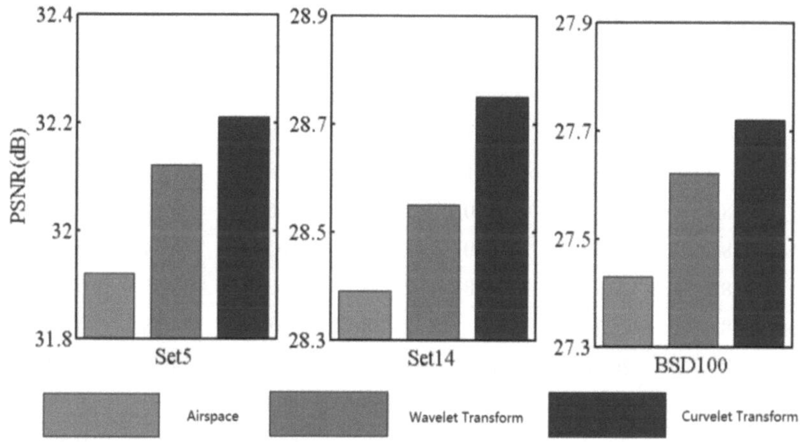

Fig. 5. Comparison results of different transform domain

3.2 Subjective and Objective Result Evaluation

The PSNR and SSIM indicators are used to objectively evaluate the proposed method. Table 2.2 shows the PSNR and SSIM of different super-resolution methods on Set5, Set14 and BSD100 datasets. The comparison methods include LapSRN, DRRN, MSLapSRN, MemNet, MSRN, IDN, and CARN. Table 2 marks the best super-resolution results in bold. It can be seen that the proposed method provides the best super-resolution reconstruction performance on the super-resolution tasks of 2×, 1× and 8× scales.

Table 2. The average PSNR(dB)/SSIM results of different methods

Methods	Scale	Set5		Set14		BSD100	
		PSNR	SSIM	PSNR	SSIM	PSNR	SSIM
Bicubic	2×	33.66	0.9299	30.24	0.8688	29.56	0.8431
LapSRN	2×	37.52	0.9591	33.08	0.9130	31.08	0.8950
DRRN	2×	37.74	0.9591	33.23	0.9136	32.05	0.8973
MemNet	2×	37.78	0.9597	33.28	0.9142	32.08	0.8978
MSRN	2×	38.08	0.9605	33.74	0.9170	32.23	0.9013
IDN	2×	37.83	0.9600	33.30	0.9148	32.08	0.8985
CARN	2×	37.76	0.9590	33.52	0.9166	32.09	0.8978
Ours	2×	38.15	0.9621	33.85	0.9192	32.29	0.9020
Bicubic	4×	28.42	0.8104	26.00	0.7027	25.96	0.6675
LapSRN	4×	31.54	0.8850	28.19	0.7720	27.32	0.7270
DRRN	4×	31.68	0.8888	28.21	0.7720	27.38	0.7284
MemNet	4×	31.74	0.8893	28.26	0.7723	27.40	0.7281
MSRN	4×	32.07	0.8903	28.60	0.7751	27.52	0.7273
IDN	4×	31.82	0.8903	28.25	0.7730	27.41	0.7297
CARN	4×	32.13	0.8937	28.60	0.7806	27.58	0.7349
Ours	4×	32.21	0.8951	28.75	0.7852	27.72	0.7371
Bicubic	8×	24.40	0.6580	23.10	0.5665	23.67	0.5480
LapSRN	8×	26.15	0.7380	24.35	0.6200	24.54	0.5860
DRRN	8×	26.22	0.7418	24.39	0.6252	24.55	0.5869
MemNet	8×	26.16	0.7414	24.38	0.6199	24.58	0.5842
MSRN	8×	26.58	0.7612	24.69	0.6315	24.59	0.5902
IDN	8×	26.39	0.7533	24.55	0.6300	24.50	0.5847
CARN	8×	26.71	0.7622	24.88	0.6342	24.69	0.5923
Ours	8×	27.21	0.7840	25.13	0.6480	24.88	0.5950

Figures 6 show the results of image super-resolution at four times the scale on the Set14 and BSD100 datasets, and the comparison methods are the results of bicubic interpolation upsampling, MSLapSRN, IDN,CARN,and MSRN. Compared with ordinary bicubic interpolation upsampling,these deep learning methods significantly improve the visual quality and restore some high-frequency details, but they cause serious distortion in areas with dense texture distribution. Obviously, compared with other methods, the edge and texture structure of the proposed curvelet domain dense method are closer to the original high-resolution image.

Fig. 6. Visual results of different image super resolution methods (Scale factor is 4)

3.3 Image Classification Task Evaluation

In order to further verify the effectiveness of the proposed method, this section will verify the performance of different super-resolution methods on image classification tasks. In the experiment, the ResNet50 network trained with 256 × 256 size images is used as the image classification evaluation model. In the performance evaluation, 1000 images are randomly selected from the ImageNet validation set as test images, and then the test images are downsampled using the linear interpolation method to become low-resolution images with a resolution of 56 × 56. Then, the test images are upsampled 4 times using the bicubic interpolation method and six super-resolution methods to obtain super-resolution images with a size of 256 × 256. The selected image super-resolution methods are DRRN, MSLapSRN, MemNet, MSRN, IDN and the proposed method. Table 3 shows the image classification results of different super-resolution methods. It can be seen from the Top-1 and Top-5 errors that the classification accuracy of the method proposed in this chapter is closer to the accuracy achieved by high-resolution images, and the classification performance is significantly better than other compared super-resolution methods, proving that the method in this chapter has greater application value in high-level tasks.

Table 3. The performance of image classification

Methods	Top-1 error	Top-5 error
Bicubic Difference	0.506	0.266

(continued)

Table 3. (*continued*)

Methods	Top-1 error	Top-5 error
DRRN	0.422	0.225
MSLapSRN	0.434	0.231
MemNet	0.402	0.203
IDN	0.388	0.196
MSRN	0.391	0.197
Ours	0.375	0.185
High-resolution images	0.260	0.072

4 Summary of this Chapter

This chapter proposes a dense convolutional neural network in the curvelet transform domain for image super-resolution. This method fully mines and utilizes image features at different levels by connecting more dense convolutional neural networks, and solves the long-term dependency problem in deep networks. On this basis, this paper introduces the curvelet transform to improve the ability to restore high-frequency details such as structure and texture. The test results on the standard test set show that this method shows the best performance in the image super-resolution task, not only obtaining better objective evaluation results, but also including more realistic structure and texture details in the subjectively reconstructed image.

References

1. Gao, X., Wang, F., Hu, X.: Research on efficient image super-resolution algorithm based on lightweight Transformer. J.of Dalian Univ. Technol. (02) (2025)
2. Matsui, Y., et al.: Sketch-based manga retrieval using manga109 dataset . Multimedia Tools Appl. (2017)
3. Dong, C., Loy, C.C., He, K., Tang, X.: Image super-resolution using deep convolutional networks. IEEE Trans. Pattern Anal. Mach. Intell. **38**(2), 295–307 (2015)
4. Zhou, F., Xu, H.: Advertising click-through rate prediction model based on improved Transformer. Comput. Appl. Res. (08) (2021)
5. Chen, L., Yu, G., Bao, H., Bian, X., Chen, C.: Multi-scale deformable 3D medical image registration based on Transformer. Comput. Syst. Appl. (01) (2025)
6. Wang, Z., Bovik, A.C., Sheikh, H.R., Simoncelli, E.P.: Image quality assessment: from error visibility to structural similarity. IEEE Trans. Image Process. **13**(4), 600–612 (2004)
7. Liu, W., Lu, X.: Research progress of Transformer based on computer vision. Comput. Eng. Appl. (06) (2022)
8. Luo, Y., Feng, T., Shao, J.: Facial expression recognition in the wild based on attention and visual Transformer. Computer Eng. Appl. (10) (2022)
9. Meng, H., Li, Y.: Anomaly detection and relationship extraction of time series data based on Transformer reconstruction. Comput. Eng. (02) (2021)

10. Jiao, S., Zhang, S.: A two-stream document image quality assessment algorithm based on Transformer. Softw. Eng. (02) (2025)
11. Cui, W., Shi, W., Shao, H.: MKGViLT: visual-and-language transformer based on medical knowledge graph embedding. High Technol. Lett. (01) (2025)
12. Wang, X., Xing, S., Mao, G.: Research on lightweight Transformer medical image segmentation with multi-scale feature fusion. J. Biomed. Eng. (02) (2025)
13. Zhou, L., Chen, H., Li, Q., Li, F.: Research progress in medical image segmentation based on Transformer deep learning model. J. Biomed. Eng. (04) (2024)
14. Ling, Z., Zhu, J., Shi, J., Weng, S.: Efficient mixed transformer for single image super-resolution. Eng. Appl. Artif. Intell. (2024)

Design and Practice of Ship Real-Time Collision Avoidance System Based on Artificial Intelligence

Yan Ma[1,2](✉), Jing Wei[1,2], Qinqian Sun[1,2], and Wei Wei[1]

[1] Weihai Ocean Vocational College, Weihai 264200, Shandong, China
632660261@qq.com

[2] Ship Control Engineering and Intelligent Systems Engineering Technology Research Center of Shandong Province, Weihai 264200, Shandong, China

Abstract. This paper focuses on ship navigation safety and conducts research on the design and practice of an artificial intelligence-based real-time ship collision avoidance system. By integrating deep learning with multi-sensor data, an intelligent collision avoidance decision-making model is constructed to realize real-time analysis of ship movement status and prediction of collision risks. Verified by simulation tests, the system significantly improves the efficiency and accuracy of collision avoidance decisions, providing an innovative solution for ensuring ship navigation safety.

Keywords: Real-time ship collision avoidance · Artificial intelligence · System design

1 Research Background and Significance

With the rapid development of the global economy and the increasing frequency of international trade, the maritime transportation industry, as an important means of facilitating economic and trade exchanges among countries, has witnessed rapid growth. The increasing number of ships and the complication of shipping routes have posed severe challenges to maritime traffic safety. Ship collision accidents not only result in huge economic losses but may also lead to environmental pollution and casualties.To ensure navigation safety, many merchant ships are currently equipped with some electronic devices, such as marine radars and the Automatic Identification System (AIS), which serve as auxiliary collision - avoidance equipment. However, due to the uneven operation levels of crew members, navigation accidents such as ship collisions, groundings, and reef - touchings still occur occasionally. In February 2024, the collision of the Lixinsha Bridge in Nansha, Guangzhou, led to the deaths of five people. The direct cause of these accidents is that subjective and objective factors cause ships to deviate from their courses, but they are also closely related to the background of the rapid development of China's transportation industry, the increasing number of cross - channel bridges, and the significant increase in the density of navigable ships.

J. Li et al. (Eds.): APWeb-WAIM 2025, CCIS 2660, pp. 164–174, 2026.
https://doi.org/10.1007/978-981-95-7996-9_15

Currently, the world has entered the era of the Fourth Industrial Revolution. The market competition in the shipbuilding industry is intensifying day by day. Actively developing intelligent shipbuilding and comprehensively transforming towards high - end "intelligence" is a major trend in the global shipbuilding industry. With the development of the transportation industry, the traffic volume of ships, cars, and trains has increased, and the number of dangerous incidents triggered by ship collisions is rising. Therefore, it is particularly important to develop an efficient and intelligent collision avoidance system. Therefore, it is particularly important to develop an efficient and intelligent collision - avoidance system.

This system is based on advanced artificial intelligence algorithms and marine data analysis techniques. It can monitor the dynamic status of ships, weather changes, and hydrological conditions in the surrounding sea areas in real-time. Through comprehensive analysis of multiple data sources, the system can quickly identify potential collision risks and provide scientific and reasonable collision avoidance decision - making suggestions for ships. The core functions of the system include real -time data collection, collision risk assessment, route optimization, and decision support, aiming to assist crew members in making more accurate judgments in complex situations.

2 Data Collection for Ship Real-Time Collision Avoidance System

2.1 Data Requirements for Collision Avoidance Decision-Making

In the process of ship collision avoidance decision-making, comprehensive and accurate data support is essential. This primarily involves data related to the ship's navigation environment, its own status, information about surrounding vessels, and obstacles, among other aspects.

Navigation and Environmental Data. On one hand, meteorological data includes wind direction, wind speed, temperature, atmospheric pressure, and visibility. Wind direction and speed affect the ship's speed and course, with strong winds potentially causing the ship to deviate from its intended route. Changes in temperature and atmospheric pressure can lead to severe weather, reducing visibility and increasing collision risks. On the other hand, sea condition data, such as wave height, wave direction, current speed, and current direction, are also critical. Waves impact the ship's stability and maneuverability, while currents can cause deviations in the ship's actual speed and course from the planned values, both of which must be considered in collision avoidance decisions.

Ship's Own Status Data. The ship's speed determines its motion state and the time required to reach the target position, while the course indicates the direction of navigation. Accurate course data is crucial for assessing the ship's movement trends and relative motion. The ship's turning rate reflects its ability to change direction, and this varies among different types of vessels. During collision avoidance operations, this rate is used to evaluate whether the ship can alter its course in time. The draft depth is related to the ship's load and the water depth of the navigation area. In shallow waters, excessive draft depth may cause the ship to run aground, so it must be considered in collision avoidance decisions to ensure navigable areas are not restricted.

Surrounding Vessel Information. Using the Automatic Identification System (AIS), static information about surrounding vessels, such as ship name, call sign, vessel type, length, and width, can be obtained. However, dynamic information, including position, speed, course, and turning rate, is even more critical as it reflects the movement state and trends of the target vessel. By analyzing this information, the future position of the target vessel can be predicted, and potential collision risks can be assessed.

Obstacle-Related Data. During navigation, ships may encounter obstacles such as islands, reefs, and drilling platforms. The position and range of these obstacles are vital for collision avoidance decisions. Such data can be acquired through electronic navigational charts, radar, and other sensors. Electronic charts provide detailed geographical information, while radar can detect surrounding obstacles in real-time and display their relative positions and distances. When approaching obstacles, the ship adjusts its route based on their position and range to ensure safe passage.

2.2 Data Collection

To achieve real-time collision avoidance decision-making, data collection must possess a high degree of timeliness and accuracy. At the same time, it is necessary to perform quality control and verification on the collected data to promptly identify and correct errors.

Navigation Environment Data Collection. Navigation environment data is of crucial importance for ship collision avoidance decision-making. Its collection involves two key areas: meteorological and sea-state data.

In terms of meteorological data, advanced equipment such as meteorological stations and satellite remote sensing are comprehensively utilized. Meteorological stations, equipped with high - precision sensors, can monitor meteorological elements such as wind direction, wind speed, air temperature, and air pressure in real-time and with high accuracy, providing local and immediate meteorological information for ships. Satellite remote sensing technology, from a macroscopic perspective, uses multi-spectral sensors on satellites to obtain large-scale meteorological data, especially key indicators such as visibility that affect navigation sight, enabling ships to have a more comprehensive perception of the meteorological situation.

The collection of sea-state data mainly relies on professional devices such as ocean monitoring buoys and sonars. Ocean monitoring buoys, acting as the "sentinels" of the marine environment, are anchored in specific sea areas to continuously measure parameters such as wave height and wave direction, providing real-time wave information for ships. Sonar technology uses the propagation characteristics of sound waves in water to accurately detect the speed and direction of ocean currents, helping ships understand the underwater environment and plan safe routes in advance.

Ship's Own State Data Collection. The ship's own state data is an important basis for collision avoidance decision - making. Its collection covers multiple aspects such as position, speed, course, turning rate, and draft. The Global Positioning System (GPS), as the core device for ship positioning, can determine the ship's geographical location in real-time and accurately by receiving signals from multiple satellites, providing basic coordinates for ship navigation and collision avoidance.

Sensors play a crucial role in the collection of the ship's own state data. By installing sensors in different parts of the ship, dynamic data such as speed, course, and turning rate can be obtained in real-time, helping crew members understand the ship's motion state. The draft is measured by a special draft sensor, which uses the principle of pressure sensing to accurately detect the ship's draft, providing an important reference for the ship's safe navigation in different waters.

Collection of Surrounding Ship Information and Obstacle Data. The acquisition of surrounding ship information mainly depends on the Automatic Identification System (AIS). The AIS system exchanges information between ships to obtain real - time static information (such as ship name, call sign, length, and width) and dynamic information (such as speed, course, and position) of surrounding ships, helping crew members comprehensively grasp the navigation situation of surrounding ships and predict potential collision risks in advance.

The collection of obstacle data requires the coordinated cooperation of electronic nautical charts and radars. Electronic nautical charts, as the digital maps for ship navigation, provide basic information about obstacles such as the location and height of reefs, shoals, and bridges. Radars use the principle of electromagnetic wave reflection to detect the relative position and distance of obstacles in real - time, providing real - time obstacle warnings for ships and ensuring safe navigation in complex waters.

After all kinds of collected data are unified and integrated into the data center, in order to ensure the accuracy and stability of data collection, detailed settings need to be made according to different data sources. For the AIS data source, the IP address and port number of the AIS receiver need to be accurately input to establish a stable data transmission channel. The setting of the radar data source requires selecting an appropriate radar model and inputting relevant connection information to ensure smooth communication between the radar and the data center. For the meteorological data source, an authoritative meteorological service API needs to be selected, and the API key needs to be input to obtain accurate and timely meteorological data. The setting of the ocean environment data source requires selecting a reliable data provider and inputting access credentials to ensure the stable acquisition of ocean environment data.

Through the above comprehensive and meticulous data collection and setting work, combined with a strict data quality control and verification mechanism, comprehensive, accurate, and real - time data support can be provided for the intelligent collision avoidance system, effectively reducing the risk of ship collisions and promoting the development of the shipbuilding industry towards intelligence and safety. The main loop code for data collection (see Fig. 1).

```
    while self.collection_status:
        start_time = time.time()

        # 执数据采集
        new_data = self._fetch_data()

        # 应用过滤条件
        filtered_data = self._apply_filters(new_data)

        # 保存数据
        self._store_data(filtered_data)

        # 等待下一个采集周期
        elapsed = time.time() - start_time
        sleep_time = max(0, self.data_sources['frequency'] - elapsed)
        time.sleep(sleep_time)

def _fetch_data(self):
```

Fig. 1. The main loop code for data collection. Execute data collection; then, apply filtering conditions, save the collected data, and finally, wait for the next collection cycle.

3 Data Processing and Fusion

3.1 Data Processing

The raw data collected from different sources often have issues such as noise, missing values, and inconsistencies. These problems can have a negative impact on subsequent collision avoidance decisions. Therefore, data pre-processing is required before data fusion.

Regarding meteorological and sea-state data, since sensors operate in a complex marine environment, they may be affected by factors such as ocean waves and electromagnetic interference, resulting in noise in the data. For this type of data, the moving average filtering method is adopted. For example, for wind speed data, a group of 5 consecutive sampling values is taken, and their average is calculated as the wind speed estimate at that time point, thereby smoothing the data and reducing the influence of random noise.

For the obstacle data collected by radar, as radar echoes may be interfered with by clutter, the median filtering algorithm is used. The obstacle distance data detected by the radar is sorted, and the middle value is taken as the estimate of the true distance, effectively removing abnormal clutter interference.

In the actual data collection process, due to equipment failures, communication interruptions and other reasons, there may be missing data. For missing values in the ship's own state data, such as missing speed or heading data, the linear interpolation method can be used for filling. Based on the valid data at adjacent time points before and after, the missing point value is estimated through linear fitting.

If there are missing parts of static information (such as the ship's name) in the information of surrounding ships provided by the AIS system, the missing information can be supplemented by querying the ship database and using other known information of the ship (such as the MMSI code).

Different types of data may have different dimensions and value ranges. To facilitate subsequent data fusion and analysis, normalization processing of the data is required. For meteorological data with different dimensions, such as temperature (unit: degrees Celsius), wind speed (unit: meters per second), and air pressure (unit: hectopascals), the minimum - maximum normalization method is adopted to map the data to the [0, 1] interval. Where X represents the original data, X_{min} and X_{max} are the minimum and maximum values of this data respectively.

$$X_{norm} = \frac{X - X_{\min}}{X_{\max} - X_{\min}} \tag{1}$$

Judge the encounter situation of ships through data pre-processing, then establish a prediction model for judging the ship collision risk, and use algorithms to accurately identify the degree of collision-avoidance risk.Schematic diagram of obstacles (see Fig. 2).

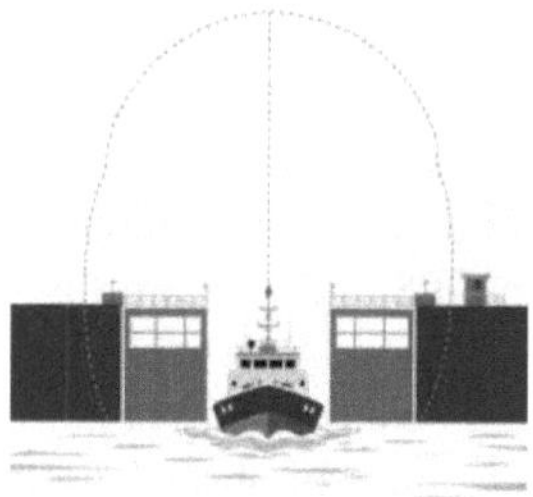

Fig. 2. The gates on both sides serve as obstacles. The successful passage of the ship's bow through the gates indicates that the obstacles have been successfully avoided.

Consider the position and motion parameters of the dynamic ship and the obstacle in a single coordinate system. Let the position coordinates of the dynamic ship be($x_{u,}y_u$), and the components of its velocity on the x - axes and y - axes be(v_{ux},v_{uy}); the position coordinates of the obstacle be (x_o, y_o), and the components of its velocity on the x - axes and y - axes be (v_{ox},v_{oy}). Calculate the motion coefficient between the dynamic ship and the obstacle. Through modeling, uniformly convert the data processing results of environmental perception sensors into the distance and azimuth of the obstacle. The schematic diagram of the obstacle is shown in Fig. 1, which presents the coordinate system for the motion coefficient between the dynamic ship and the obstacle. Coordinate system for the motion coefficient between the dynamic ship and the obstacle (see Fig. 3).

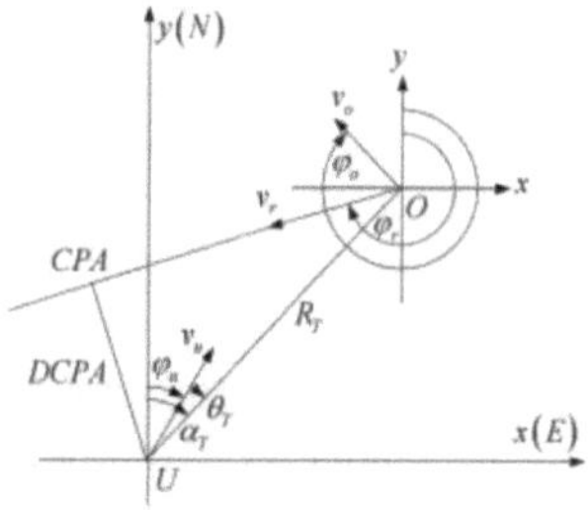

Fig. 3. Coordinate system for the motion coefficient between the dynamic ship and the obstacle.

3.2 Data Fusion

For the dynamic data of ships, such as position, speed, and heading, since there are certain errors in GPS positioning, and the ship's dynamic information provided by the AIS system may also have delays and inaccuracies, the Kalman filtering method can be used for data fusion. In the prediction step, according to the ship's motion model (such as the uniform linear motion model), the state estimate of the previous moment and the system noise covariance are used to predict the state at the current moment. In the update step, the measurement values provided by GPS and AIS are compared with the predicted values. Based on the measurement noise covariance, the Kalman gain is calculated to update the state estimate. Through Kalman filtering, more accurate ship dynamic information can be obtained, and the reliability of collision avoidance decisions can be improved.

To achieve more efficient data fusion, a hierarchical structure based on multi - sensor information fusion is adopted. The data is divided into three levels: data - layer fusion, feature-layer fusion, and decision-layer fusion.

In data-layer fusion, the raw data from different sensors are directly fused. For example, the position data from GPS and AIS are fused to obtain more accurate ship position information.

In feature-layer fusion, features are first extracted from the raw data, and then these features are fused. For instance, features of obstacles (such as position, size, and shape) are extracted from radar and electronic chart data and then fused.

In decision-layer fusion, comprehensive decisions are made based on the decision - making information provided by different sensors (such as risk assessment based on meteorological data and collision warnings based on ship dynamic data). Ultimately, a unified decision - making recommendation is provided for ship collision avoidance.

Through the above data processing and fusion methods, the information from different data sources can be effectively integrated, providing accurate, reliable, and real - time data support for the ship's intelligent collision avoidance system. Consequently, the safety and efficiency of ship navigation can be improved.

4 System Practice and Optimization

4.1 System Practice

Hardware selection includes various sensors such as LiDAR, sonar, cameras, and meteorological sensors, which are used to monitor the surrounding environment in real time. All these sensors are connected to the on-board data acquisition unit. This unit is responsible for collecting data from each sensor, converting it into a digital format, and transmitting it to the central processing unit for further processing. The acquired sea condition information is preprocessed and organized into a data format suitable for analysis and processing [4] using a water area mathematical model.

For software, a suitable experimental operating environment is chosen—a computer configuration with the Windows 10. Under the Windows 10, the network model is built and trained based on the Matlab deep learning framework. Design and Practice of Ship Real-time Collision Avoidance System Based on Artificial Intelligence (see Fig. 4).

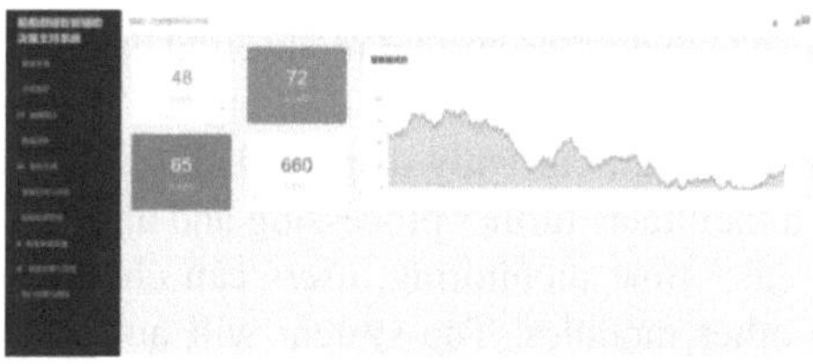

Fig. 4. Design and Practice of Ship Real-time Collision Avoidance System Based on Artificial Intelligence.

Real-time monitoring module:

On the system's main interface, users can select the "Real-time Monitoring" module to enter the monitoring interface and start related operations. The system automatically loads the real - time status of all current ships, including:

- Ship name
- Ship position (latitude and longitude)
- Ship heading
- Ship speed
- Ship type

Users can also use filtering conditions such as ship type and status to narrow down the display scope, so as to view the required ship information more accurately.

The system automatically updates the real - time data of ships at regular intervals (e.g., every 5 s). Meanwhile, users can manually refresh the data to ensure they obtain the latest ship information.

Users can select any ship and click the "Track Replay" button. At this time, the system will display the navigation track of the selected ship within a specified time period. Moreover, users can independently choose the replay speed (e.g., normal, accelerated, decelerated) and the time range (e.g., the past 1 h, the past 24 h), which facilitates a detailed analysis of the ship's navigation situation.

The system automatically calculates the potential collision risk based on the real - time data of ships. When a collision risk is detected, the system will immediately issue an alarm and highlight the relevant ships on the interface. Users can click on the alarm information to view a detailed collision risk analysis, which includes important contents such as the risk level (high, medium, low), the estimated collision time, and the estimated collision position.

Below the collision warning information, the system will provide intelligent decision - making support suggestions, such as suggesting a change in heading, suggesting deceleration or acceleration, and suggesting communication with other ships. Users can perform corresponding operations according to these suggestions to reduce the collision risk.

Users can enter the ship name or ID in the search box to quickly find the information of a specific ship. After clicking on the ship information, they can view detailed information, including the ship's basic information, historical navigation records, and safety records, which helps to comprehensively understand the ship's situation.

Users can also choose to export the current monitoring data, and the system supports multiple formats such as CSV and Excel. After clicking the "Export Data" button, users can select the export scope (e.g., currently displayed ships, all ships) and confirm the export operation, which facilitates further processing and analysis of the data.

After completing real - time monitoring, users can choose to return to the main interface or switch to other modules. The system will automatically save the user's monitoring settings, making it convenient to quickly restore to the previous monitoring state for the next use.

The collision avoidance system operation module consists of four main steps: data input, collision avoidance algorithm selection, calculation of collision avoidance paths, and decision - making support. Its aim is to provide intelligent collision avoidance auxiliary decision - making for ship navigation.

Collision avoidance system operation module:

First, input the detailed status information of the current ship that has been collected, including the ship's position (latitude and longitude), ship heading, ship speed, ship type (such as cargo ships, oil tankers, passenger ships, etc.), and ship dimensions (length, width, draft). Also, input relevant data information about the surrounding environment, such as water flow speed and direction, wind speed and direction, visibility (weather conditions like fog and rain), and the status information of other ships (position, heading, speed, etc.).

Second, select a suitable collision avoidance algorithm according to specific needs and set the corresponding parameters. For example, set algorithm parameters such as the collision avoidance distance threshold, reaction time, and ship priority. Finally, conduct path planning and optimization based on the input data and the selected algorithm.

Path planning: Combine the input ship status and environmental information, and use the selected collision avoidance algorithm to plan a safe navigation path to avoid other ships and obstacles.

Path optimization: Optimize the calculated path, comprehensively considering the ship's dynamic characteristics, environmental changes (such as water flow and wind speed), and the ship's navigation efficiency.

Finally, generate specific collision avoidance decision-making suggestions based on the calculation results, such as changing the heading, adjusting the speed, and selecting a safe navigation area. Present the collision avoidance path and decision - making suggestions to the user in a visual form, and simultaneously display the current ship's position, the positions of other ships, and their motion trajectories to facilitate the user's intuitive understanding of the navigation situation.

4.2 System Optimization

The optimized real-time ship collision avoidance system is applied to various types of vessels, such as container ships, oil tankers, and bulk carriers. Different types of vessels have distinct characteristics and navigation requirements, necessitating appropriate adjustments and optimizations to the system. For example, container ships typically travel at higher speeds, demanding greater real-time responsiveness in collision avoidance decisions. Oil tankers, with their large carrying capacity and poorer maneuverability, require more cautious collision avoidance strategies.

In addition to the original environmental data monitoring, the system can also achieve real - time detection of the activities of marine organisms. Through underwater sonar and optical sensors, it can identify the species, quantity, distribution area and movement direction of marine organisms. For example, when a large group of whales is detected swimming on the ship's route, the system will promptly issue a prompt to remind the crew to adjust the route, avoiding collisions with marine organisms and protecting the marine ecological environment at the same time.

When sailing in cold waters, the system can utilize satellite remote-sensing data and the ship's own radar equipment to monitor the position, size and movement trend of sea ice and icebergs in real - time. For sea ice areas and icebergs that may affect the ship, s navigation safety, the system will clearly mark them on the map and provide detailed avoidance suggestions.

The water quality monitoring equipment installed on the ship will detect the pollution status of seawater in real-time, including oil spills and chemical substance leaks. Once the pollution exceeds the standard, the system will immediately issue an alarm and record information such as the time, location and pollution level of the pollution occurrence, providing a basis for subsequent environmental protection treatment and liability determination.

The intelligent alarm function has been upgraded. Besides displaying alarm information on the system interface, the system will also send alarm notifications to the crew and relevant management personnel through multiple channels such as text messages, emails and ship - borne broadcasts. This ensures that even if the crew is not near the operation interface, they can receive the alarm information in a timely manner.

During the application process, feedback from the use of different types of vessels is collected to further improve the system's performance and applicability. By deploying the system on a variety of vessel types, its universality and reliability are validated, laying the groundwork for its broader adoption. In the future, the AI-based real-time ship collision avoidance system will be promoted across a wider range of maritime domains, providing robust support for the safety of ship navigation.

Funded by Ship Control Engineering and Intelligent Systems Engineering Technology Research Center of Shandong Province (Grant Number:SSCC20250001).

References

1. Zhang Xiaoyuan, F., Fu Jianping, S.: Safety monitoring system for hoisting machinery equipment based on ZigBee wireless sensor network. Chin. J. Sens. Actuators **35**(12), 1722–1727 (2022)
2. Shang Chuan, F., Gao Junwei, S.: Design of train axle temperature monitoring system based on ZigBee and cloud. Transducer Microsyst. Technol. **40**(8), 110–112 + 120 (2021)
3. Wu Haixin, F., Xiao Lei, S.: Wireless IoT sensor system based on ZigBee. Comput. Technol. Dev. Appl. **39**(6) (2024)
4. Shu Linzhuo, F.: Analysis of motion characteristics and collision risk of ship - hitting riverside overhead box culverts. Doctoral Dissertation, Chongqing Jiaotong University (2022)

5. Teng Qimou, F.: Generation method of test scenarios for intelligent ship navigation. Ship Sci. Technol. **21**, 93–96 (2023)
6. Xia Yuanqing, F.: Cross - domain collaborative control and intelligent decision - making of multi - moving body systems in the integrated sky - earth - space network environment. Control Decis. **5**, 1176–1199 (2023)

A Chinese Named Entity Recognition Approach with Multi-head Attention

Guanfeng Li(✉), Yuhang Xiao, and Dezhou Hu

Ningxia University, Yinchuan 750021, China
ligf@nxu.edu.cn

Abstract. Named entity recognition is the basis of many downstream tasks in the field of natural language processing, and its accuracy directly affects the work effect of subsequent tasks. In order to solve the problems of insufficient feature extraction ability and difficulty in capturing long-distance dependence of existing Chinese named entity recognition models, we propose a Chinese named entity recognition method that integrates multi-head attention and bidirectional gated recurrent unit. Firstly, the ALBERT pre-trained model is used to generate dynamic word vectors, and the vector sequences are input into BiGRU to extract global semantic features, and then the multi-head attention mechanism is used to capture long-distance dependent information to enhance semantic features, and finally the optimal sequence is obtained by conditional random field decoding. Experiments show that the proposed method achieves 95.84% and 94.93% F1 on the People's Daily and MSRA Chinese datasets, respectively, which is better than other models, and the training time of the proposed method is reduced by about 14% compared with the BERT-BiLSTM-CRF model, which proves the effectiveness and versatility of our model.

Keywords: Named Entity Recognition · Multi-head Attention · Bidirectional Gate Recurrent Unit · Conditional Random Field

1 Introduction

Named Entity Recognition (Named Entity Recognition, NER) aims to extract entities with specific meanings from text. NER plays a significant role in applications such as machine translation, text understanding, and question-answering systems [1]. English NER has an early development and has achieved remarkable results. However, Chinese NER is more complex due to its high sensitivity to word segmentation outcomes and the widespread nature of Chinese word polysemy [2]. Traditional NER methods are mainly rule-based and statistical machine learning-based. While traditional methods are interpretable, they are labor-intensive and have poor transferability.

Deep learning-based NER methods [8] now dominate the field, thanks to their strong feature extraction capabilities. You et al. [3] used a BiLSTM-CRF model for entity recognition in medical imaging reports and achieved good results. However, Word2vec cannot solve the problem of word polysemy. Song et al. [4] employed attention mechanisms to

J. Li et al. (Eds.): APWeb-WAIM 2025, CCIS 2660, pp. 175–186, 2026.
https://doi.org/10.1007/978-981-95-7996-9_16

semantically enrich the text information extracted by the encoding layer. They achieved an F1 score of 91.35% on the CONLL-2003 dataset, proving the effectiveness of attention mechanisms in NER, yet they did not utilize pre-trained models like BERT to generate dynamic semantic vectors. Zhang et al. [5] proposed the Lattice-LSTM model for Chinese NER, which avoids Chinese word segmentation errors and can automatically match word information from context. However, this model cannot be trained in parallel and has poor transferability. Guan et al. [6] used BERT to generate word embedding vectors that integrate left and right context information, enhancing the semantic representation of characters and effectively alleviating the problem of word polysemy. Nevertheless, the BERT model has a large number of parameters and a long training time, limiting its application in specialized fields. Zhang et al. [7] applied a BERT-IDCNN-CRF model to entity recognition in the military domain and achieved better results than the Lattice-LSTM model. Existing NER research has not fully utilized the advantages of pre-trained models, deep neural networks and attention mechanisms, and the feature extraction capabilities of NER models need to be further improved.

Based on the above research, this paper proposes an ALBERT-BiGRU-MA-CRF NER model. The BERT model can solve the problem of word polysemy but has a large number of parameters. Therefore, this study uses ALBERT, a lightweight version of BERT, to generate dynamic text sequence vectors. The BiLSTM network structure is relatively complex and has many parameters. Thus, this study uses BiGRU, which has fewer parameters, to extract global text sequence features and further reduce model training time. This study also introduces a multi-head attention mechanism to address long-distance dependence problems and enhance model feature extraction capabilities. Finally, CRF is used to add rule constraints and decode the optimal sequence.

2 Related Work

Deep learning-based NER methods now dominate the field, and recent models have also been built on deep learning foundations. Related work categorizes these methods into three types for detailed analysis.

2.1 CNN-Based NER Methods

Although Convolutional Neural Network (CNN) are more commonly used in the field of computer vision, their strong ability to capture local features also makes them widely used for feature extraction from text sequences. Collobert et al. [9] adopted a multi-channel CNN approach, which utilizes multiple convolutional kernels to extract diverse features from text sequences. The extracted features are concatenated into multi-channel feature maps and then fed into subsequent classifiers for NER tasks. Aguilar et al. [10] proposed a multi-task NER method that leverages CNNs to capture orthogonal features and word shapes at the character level. On the other hand, Wu et al. [11] used convolutional layers to generate global features represented by multiple global hidden nodes. These local and global features are then input into a standard affine network to identify named entities in clinical texts.

2.2 Recurrent Neural Network-Based Named Entity Recognition Techniques

It has been proven that Long Short-Term Memory (LSTM) network models are effective in extracting contextual semantic information for NER tasks, enabling deeper text understanding and improving model accuracy. Kuru et al. [12] proposed CharNER, an LSTM-based character-level NER method. This model treats sentences as character sequences, uses LSTMs to extract character-level representations, and outputs label distributions for each character rather than each word, deriving word-level labels from character-level tags. Wan et al. [13] replaced unidirectional LSTM with a BiLSTM model to better capture character features. By incorporating a SoftMax layer, data features can be transformed into annotation results, facilitating sequence labeling tasks and their application in NER. The combination of BiLSTM and Conditional Random Field (CRF) [14] is one of the most commonly used models for NER tasks. It uses BiLSTM to extract sentence features and CRF to model label dependencies. Ma and Hovy [15] proposed a neural network model integrating BiLSTM, CNN, and CRF. This model uses GloVe [16]-pretrained 100-dimensional word embeddings to automatically extract word and character-level features. It employs CRF for global modeling of tag sequences, alleviating potential information loss issues encountered by BiLSTM when processing long sequences and thus enhancing sequence labeling performance.

Rei et al. [18] utilized Gated Recurrent Unit (GRU) mechanisms to integrate character-level representations with word embeddings, producing better long-term memory capabilities compared to RNN models. With fewer parameters than LSTM, GRU is easier to train. Yang et al. [19] employed a deep GRU model to encode morphological and contextual information at both character and word levels. Their model can dynamically determine how much information to utilize from character-level or word-level representations. Additionally, pre-trained models based on the Transformer model have become a new paradigm for NER tasks. The Generative Pre-trained Transformer (GPT) and Bidirectional Encoder Representations from Transformers (BERT) are both pre-trained models based on the Transformer model. Models such as RoBERTa [20], ALBERT [21], and XLNet [22] further refine the BERT model series.

2.3 Graph Neural Network-Based Named Entity Recognition Methods

Graph Neural Networks (GNNs) transform NER tasks into node classification problems on graph structures. By modeling relationships and context between nodes, GNNs learn more accurate node representations. Common GNN methods include Graph Convolutional Network (GCN)-based and Graph Attention Network (GAT)-based NER models.

GCN-based models use convolutional operations to aggregate information from neighboring nodes and learn node representations through stacked GCN layers. Hanh et al. [23] proposed combining context features with graph convolutional networks to enhance NER performance, supported by extensive experiments. Tang et al. [24] used cross GCNs to process character-oriented directed acyclic graphs in two directions simultaneously and introduced global attention GCN blocks to learn node representations under global context conditions.

GAT-based models employ self-attention mechanisms to dynamically aggregate features of neighboring nodes based on their importance. Chen et al. [25] introduced additional graph attention network layers to strengthen the representation of phrase-level dependencies. Wang et al. [26] proposed a polysemous graph attention network that dynamically adjusts fine-grained correlations between matching characters and words from multiple dimensions to enhance character representations. Tian [27] presented a multi-task learning method with a simple and effective word selection strategy for lexicon-enhanced Chinese NER. One task score matched words and selects the top K most useful ones, while another integrates the selected words through a multi-head attention network and achieves Chinese NER through character-level sequence labeling.

In summary, each model has its own advantages and disadvantages when processing entities. For example, CNNs can process data in parallel to improve computational speed but lack contextual information retention. As an RNN variant, LSTM effectively learns long-term dependencies but still faces gradient issues. GNNs can more effectively reveal relationships between entities by continuously mining graph data, but their complex model structures can compromise flexibility and scalability. Transformers, usually pre-trained with BERT, generate profound linguistic features at the cost of substantial computational resources.

3 Methodology

Figure 1 illustrates the overall model architecture, which consists of four modules: the ALBERT layer, the BiGRU layer, the multi-head attention layer, and the CRF layer. The ALBERT layer uses a pre-trained ALBERT model to obtain dynamic, contextualized representations. The BiGRU layer extracts global semantic features from the text sequence. The multi-head attention layer identifies internal relationships within the sequence by considering dependencies among features. Finally, the CRF layer decodes the sequence to yield the optimal output. To enhance the theoretical rigor of the proposed architecture, this study integrates ALBERT and BiGRU to reduce model complexity and training costs and improve semantic representation. Specifically, a multi-head attention mechanism is introduced after the BiGRU encoding stage. Theoretically, this approach leverages the complementary strengths of local sequential modeling and global dependency capture. BiGRU effectively encodes contextual information, and multi-head attention allows the model to focus on various long-range semantic relationships across different subspaces. This hierarchical structure, which involves local encoding followed by global refinement, offers a more structured and computationally efficient alternative to traditional BiLSTM-attention frameworks.

3.1 ALBERT Layer

The BERT model uses a bidirectional Transformer for encoding, offering strong semantic representation. It learns rich language representations through unsupervised training and can be adapted to NER tasks with a label classification layer and fine-tuning. ALBERT, like BERT, uses a Transformer encoder but optimizes BERT's large parameter size in two ways:

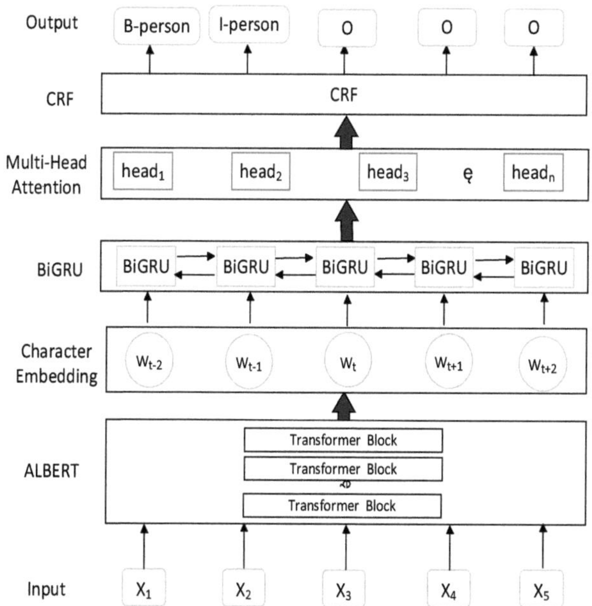

Fig. 1. The Framework of Model

Embedding Factorization: In BERT, the embedding and hidden layer dimensions are the same, increasing the model's parameter size. ALBERT decouples these dimensions, introducing an auxiliary matrix to transform the embedding layer's dimensions. This reduces time complexity, making ALBERT more parameter-efficient than BERT.

Parameter Sharing: BERT uses multiple Transformer encoder layers without parameter sharing. ALBERT shares parameters across attention and fully connected layers, reducing parameter size while improving model stability.

3.2 BiGRU Layer

Gated Recurrent Units (GRUs) are a type of RNN that, like LSTMs, use gating mechanisms to control information flow. While LSTMs have three gates (forget, input, and output), GRUs have two: update and reset, as shown in Fig. 2. This simpler structure means GRUs have fewer parameters and are easier to train. To address the limitation of unidirectional GRUs in capturing full context, bidirectional GRUs (BiGRUs) are used. BiGRUs employ both forward and backward GRUs to extract contextual features, combine their outputs through weighted sums, and pass the result through a linear layer to produce the final output label vector list, as detailed in Eqs. (1)–(3).

$$\overrightarrow{\mathrm{h_t}} = GRU\left(\mathrm{x_t}, \overrightarrow{\mathrm{h_{t-1}}}\right) \tag{1}$$

$$\overleftarrow{\mathrm{h_t}} = GRU\left(\mathrm{x_t}, \overleftarrow{\mathrm{h_{t-1}}}\right) \tag{2}$$

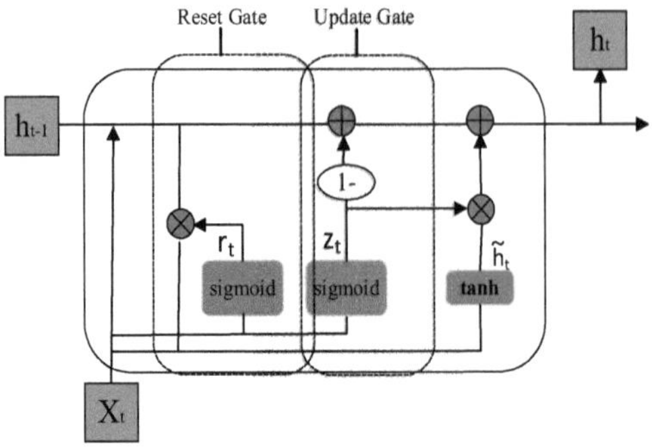

Fig. 2. GRU Network Architecture Diagram.

$$h_t = \left[\overrightarrow{h_t}; \overleftarrow{h_t}\right] \tag{3}$$

3.3 Multi-head Attention Layer

After encoding by the BiGRU layer, text vectors have equal weights. However, words in a sentence often vary in importance for NER. For example, in "Lei Jun founded Xiaomi Corporation in 2010", "2010" is less important for identifying "Xiaomi" as a company entity and is assigned a lower weight. In contrast, "Lei Jun" is more important and gets a higher weight. This weight allocation improves the accuracy of entity recognition for "Xiaomi".

Attention mechanisms enhance the extraction of relevant semantic features and boost the model's local feature extraction ability. Attention weights depend on word vectors themselves, not word distances, which helps solve long-distance dependency issues.

For the BiGRU output matrix H, the attention mechanism projects it into query (Q), key (K), and value (V) matrices. In multi-head attention, each head independently computes attention matrices, and the results are concatenated to capture text context. W_i^Q, W_i^K, W_i^V are weight parameters, and n is the number of attention heads. The specific calculation formula is as follows:

$$Attention = softmax\frac{QK^T}{\sqrt{d_k}}V \tag{4}$$

$$head_i = Attention\left(QW_i^Q, KW_i^K, VW_i^V\right) \tag{5}$$

$$Multi(Q, K, V) = Concat[head_1, \cdots\cdots, head_n] \tag{6}$$

3.4 CRF Layer

Conditional Random Fields (CRF) is a common machine-learning algorithm for NER tasks that models label dependencies, taking into account constraints between adjacent words, such as "B-ORG" not being followed by "I-LOC", "B-PER" being followed only by "I-PER" or "O", and entities starting only with "B-". For the vector matrix generated by the multi-head attention layer, CRF calculates the score of the predicted label sequence, uses the softmax function to compute the probability of the output label sequence, and then applies the Viterbi algorithm to determine the optimal global label sequence for the text.

4 Experimental Results and Analysis

4.1 Experimental Data and Evaluation Criteria

The experiments use two authoritative Chinese datasets: the 1998 People's Daily and MSRA datasets. They focus on identifying three entity types: persons, locations, and organizations. Each dataset is randomly split into training and testing sets in an 8:2 ratio. Table 1 below shows the scale of these datasets. The datasets are annotated using the BIO tagging scheme, where B-X denotes the beginning of entity X, I-X denotes the middle or end of entity X, and O indicates segments that do not belong to any named entity type.

Table 1. Dataset Information

Dataset	Training Set	Testing Set	Total
《People's Daily 》	36257	9059	45316
MSRA	64712	16172	80884

Precision (P), recall (R), and F1 score are used as evaluation metrics. TP refers to correctly predicted positive samples, FP refers to incorrectly predicted positive samples, and FN refers to missed positive samples in the prediction. The specific formulas for these metrics are shown in Eqs. (7)–(9).

$$P = \frac{\mathrm{TP}}{\mathrm{TP+FP}} \tag{7}$$

$$R = \frac{\mathrm{TP}}{\mathrm{TP+FN}} \tag{8}$$

$$F1 = \frac{2\times \mathrm{P} \times \mathrm{R}}{\mathrm{P+R}} \tag{9}$$

4.2 Experimental Environment and Parameter Settings

The experiments were conducted on a Windows 10 operating system using the PyTorch framework for model construction, with specific parameter settings detailed in Table 2.

Table 2. Parameter Settings

Parameter	Value
Number of ALBERT-Base Layers	12
Dimension of ALBERT-Base Hidden Layers	768
Maximum Sequence Length	128
Dimension of GRU Hidden Layers	128
Dropout	0.5
Batch Size	32
Number of Attention Heads	8
Optimizer	Adam

4.3 Experimental Results and Analysis

To validate our model's effectiveness, we conducted comparative experiments on two datasets using six models: LSTM-CRF, BiLSTM-CRF, BERT-IDCNN-CRF, BERT-BiLSTM-CRF, RoBERTa-BiGRU-CRF and XLNet-BiGRU-CRF. The results are shown in Tables 3 and 4.

On the 1998 People's Daily dataset, our model outperformed the others across all metrics. BiLSTM-CRF's F1 score was 4.19% higher than LSTM-CRF's, highlighting the advantages of bidirectional encoding in capturing more semantic information. BERT-IDCNN-CRF's F1 score was 2.26% lower than BERT-BiLSTM-CRF's, indicating BiLSTM's superior semantic capture ability compared to CNN.

Compared to BiLSTM-CRF and BERT-BiLSTM-CRF, our model achieved significant improvements of 7.37%, 9.85%, and 8.62% in P, R, and F1, respectively. These gains are mainly due to the use of dynamic word vectors generated by pre-trained models (BERT or ALBERT), which outperform the static vectors from Word2vec. Our model's slight edge over BERT-BiLSTM-CRF stems from the integration of multi-head attention, enabling the model to focus more on key content and compensating for the local feature capture limitations of neural networks.

Furthermore, we conducted a comparative evaluation of our proposed approach against two additional pre-trained encoder-based baselines: RoBERTa-BiGRU-CRF and XLNet-BiGRU-CRF. While both baseline models demonstrated competitive overall performance, our method achieved superior results, surpassing RoBERTa-BiGRU-CRF by 5.19% and XLNet-BiGRU-CRF by 3.58% in terms of F1 score on the People's Daily dataset. These findings underscore the effectiveness of integrating multi-head attention into the ALBERT-BiGRU architecture, which not only enhances global semantic representation but also improves generalization capabilities for identifying boundary-sensitive and long-range dependent entities.

Table 3. Experimental Results on the People's Daily Dataset.

Model	P	R	F1
LSTM-CRF	84.56	80.36	82.41
BiLSTM-CRF	87.76	85.47	86.60
BERT-IDCNN-CRF	92.43	91.67	92.05
BERT-BiLSTM-CRF	94.26	94.37	94.31
ROBERTa-BiGRU-CRF	89.23	90.90	90.04
XLNet-BiGRU-CRF	91.46	91.83	91.64
OUR MODEL	95.13	95.32	95.22

On the MSRA dataset, our model achieved state-of-the-art (SOTA) performance in entity recognition. Specifically, it surpassed the BiLSTM-CRF model by 6.5%, the BERT-IDCNN-CRF model by 2.9%, and the BERT-BiLSTM-CRF model by 0.8% in terms of F1 score. These results provide further empirical evidence of the effectiveness and robust generalizability of our proposed NER framework. Notably, our model not only outperformed RoBERTa-BiGRU-CRF by 5.93% and XLNet-BiGRU-CRF by 3.06% in F1 score but also demonstrated significant efficiency advantages, requiring 18% less training time and 23% fewer parameters compared to the strongest baseline (RoBERTa-BiGRU-CRF).

Table 4. Experimental Results on the MSRA Dataset.

Model	P	R	F1
LSTM-CRF	86.79	81.31	83.96
BiLSTM-CRF	89.67	87.26	88.45
BERT-IDCNN-CRF	92.45	91.76	92.11
BERT-BiLSTM-CRF	94.86	93.51	94.18
ROBERTa-BiGRU-CRF	88.56	89.66	89.09
XLNet-BiGRU-CRF	92.69	91.25	91.96
OUR MODEL	94.98	95.07	95.02

Among the comparison models, the BERT-BiLSTM-CRF model performed the best. Using the same training samples from the People's Daily dataset, we trained the above models and recorded the training time for each model, as shown in Table 5.

The experimental results demonstrate that the BERT-BiLSTM-CRF model required 2,133 s for training, whereas our proposed model completed training in 1,832 s, achieving a 14.1% reduction in training time relative to the baseline. This efficiency gain primarily stems from two architectural optimizations: (1) replacing BERT with ALBERT, which reduces parameter count through factorized embedding parameterization and cross-layer parameter sharing, while preserving competitive contextual representation capabilities;

and (2) substituting BiLSTM with BiGRU, which simplifies the recurrent unit architecture by eliminating redundant gates, thereby lowering computational complexity without compromising bidirectional semantic dependency modeling.

Table 5. Model Runtime

Model	Training Time (seconds)
BERT-BiLSTM-CRF	2133
ALBERT-BiGRU-MA-CRF	1832

Compared to traditional BERT-BiLSTM-CRF models, we opt for ALBERT and BiGRU due to performance and efficiency. ALBERT, a lightweight BERT variant, cuts parameters via factorized embedding and cross-layer sharing, yet keeps strong contextual encoding. BiGRU, simpler than BiLSTM with fewer gates, reduces computation while capturing bidirectional semantics.

4.4 Ablation Study Results and Analysis

To investigate the impact of each module (ALBERT, BiGRU, MA, CRF) on the model, comparative experiments were conducted on the People's Daily dataset, with results presented in Table 6.

Table 6. Ablation Study

Model	P	R	F1
ALBERT	90.16	90.79	90.48
ALBERT-CRF	91.78	91.37	91.57
ALBERT-BiGRU-CRF	92.96	93.76	93.35
ALBERT-BiGRU-MA-CRF	95.13	95.32	95.22

As shown in Table 5, the ALBERT-CRF model outperforms the ALBERT-only model across all metrics. This indicates that adding a CRF layer imposes more constraints on label prediction, thereby enhancing model performance. Building on the ALBERT - CRF model and incorporating the BiGRU module further improves precision, recall, and F1 score by 1.18%, 2.39%, and 1.78% respectively. This is because BiGRU's ability to capture global bidirectional features positively impacts named entity recognition. Furthermore, adding the MA module to the ALBERT- BiGRU-CRF model results in additional gains of 2.17%, 1.56%, and 1.87% in precision, recall, and F1. The multi-head attention mechanism in this module effectively captures text features from multiple dimensions and addresses long- distance dependency issues.

5 Conclusion

This paper introduces the ALBERT-BiGRU-MA-CRF model to overcome the shortcomings of existing Chinese NER models in feature extraction and long-distance dependency capture. ALBERT generates dynamic vectors to tackle word polysemy, while BiGRU extracts global semantics, cutting training time. The multi-head attention mechanism improves accuracy by capturing long-distance dependencies, and CRF enhances results by handling label interdependencies. Experiments demonstrate that our model achieves over 95% F1 on the 1998 People's Daily and MSRA datasets, outperforming others with reduced training time.

Despite its strong performance, the model's generalization to domain-specific texts like medical or legal texts is uncertain, and nested entity recognition remains suboptimal. Future research could explore domain adaptation, incorporate external resources, extend to multilingual tasks, and refine boundary detection.

Acknowledgments. This study was funded in part by the Natural Science Foundation of Ningxia under Grant 2024AAC03098, in part by the Key Research and Development Program of Ningxia under Grant 2023BSB03066, and in part by the National Natural Science Foundation of China under Grant 62066038.

Disclosure of Interests. It is now necessary to declare any competing interests or to specifically state that the authors have no competing interests.

References

1. Li, J., Sun, A.X., Han, J.L., et al.: A survey on deep learning for named entity recognition. IEEE Trans. Knowl. Data Eng. **34**(1), 50–70 (2022)
2. Peng, X., Zhao, H., Zheng, Z., et al.: Fusion of multiple embedding representations for Chinese named entity recognition. J. Changchun Univ. Technol. **43**(01), 81–90 (2022)
3. You, L.J., Yin, Y.F.: Medical imaging report information entity recognition based on BiLSTM-CRF model. Microcomput. Appl. **39**(10), 134–137 (2023)
4. Song, J.R., Chen, Y.P., Wang, K., et al.: Named entity recognition semantic supplement method based on affix-attention. J. Shandong Univ. (Engineering Edition) **53**(02), 70–76 (2023)
5. Zhang, Y., Yang, J.: Chinese NER using lattice LSTM. In: Proceedings of the 56th Annual Meeting of the Association for Computational Linguistics, pp. 1554–1564 (2018)
6. Guan, S.Q., Dong, T.T., Wan, Z.J., et al.: Research on fire case entity recognition based on BERT-CRF model. Fire Sci. Technol. **42**(11), 1529–1534 (2023)
7. Zhang, Q., Li, C.J., Liu, J.S.: Research on named entity recognition in military field based on BERT_IDCNN_CRF. Aerosp. Electr. Warfare **37**(05), 56–60 (2021)
8. Hu, Z., Hou, W., Liu, X.: Deep learning for named entity recognition: a survey. Neural Comput. Appl. **36**(16), 8995–9022 (2024)
9. Collobert, R., Weston, J., Bottou, L., et al.: Natural language processing (almost) from scratch. J. Mach. Learn. Res. **12**, 2493–2537 (2011)
10. Aguilar, G., Maharjan, S., López-Monroy, A. P., et al.: A multi-task approach for named entity recognition in social media data. In: Proceedings of the 3rd Workshop on Noisy User-Generated Text, pp. 48–153 (2019)
11. Wu, Y., Jiang, M., Lei, J., et al.: Named entity recognition in Chinese clinical text using deep neural network. In: MEDINFO 2015: eHealth-enabled Health, pp. 624–628. IOS Press (2015)

12. Kuru, O., Can, O.A., Yuret, D.: CharNER: character-level named entity recognition. In: Proceedings of COLING 2016, the 26th International Conference on Computational Linguistics: Technical Papers, pp. 911–921 (2016)
13. Wan, Q., Liu, J., Wei, L., et al.: A self-attention based neural architecture for Chinese medical named entity recognition. Math. Biosci. Eng. **17**(4), 3498–3511 (2020)
14. Wallach, H.M.: Conditional random fields: an introduction, vol. 22, pp. 2–6. Technical reports (CIS) (2004)
15. Ma, X.: End-to-end sequence labeling via bi-directional LSTM-CNNs-CRF. In: Proceedings of the 54th Annual Meeting of the Association for Computational Linguistics, pp. 1064–1074 (2016)
16. Pennington, J., Socher, R., Manning, C.D.: GloVe: global vectors for word representation. In: Proceedings of the 2014 Conference on Empirical Methods in Natural Language Processing (EMNLP), pp. 1532–1543 (2014)
17. Bose, P., Srinivasan, S., Sleeman, W.C., IV., Palta, J., Kapoor, R., Ghosh, P.: A survey on recent named entity recognition and relationship extraction techniques on clinical texts. Appl. Sci. **11**(18), 8319 (2021)
18. Rei, M., Crichton, G.K.O., Pyysalo, S.: Attending to characters in neural sequence labeling models. In: Proceedings of COLING 2016, the 26th International Conference on Computational Linguistics, pp. 309–318 (2016)
19. Yang, Z., Salakhutdinov, R., Cohen, W.: Multi-task cross-lingual sequence tagging from scratch. arXiv preprint: arXiv:1603.06270 (2016)
20. Liu, Z., Lin, W., Shi, Y., Zhao, J.: A robustly optimized BERT pre-training approach with post-training. In: China National Conference on Chinese Computational Linguistics, pp. 471–484. Springer International Publishing, Cham (2021)
21. Lan, Z., Chen, M., Goodman, S., Gimpel, K., Sharma, P., Soricut, R.: ALBERT: a lite BERT for self-supervised learning of language representations. In: International Conference on Learning Representations, pp. 102–108 (2020)
22. Yang, Z., Dai, Z., Yang, Y., Carbonell, J., Salakhutdinov, R.R., Le, Q.V.: XLNet: generalized autoregressive pretraining for language understanding. In: Advances in Neural Information Processing Systems, vol. 32 (2019)
23. Hanh, T.T.H., Doucet, A., Sidere, N., Moreno, J.G., Pollak, S.: Named entity recognition architecture combining contextual and global features. In: International Conference on Asian Digital Libraries, pp. 264–276. Springer International Publishing, Cham (2021)
24. Tang, Z., Wan, B., Yang, L.: Word-character graph convolution network for Chinese named entity recognition. IEEE/ACM Trans. Audio, Speech, Lang. Process. **28**, 1520–1532 (2020)
25. Chen, C., Kong, F.: Enhancing entity boundary detection for better Chinese named entity recognition. In: Proceedings of the 59th Annual Meeting of the Association for Computational Linguistics and the 11th International Joint Conference on Natural Language Processing, Volume 2: Short Papers, pp. 20–25 (2021)
26. Wang, Y., Lu, L., Wu, W., Chen, Y.: Polymorphic graph attention network for Chinese NER. Expert Syst. Appl. **203**, 117467 (2022)
27. Tian, X., Bu, X., He, L.: Multi-task learning with helpful word selection for lexicon-enhanced Chinese NER. Appl. Intell. **53**(16), 19028–19043 (2023)

A Feature-Clustering Algorithm Based on K-Medoids

Xuehao Zhang, Yingjun Ji(✉), Xiwang Guo, and Shujin Qin

Liaoning University, Shenyang 110036, Liaoning, China
jiyingjun@lnu.edu.cn

Abstract. Clustering is an important problem in machine learning. Existing clustering methods mainly focus on sample clustering, while research on feature clustering remains limited. Unlike sample clustering, features exhibit substantial heterogeneity in their types and distributions, rendering traditional clustering algorithms based on homogeneity assumptions ineffective. To address this issue, this work proposes a novel feature-clustering algorithm based on k-medoids (FCKM). It adapts three distance metrics according to different types of features and employs an improved k-medoids algorithm on feature clustering. Besides, this work proposes a novel metric to evaluate the performance of feature clustering. Experimental results on real-world datasets show that our proposed algorithm has better performance than its competitor, showcasing strong practical applicability.

Keywords: Feature Clustering · Mixed-type Data · Clustering Measure

1 Introduction

Clustering is a common unsupervised learning technique designed to discover patterns inherent in unlabeled data. It can organize data into meaningful groups without any predefined category labels by grouping similar data instances into the same clusters based on a defined similarity metric. Clustering algorithms display an important role in scenarios where manual annotation is not feasible or practical.

Researchers have proposed various clustering validity indexes to evaluate the unsupervised clustering results. Compact-Separate Proportion Index (CSP) proposed by Zhou et al. [1] simultaneously measures the compactness and separateness of the clustering structure, which effectively improves the robustness. However, it requires a predetermined number of clusters and is more complex to compute. Cheng et al. [2] proposes a Local Cores-based Cluster Validity Index (LCCV) that performs well when dealing with noisy and different density data, although its sensitivity to the core density threshold may limit the practical application. Density-Core-based Clustering Validation Index (DCVI) proposed by Xie et al. [3] focuses on the evaluation based on the density of the core members, which may be less effective in datasets that are irregular or have large density differences. Şenol [4] proposes a Validity Index for Arbitrary-Shaped Clusters based on the Kernel Density Estimation (VIASCKDE) that can effectively recognize arbitrarily

J. Li et al. (Eds.): APWeb-WAIM 2025, CCIS 2660, pp. 187–198, 2026.
https://doi.org/10.1007/978-981-95-7996-9_17

shaped clusters and avoids the pre-setting of parameters. However, its selection of kernel function parameters is difficult and the computational overhead is large. Augmented Non-shared Nearest Neighbors (ANCV) proposed by Duan et al. [5] performs well in coping with the noisy data and improving the scalability, however, the performance may be degraded on datasets with high dimensional space or high cluster overlap.

The Clustering Validation based on Nearest Neighbor (CVNN) proposed by Liu et al. [6] combines the nearest neighbor-based separation with the traditional compactness term to capture arbitrarily shaped clusters and complex cluster structures. Their experiments show that CVNN outperforms many classical metrics, especially when the data contains clusters with non-standard shapes. However, to the best of the authors' knowledge, most of the existing studies are based on homogeneous clustered data points, and studies have yet to be conducted on feature clustering with the inapplicability of a single distance metric used in existing clustering algorithms when measuring different types of features.

Feature clustering plays a crucial role in many real-world applications, such as identifying clusters of interrelated features in high-dimensional datasets, or achieving dimensionality reduction by merging redundant variables. Feature clustering presents many challenges, especially when dealing with datasets containing multiple data types. In such scenarios, defining a mathematically sound and statistically valid similarity or distance measure between features is a complex problem. Traditional clustering algorithms typically assume a uniform data type when measuring data similarity or calculating distances. They lack a generalized similarity metric that is equally valid among heterogeneous feature types. To address this problem, this work constructs a more general and robust feature clustering algorithm that can adapt to mixed data types and provide a reliable clustering result. The main contributions of this work are as follows.

1) A similarity measure oriented between different types of features is proposed. We propose a feature-oriented clustering method for the first time. The method defines the corresponding feature similarity metrics according to the variable types (e.g., continuous or categorical), ensuring that different types of features can be subject to a uniform similarity metric.
2) An improved k-medoids clustering algorithm is proposed. Using the above feature similarity metric, we design a feature clustering algorithm based on improved k-medoids to make it well-suited for feature clustering scenarios. The centroid selection and updating steps are optimized to converge to a stable segmentation of feature clusters.
3) A new metric used to evaluate the effectiveness of feature clustering is designed. The metric measures the effectiveness of feature clustering based on the accuracy of the subset corresponding to the feature clustering result in the classification task.

2 Proposed Algorithm

The proposed FCKM aims at clustering features of mixed-type data. The method first constructs the statistical distance matrix between features and generates the clustering divisions under different numbers of clusters using the improved k-medoids algorithm, and then determines the optimal number of clusters by an internal metric based on the nearest neighbor structure. To evaluate the effectiveness of the clustering results, Average Classification Accuracy Across Features (ACAF) is introduced in this work to measure the ability of feature clustering to support downstream classification tasks. The pseudo-code of the proposed algorithm is shown in Algorithm 1. Its specific steps are as follows.

Algorithm 1: FCKM

Input: Dataset with n instances and d features; maximum clusters $K_{\max}$; nearest-neighbor parameter k.
Output: Optimal feature clustering P_{K^*}; optimal number of clusters K^*.
1: Calculate raw distances δ_{pq} for all feature pairs using Eq. (1).
2: Apply global normalization to obtain the distance matrix $D' = [\tilde{\delta}pq]$ using Eq. (2).
3: **for** $K = 2$ **to** $K_{\max}$ **do**
    4: Perform k-medoids clustering on D' to obtain partition $P_K = \{C_1, \ldots, C_K\}$.
    5: Compute intracluster compactness $\mathrm{Comp}(P_K)$ and intercluster separation $\mathrm{Sep}(P_K)$ using Eqs. (4) and (5).
6: Calculate structural score $FCKM(P_K)$ using Eq. (6).
7: **end for**
8: Determine $K^* = \mathrm{argmin}_{2 \le K \le K_{\max}} FCKM(P_K)$ using Eq. (7).
9: **Return** the feature clusters P_{K^*} and the optimal number K^*.

2.1 Distance Matrix Construction

In feature clustering, a key prerequisite is to quantify the similarity between each pair of features. We construct a $d \times d$ distance matrix $D = [\delta_{pq}]$, where d denotes the number of features and each element δ_{pq} denotes the distance between the features x_p and x_q. In order to accommodate mixed data types and to reflect the statistical relationships between features, we employ appropriate statistical hypothesis testing methods to construct δ_{pq} rather than using simple Euclidean distances. Specifically, for each pair of features (x_p, x_q).

1) If x_p and x_q are both continuous variables, the Kolmogorov-Smirnov test (KS) [7] is used to quantify the difference in their distributions. The corresponding test statistic D_{pq} is used as a measure of the distance between the two, with smaller values of D_{pq} indicating that the distributions of x_p and x_q are more similar, and therefore less distant.

2) If one variable is continuous and the other is discrete, the Kruskal-Wallis test (KW) is used to determine whether there is a difference in the distribution of the continuous variable under different categories. The statistic obtained from this test H_{pq} is used as an indicator of distance.H_{pq} The smaller it is, the more consistent the distribution is between different categories, indicating a weaker dependence between variables and a smaller distance.

Table 1. Test Statistics For Mixed-Type Feature Pairs

Feature Types	Method	Formula
Continuous - Continuous	KS	$D_{pq} = \sup_v \lvert F_p(v) - F_q(v) \rvert$
Continuous - Categorical	KW	$H_{pq} = \frac{12}{n(n+1)} \sum_{j=1}^{c} n_j \left(\bar{R}_j - \frac{n+1}{2}\right)^2$
Categorical - Categorical	χ^2	$\chi^2_{pq} = \sum_{i=1}^{r} \sum_{j=1}^{c} \frac{\left(O_{ij} - E_{ij}\right)^2}{E_{ij}}$

n: total number of samples;$F_p(v), F_q(v)$: empirical CDFs of x_p and x_q; $\sup_v$: supremum over all ;vc : number of groups;n_j : number of samples in group;$j\bar{R}j$: average rank in group;$jOij, E_{ij}$: observed and expected frequencies; rc : number of rows and columns in the contingency.

3) If both x_p and x_q are discrete variables, the Chi-Square test (χ^2) [8] is performed through a column table and the statistic is recorded as χ^2_{pq}.χ^2_{pq} The smaller the value, the weaker the statistical correlation between the two variables, the more similar the distributions, and therefore the smaller the distance.

In all cases, we define the raw distance δ_{pq} as the test statistic returned by the corresponding statistical test, the form of which is given in Table 1. Test Statistics For Mixed-Type Feature Pairs. The test statistic is the same as that returned by the statistical test. This method provides a uniform similarity measure for pairs of heterogeneous variables, where a smaller statistic denotes stronger similarity between the two features. The process of constructing δ_{pq} in the above three cases is summarized in Eq. (1).

$$\delta_{pq} \begin{cases} D_{pq}, & \textit{if } x_p \textit{ and } x_q \text{ are continuous} \\ H_{pq}, & \text{if one is continuous and one categorical} \\ \chi^2_{pq}, & \textit{if } x_p \textit{ and } x_q \text{ are categorical} \end{cases} \tag{1}$$

where $D = \left[\delta_{pq}\right]$ is a symmetric matrix satisfying $D_{pq} = D_{qp}$ and $D_{pp} = 0$. The smaller the value of D_{pq}, the more closely the features x_p and x_q are related.

The raw distance values obtained from different statistical tests may differ in distribution and scale. To ensure the comparability of distances among all feature pairs, we apply min-max normalization to the distance matrix. Let $\delta_{\min}$ and $\delta_{\max}$ be the minimum and maximum raw distances among all non-diagonal terms, respectively, then each distance value δ_{pq} is linearly transformed to the normalized distance $\overset{\sim}{\delta}_{pq}$, which is defined as.

$$\overset{\sim}{\delta}_{pq} = \frac{\delta_{pq} - \delta_{min}}{\delta_{max} - \delta_{min}} \tag{2}$$

After min-max normalization, it satisfies $\overset{\sim}{\delta}_{pq} \in [0,1]$ and yields the normalized distance matrix $D' = \left[\overset{\sim}{\delta}_{pq}\right].\overset{\sim}{\delta}\ pq = 0$ This corresponds to the most similar pair of features and $\overset{\sim}{\delta}\ pq = 1$ corresponds to the least similar pair of features. In the subsequent steps, all distances used are normalized to $\overset{\sim}{\delta}_{pq}$ unless otherwise stated.

2.2 The k-Nearest Neighbor Consistency Principle in Feature Clustering

A core principle of FCKM is K-Nearest Neighbor (kNN) consistency in clustering, i.e., in a well-structured cluster, the nearest neighbor of each object should be in the same cluster as that object. If multiple nearest neighbors of an object are located in other clusters, this indicates that the kNN consistency is broken, suggesting that there is poor separation between cluster.

FCKM extends the kNN consistency principle from sample space to feature space. If two features are similar to each other, they should be clustered in the same cluster, and they should also have similar neighbor features. For feature clustering, kNN consistency means that for any feature in a cluster, its k nearest neighbors in the feature space should also belong to the same cluster. FCKM bases its "separability" criterion for clustering on this principle - it tends to look for a clustering structure such that each feature's neighbors are mainly concentrated in the cluster to which it is assigned. By introducing kNN, the algorithm is able to reflect the intrinsic structure of the feature space, which is difficult to achieve with traditional Euclidean distance-based methods.

2.3 Feature Clustering Based on k-medoids

After obtaining the normalized distance matrix D', the FCKM algorithm starts clustering the features. Compared with the k-means algorithm used in CVNN which requires the input to be numerical feature vectors, the improved k-medoids method we use to group similar features has better adaptability and robustness to different types of features. Because k-medoids selects actual samples as clustering centers and it is able to measure the similarity between features directly using D', it improves the execution efficiency of the algorithm.

For a given number of clusters K, the k-medoids algorithm divides the d features into a set of K clusters,$\mathcal{P}_K = \{C_1, C_2, \ldots, C_K\}$, with the objective of minimizing the sum of intra-cluster distances. For each clusterC_k, a centroid m_k is chosen. The objective function of k-medoids can be formulated as minimizing the total distance from all features in the cluster to their centroids.

$$\min_{m_1,\ldots,m_K} \sum_{i=1}^{K} \sum_{x_p \in C_i} \overset{\sim}{\delta}\left(x_p, m_i\right) \tag{3}$$

where $\overset{\sim}{\delta}\left(x_p, m_i\right)$ is the distance between the feature x_p and the center element m_k of the cluster it belongs to. The centroid m_k denotes the feature in the cluster that minimizes the total distance to other features. The clustering process proceeds as follows: first K initial centroids are selected. The algorithm then iterates between the following two steps: 1) assigning each feature to the cluster to which its closest centroid belongs; and 2) re-selecting a feature from all the features in that cluster to minimize the total distance between it and the other features in the same cluster. This iterative process continues until the cluster assignment converges. The final output is the cluster $\mathcal{P}_K$ and the set of centroid elements for each cluster $\{m_1, \ldots, m_K\}$.

The above clustering process will be iterated at multiple K values to obtain multiple levels of feature clustering divisions. For each clustering number K, a feature clustering result C_K is obtained. The next subsection presents the FCKM algorithm for selecting the most appropriate number of clusters from these results.

2.4 FCKM Algorithm with Optimal Cluster Number Selection

To select the appropriate number of feature clusters, we define two internal metrics: intracluster compactness and intercluster separation. These two metrics are inspired by the nearest neighbor-based clustering effectiveness metric CVNN and are adapted to the task of feature clustering in this study. We set the number of nearest neighbors k to be rounded up to the square root of the number of features d, i.e.,$k = \lceil \sqrt{d} \rceil$, which is used to calculate the k nearest neighbors of each feature in the clustering process.

Intracluster compactness measures the degree to which features (feature columns) within the same cluster remain connected to each other in distance space. Defined as the sum of the average pairwise distances between all feature columns within each cluster:

$$Comp(P) = \sum_{k=1}^{K} \left[\frac{2(|C_k|-1)}{|C_k|} \sum_{x_p, x_q \in C_k, p<q} \tilde{\delta}_{pq} \right] \tag{4}$$

where $|C_k|$ denotes the number of features contained in the cluster. For each cluster, we calculate distances for all member features in the cluster two by two, and average these distances after summing them up, and finally sum up the average internal distances of all the clusters as a compactness measure for the overall clustering classification. This metric reflects the degree of similarity between the members of a cluster, with smaller values indicating that the cluster features are closer to each other and the clustering structure is more compact. Ideally, if the distance between features in all clusters is extremely small, i.e., Comp(P) $= 0$, it indicates that the feature clustering division is highly structurally concentrated and has good internal consistency.

Inter-cluster separation is used to measure the degree of separation between different clusters. It focuses on the single nearest neighbor feature of each feature and determines whether that nearest neighbor belongs to the same cluster. It is defined as the proportion of features whose nearest neighbors belong to different clusters.

$$Sep(P) = \max_{1 \le k \le K} \left[\frac{1}{|C_k|} \sum_{x_p \in C_k} \frac{h_p}{k} \right] \tag{5}$$

where $h_p = \sum_{r=1}^{k} \mathbb{I}C(NN_r(x)) \neq C(x_p)$ indicates how many neighbors of the first k nearest neighbors of the feature x_p are from other clusters,$NN_r(x_p)$ is its r nearest neighbor, and $\mathbb{I}\cdot$ is an indicator function, which takes the value of 1 when the condition is valid, otherwise it is 0. This indicator first calculates the proportion of neighbors from other clusters for all features in each cluster, then averages the clusters, and finally takes the maximum value from all clusters as the overall separateness.$Sep(P)$ The smaller it is, the clearer the boundaries between the clusters are. Ideally, if the nearest neighbors of all the features fall within the clusters to which they belong, there is $Sep(P) = 0$, which indicates that the feature clustering results have good separability.

The FCKM algorithm is proposed to select the optimal number of clusters based on intra-cluster tightness $Comp(P)$ and inter-cluster separation$Sep(P)$. FCKM calculates the following function values for each feature cluster number K corresponding to the cluster division P_K:

$$FCKM\,(P) = \frac{Comp(P)}{\max\limits_{K}(Comp(P))} + \frac{Sep(P)}{\max\limits_{K}(Sep(P))} \tag{6}$$

The two components in Eq. (6) converge to 0 in the ideal case, at this time $Comp(P) = 0$ and $Sep(P) = 0$, which means that the features are clustered most compactly and clearly separated from each other, and $FCKM\,(P)$ also reaches the minimum value, which represents the optimal clustering structure.

The FCKM algorithm will run on the number of clusters $K \in [2, K_{\max}]$ and compute the value of the structure function $FCKM\,(P_K)$ for the division P_K corresponding to each number of clusters and select the optimal number of clusters accordingly:

$$K^* = \arg \min_{2 \le K \le K_{max}} FCKM\,(P_K) \tag{7}$$

K^* is the number of clusters that minimizes the FCKM value of the cluster partition P_K. The model selection process is completely unsupervised, relying only on the structure of the distances between features, without any external labeling or human intervention.

In practice, in order to improve the stability and reliability of the selection results, we can also set the range of the number of clusters to $\left[2, \lceil\sqrt{d}\rceil\right]$, i.e., the number of clusters does not exceed the square root of the sample size, so as to avoid overfitting. Combined with the "1-SE rule" (one-standard-error rule) [9], we can select the smallest K as the final number of clusters K^* within the tolerance interval formed by the minimum value of $FCKM\,(P_K)$ and its double standard error. This strategy can improve model simplicity while maintaining structural rationality.

2.5 Average Feature Classification Accuracy Metric

Traditional clustering algorithms are hardly to evaluate the performance of feature clustering. We introduce a new evaluation metric for feature clustering called Average Classification Accuracy Across Features (ACAF). This metric measures the quality of clustering delineation by the magnitude of the accuracy of the subsequent classification tasks. Note that the metrics we construct are based on the classification problem as an example, and can be replaced by other metrics if it is a regression problem. The specific calculation process of ACAF is as follows.

(1) Selecting representative features. According to the K clustering division divided by ACIC algorithm randomly selected as the representative feature of that cluster. In t his way, a total of K representative features are selected to form the feature subset.

(2) Training a classification model. Train a classification model to predict the sample categories of the dataset using the subset of representative features selected above, and calculate the classification accuracy. Due to the uncertainty of randomly selecting the representative features, this step should be repeated several times under the same number of clusters K, and the average value should be taken to get the average accuracy under the number of clusters, in order to reduce the influence of randomness.

(3) Calculating the ACAF metrics. We use three classification models, i.e., Logistic Regression (LR), Random Forest (RF) and XGBoost (XGB) to perform (1) and (2), respectively. For each model, the average accuracy obtained for different number of clusters K is recorded. The classification accuracies for all the clusters in the range of $K = 2$ to $K_{\max}$ are then averaged for each classifier to obtain the final ACAF metric value. For a given classifier $M \in$ LR, RF, XGB, ACAF is defined as:

$$\mathrm{ACAF}^{(M)} = \frac{1}{K_{max}-1} \sum_{k=2}^{K_{max}} Acc_k^{(M)} \tag{8}$$

where $Acc_k^{(M)}$ denotes the average classification accuracy of the classifier M on the selected subset of features when the number of clusters is K.

The above ACAF metric utilizes the average classification accuracy to evaluate the effectiveness of a feature clustering scheme in a supervised learning task. It is designed on the principle that if a certain feature clustering delineation can better distinguish the sample categories, then classification using representative features from each cluster will yield higher accuracy. By averaging the classification performance under different number of clusters, ACAF integrates the effect of the variation in the number of clusters on the classification performance, thus providing an evaluation of the overall quality of the clustering scheme. Compared with only selecting the performance under a fixed number of clusters, the averaging approach can reduce the interference of chance factors and obtain more robust metrics. The advantage of ACAF is that it introduces a task-oriented evaluation criterion, which puts the clustering results into the classification task for testing, and directly reflects the size of the contribution of the feature clustering to the subsequent classification task.

3 Experimental Results

In this section, we experimentally validate the effectiveness of the proposed FCKM method and compare it with the original CVNN-based method in terms of feature clustering quality and subsequent classification performance. We use five real datasets as described below.

3.1 Dataset and Experimental

The five real-world data sets used in this work are from the University of California, Irvine (UCI) machine learning repository [10], and the specific information is shown in Table 2. Summary of datasets the target variables of each of the above datasets are converted to binary classification tasks by binarization.

Table 2. Summary Of Datasets

Dataset	Instances	Features	Continuous	Categorical
HD	303	13	5	8
HF	299	12	7	5
MI	1700	123	13	100
CP	418	17	10	7
CK	400	24	12	12

HD = Heart Disease; HF = Heart Failure Clinical Records; MI = Myocardial Infarction Complications; CP = Cirrhosis Patient Survival Prediction; CK = Chronic Kidney Disease.

Table 3. CVNN vs. FCKM Optimal Clustering Number

Dataset	CVNN	FCKM
HD	13	13
HF	12	12
MI	123	123
CP	17	17
CK	24	24

In FCKM, we set the parameters according to the empirical formulas: the number of nearest neighbors $k = \sqrt{d}$ and the maximum number of clusters $c_m = \sqrt{d}$. For the above five datasets, the initial number of clusters k is about$\sqrt{13} \approx 4$, $\sqrt{12} \approx 4$, $\sqrt{123} \approx 11$, $\sqrt{17} \approx 4$, and$\sqrt{24} \approx 5$, respectively. FCKM starts from this initial number of clusters and performs feature clustering step by step by combining the ACAF metrics described in Section II. To evaluate the clustering results and their impact on the classification task, we used three classification models (logistic regression, random forest, and XGBoost), which are described in Section II-A. For each dataset, we sequentially increased the number of clusters from 2 to the maximum number of clusters for that dataset c_m and recorded the trend of the ACAF metrics and the CVNN index with the number of clusters. This allows each method to suggest an optimal number of clusters based on its own guidelines. We perform five folds of cross-validation on each dataset. In each fold, the feature selection process is consistent for both the training and validation sets and the evaluation process is unbiased.

3.2 Results and Analysis

1. **Optimal Cluster Number Comparison.** Table 3 CVNN vs. FCKM Optimal Clustering Number summarizes the optimal number of feature clusters obtained based on CVNN and FCKM methods on five real datasets (HD, HF, MI, CP, and CK). From the results, it can be observed that the optimal clustering numbers of the two methods on all the datasets are identical, which indicates that FCKM has a high degree of consistency while maintaining the original structural judgment ability of CVNN. However, the optimal clustering number is almost equal to the original feature dimension. This trend reflects that CVNN tends to subdivide the clusters to obtain a better structural separation when calculating the optimal number of clusters. Although FCKM is consistent with CVNN in the selection of optimal values, we will further discuss its performance advantage in small cluster scenarios in the subsequent accuracy analysis.

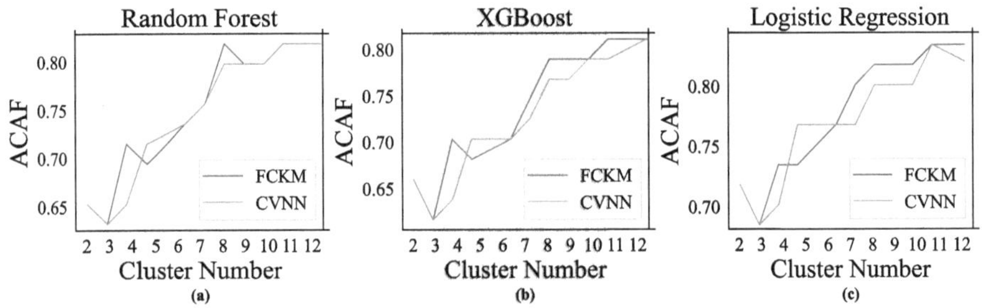

Fig. 1. Classification Accuracy Under Different Cluster Numbers

2. **Classification Accuracy Trend Under Different Cluster Numbers.** In order to compare more intuitively the performance of the two feature clustering methods, FCKM and CVNN, under different numbers of clusters, we plot Fig. 1. Classification Accuracy Under Different Cluster Numbers, which shows the trend curves of the classification accuracies obtained by the different methods with the increase of the number of clusters from 2 to the upper limit on the HD dataset.

From the figure, it can be observed that the classification performance of both methods shows a gradual improvement as the number of clustering clusters increases. This is because more cluster divisions usually provide finer feature grouping and increase the size of the feature subset, thus improving the accuracy of the downstream classifier.

When the number of feature clusters is small (e.g., only 2–4 clusters), the classification accuracy of FCKM significantly outperforms that of the CVNN method, indicating that it retains more information useful for the classification task despite the larger feature compression. In this experiment, when the number of clusters is 4, the feature grouping of FCKM with the XGBoost model can achieve an accuracy of 70.7%, while the corresponding accuracy of the CVNN method is only about 63.8%, which is a difference of 6.9%.

As the number of clusters approaches the total number of features (i.e., each feature is almost individually clustered), the performance of the two methods converges, approaching the upper limit of classification that can be achieved using all features. This trend matches the results for the optimal number of clusters in Table 3. CVNN vs. FCKM Optimal Clustering Number, validating the effectiveness of ACAF in assessing the quality of clustering.

3.3 Overall Accuracy Comparison Based on ACAF

Table 4 comparison results of CVNN and FCKM in ACAF summarizes the average classification accuracy (ACAF) for five datasets and three classes of classifiers under two feature clustering methods. The number of clusters ranges from 2 to $\sqrt{d}$. Experimental results show that FCKM-based feature partitioning outperforms the CVNN approach in all cases, demonstrating consistent benefits in supporting downstream classification tasks. On the HD dataset, the XGBoost classifier shows the most significant performance improvement, with an average accuracy of 0.721 for FCKM compared to 0.665 for CVNN, an improvement of 5.6%, indicating that FCKM is able to effectively retain the important task-relevant information and thus achieve better prediction performance. In the MI dataset, FCKM also achieves a higher average classification accuracy among all classifiers. In the case of RF, the accuracy improves from 0.821 of CVNN to 0.832 of FCKM, indicating that FCKM is able to extract a more compact subset of features in the high-dimensional feature space while improving the predictive performance of the model.

Table 4. Comparison Results Of CVNN And FCKM In ACAF

Dataset	Classifier	CVNN	FCKM	Gap (%)
HD	LR	0.721	**0.749**	2.8
	RF	0.677	**0.730**	5.3
	XGB	0.665	**0.721**	5.6
HF	LR	0.704	**0.716**	1.2
	RF	0.648	**0.666**	1.8
	XGB	0.638	**0.661**	2.3
MI	LR	0.845	**0.846**	0.1
	RF	0.821	**0.832**	1.1
	XGB	0.829	**0.837**	0.8
CP	LR	0.646	**0.651**	0.5
	RF	0.644	**0.653**	0.9
	XGB	0.692	**0.703**	1.1
CK	LR	0.867	**0.890**	2.3
	RF	0.865	**0.890**	2.5
	XGB	0.836	**0.840**	0.4

Gap = FCKM – CVNN

Moreover, FCKM performs better than CVNN when the number of feature clusters is small. In some practical applications, e.g. control systems for production lines, the user wants to control as few key features as possible that have effects on the output variable. CVNN, an unsupervised clustering validation metric for the sample space, is not optimized for the classification task and thus requires features to be subdivided into more clusters in order to achieve comparable performance to FCKM. Experimental results on five real datasets demonstrate that FCKM is able to achieve higher classification accuracy while reducing feature complexity.

4 Conclusion

The FCKM algorithm proposed in this work effectively solves the difficult problem of feature clustering for mixed-type data and achieves better classification performance than the traditional CVNN method by adopting a statistical test method for different feature types and an improved k-medoids clustering algorithm.

However, the clustering evaluation indexes used in this study still have some limitations. The design of evaluation indexes can be further optimized in future work, such as intelligent optimization algorithms [11] or reinforcement learning.

Acknowledgments. This work is supported in part by the Science and Technology Project of Liaoning Province (Grant No. 2024-BSLH-205), and in part by the Youth Program of the Liaoning Provincial Department of Education (Grant No. JYTQN2023366).

References

1. Zhou, S., Xu, Z., Liu, F.: Method for determining the optimal number of clusters based on agglomerative hierarchical clustering. IEEE Trans. Neural Netw. Learn. Syst. **28**(12), 3007–3017 (2017)
2. Cheng, D., Zhu, Q., Huang, J., Wu, Q., Yang, L.: A novel cluster validity index based on local cores. IEEE Trans. Neural Netw. Learn. Syst. **30**(4), 985–999 (2019)
3. Xie, J., Xiong, Z.Y., Dai, Q.Z., Wang, X.X., Zhang, Y.F.: A new internal index based on density core for clustering validation. Inf. Sci. **506**, 346–365 (2020)
4. Şenol, A.: VIASCKDE index: a novel internal cluster validity index for arbitrary-shaped clusters based on the kernel density estimation. Comput. Intell. Neurosci. **2022** (2022)
5. Duan, X., Ma, Y., Zhou, Y., Huang, H., Wang, B.: A novel cluster validity index based on augmented non-shared nearest neighbors. Expert Syst. Appl. **223**, 119784 (2023)
6. Liu, Y., Li, Z., Xiong, H., Gao, Y., Wu, J.: Understanding and enhancement of internal clustering validation measures. IEEE Trans. Cybern. **43**(3), 982–994 (2013)
7. Massey, F.J., Jr.: The Kolmogorov-Smirnov test for goodness of fit. J. Amer. Statist. Assoc. **46**(253), 68–78 (1951)
8. Pandis, N.: The chi-square test. Am. J. Orthod. Dentofac. Orthop. **150**(5), 898–899 (2016)
9. Bickel, P., Diggle, P., Fienberg, S., et al.: Principles and Theory for Data Mining and Machine Learning. Springer Series in Statistics. Springer, Cham (2009)
10. Frank, A., Asuncion, A.: UCI machine learning repository. University of California, Irvine (2010). http://archive.ics.uci.edu/ml
11. Ji, Y., Liu, S., Zhou, M., et al.: A machine learning and genetic algorithm-based method for predicting width deviation of hot-rolled strip in steel production systems. Inf. Sci. **589**, 360–375 (2022)

Tilt Correction of Italic Character Images

Cuiping Liu, Cunrui Wang(✉), and Bo Lu

School of Computer Science and Engineering, Dalian Minzu University, Dalian, China
wcr@dlnu.edu.cn

Abstract. Skewing of Chinese character images interferes with structure perception and feature extraction, and is a key factor affecting recognition and generation accuracy. Existing methods focus on text line level correction, which is difficult to cope with single character level images, and especially challenging when dealing with characters with italicized styles. Italicized styles tend to lead to systematic tilting, exacerbating structural deformation and further interfering with recognition. To this end, this paper proposes a single-character tilt correction method based on image alignment algorithm, which introduces multi-angle rotation and scale normalization in the feature search stage, and minimum outer rectangle analysis in the geometric transformation stage, to achieve correction of italicized tilted character images.

Keywords: Tilt correction · Rotation and scaling matching · Geometric transformation · Minimum outer rectangle analysis

1 Introduction

In computer vision tasks, the presence of tilt in the target image can cause large errors in the detection results. Existing image correction methods for Chinese characters target text documents [1], detecting straight lines in the image by Hough transform [2] and estimating the rotation angle for correction, which relies on the straight line characteristics of text lines and is suitable for images with obvious and continuous text line structures. Individual Chinese characters usually do not have obvious horizontal line structures, and the strokes are complex and have various directions, so the image correction of individual Chinese characters is not universal. Traditional image registration methods [3] are widely used for image registration, but single-character images often have complex structures and lack repeatable and describable salient feature points, especially when font styles vary greatly, and it is difficult for region-based and feature-based registration [4] methods to establish effective correspondences. Meanwhile, deep learning has been more and more widely used in image processing, and people are gradually adopting learning-based features for image registration [5], and a feature alignment model based on the Transformer architecture [6], which uses a self-attention mechanism to model the structural mapping relationship between character images of different fonts, to achieve high-precision character region alignment and matching. However, Chinese character

J. Li et al. (Eds.): APWeb-WAIM 2025, CCIS 2660, pp. 199–210, 2026.
https://doi.org/10.1007/978-981-95-7996-9_18

design styles are continuously updated, which makes the learned features lose their real-time nature. Therefore, this strategy of using tilted Chinese character images with fixed font styles as the training dataset has some limitations in individual character tilt correction.

Different from the traditional image registration methods, the proposed method in this paper no longer relies on the local contour features [7] or the precise correlation matching between pixels, but constructs a reference template with the help of standard fonts, and estimates the optimal matching direction by multi-angle rotation and scale normalization. The method inherits the core idea of structural alignment in image registration, but focuses more on the similarity analysis of the overall shape and structure in the feature selection and matching strategy, and further combines the minimum outer rectangle [8] to extract the main direction of the character, which can be effectively applied to the task of italicized style single-character image correction.

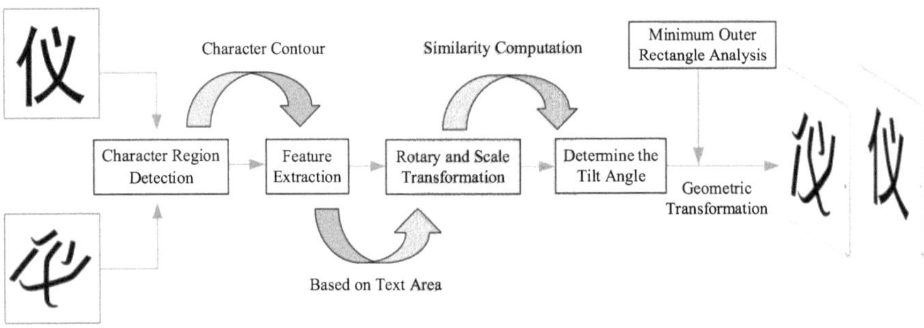

Fig. 1. Image registration process.

2 Research Methods

Among the many image registration methods, each is designed for a specific problem. For the task of aligning Chinese character images, it is necessary to consider factors such as font style and stroke morphology, and to choose a strategy that meets the actual needs in each step. In this paper, we calculate the similarity between the character image and the standard bold in the feature space and estimate the optimal tilt angle through multi-angle rotation and scale normalization; subsequently, we implement the geometric transformation in combination with the minimum outer rectangle to realize the correction of the italicized tilted character image. The correction process is shown in Fig. 1, which mainly goes through the following steps for image alignment: character region detection, feature extraction, similarity calculation, exact tilt angle, minimum outer rectangle analysis and geometric transformation [9].

2.1 Tilt Angle Detection Basedi on Rotation and Scaling

Character Area Detection. The existing extraction method is for all the contents of the image [10], in order to reduce the invalid pixel region involved in the calculation, extracting the outline of the Chinese character in the image fixes the alignment body as the Chinese character region and serves as the feature space. Firstly, the benchmark image and the image to be detected are binarized [11], and the outline extraction algorithm [12] is used to obtain the outlines of all the strokes of the Chinese characters with a set of pixel points $\boldsymbol{P}$. Then the coordinates of the extreme points of the outlines in the x and y coordinate system are calculated by taking the extreme values of $\boldsymbol{P}$. The coordinates of the extreme points are calculated by taking the extreme values of M in the coordinate system. From the extreme point coordinates, the coordinates of the four rectangular vertices of the outline are obtained, and the region enclosed is the character region of the Chinese character.

Extraction Characteristics. The image to be aligned is sequentially rotated and scaled to first align its text region with that of the reference image in terms of size, and then the scaling ratio is further adjusted to realize the alignment of the two in terms of position. For each specific rotation angle, a new image is generated and feature extraction is performed to obtain the set of pixel points of the text region under the current angle is $\boldsymbol{R}$. The set $\boldsymbol{R}$ reflects the character structure characteristics under the rotation angle, which provides the necessary data basis for the subsequent similarity analysis.

Rotary Transformation. The image coordinate system in computer vision takes the upper-left corner of the image as the origin, and a direct application of the rotation matrix, as in Eq. (1), will result in the rotation of the image around the upper-left corner, which will cause a shift in the center of the image, and may affect the complete presentation of the text area. For this reason, it is necessary to introduce a translation of the image in the x-axis and y-axis directions in order to achieve a reset of the center point after image rotation. The translations corresponding to image rotation are calculated as shown in Eq. (2) and Eq. (3).

$$\begin{bmatrix} \cos\theta - \sin\theta \\ \sin\theta \ \ \cos\theta \end{bmatrix} \tag{1}$$

$$off_x = (1 - \cos(\theta)) \cdot \frac{I_w}{2} + \sin(\theta) \cdot \frac{I_h}{2} \tag{2}$$

$$off_y = -\sin(\theta) \cdot \frac{I_w}{2} - (1 - \cos(\theta)) \cdot \frac{I_h}{2} \tag{3}$$

where I_w and I_h are the width and height of the image, θ is the rotation angle, off_x is the x-axis translation, and off_y is the y-axis translation.

The maximal outer rectangle [13] is the one that completely encloses the set of points and is parallel to the coordinate axes. Rotation changes the size of the outer rectangle of the image, and the values of its width and height may be larger. In order for the rotated image to be fully displayed on the canvas, the maximum width and height of the outer rectangle of the rotated image need to be recalculated. According to the basic principles of geometry and linear algebra, the width of the rotated image is the sum of the projection of the width of the original image in the x-axis direction and the projection of the height of the original image in the x-axis direction. The height of the rotated image is the sum of the projection of the height of the original image in the y-axis direction and the projection of the width of the original image in the y-axis direction. The principle is shown in Fig. 2, and the formula for the width and height of the new image after rotation is shown in Eq. (4) and Eq. (5).

$$I'_w = I_w \cdot |\cos(\theta)| + I_h \cdot |\sin(\theta)| \tag{4}$$

$$I'_h = I_w \cdot |\sin(\theta)| + I_h \cdot |\cos(\theta)| \tag{5}$$

where I_w and I_h are the width and height of the original image, respectively, $I_w \cdot |\cos(\theta)|$ and $I_h \cdot |\sin(\theta)|$ are the projections of the original image's width and height along the x-direction, $I_w \cdot |\sin(\theta)|$ and $I_h \cdot |\cos(\theta)|$ are the projections of the original image's width and height along the y-direction, I'_w and I'_h are the width and height of the new image.

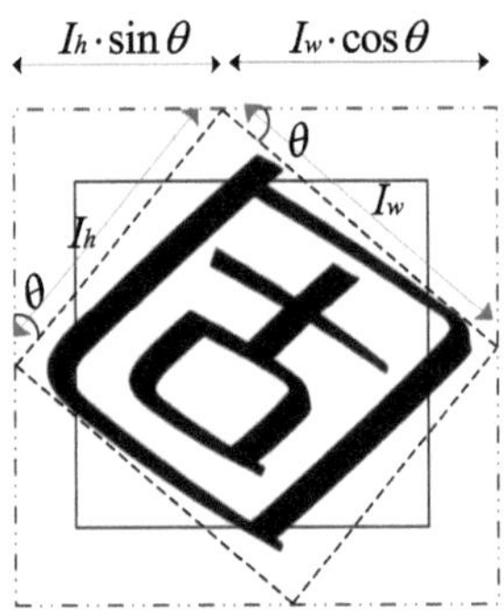

Fig. 2. Principle of image rotation.

After the image size is transformed, the center point may also be shifted, in order to ensure the accuracy of the image rotation, the parameters of the center point of the rotated image are updated by modifying the amount of translation, and the rotated translation of the new image is shown in Eq. (6) and Eq. (7), and the final image rotation matrix is obtained as Eq. (8).

$$off'_x = off_x + \frac{I'_w - I_w}{2} \tag{6}$$

$$off'_y = off_y + \frac{I'_h - I_h}{2} \tag{7}$$

$$\begin{bmatrix} \cos\theta & -\sin\theta & off'_x \\ \sin\theta & \cos\theta & off'_y \end{bmatrix} \tag{8}$$

where off_x is the x-axis translation of the original image, off_y is the y-axis translation, $(I'_w - I_w)/2$ and $(I'_h - I_h)/2$ are the offsets of the center point of the new image relative to the center point of the original image, off'_x is the x-axis translation of the new image, and off'_y is the y-axis translation of the new image.

Scaling Transformation. The scaling operation is only for the text region, and in the character region detection stage, the coordinates of the four rectangular vertices of the character region of the Chinese characters are derived using the coordinates of the extreme points, and then the width and height values of the character region are calculated. The scaling factors for width and height are shown in Eq. (9) and the scaling matrix is shown in Eq. (10).

$$\lambda_w = \frac{T'_m}{T_m}, \quad \lambda_h = \frac{T'n}{T_n} \tag{9}$$

$$\begin{bmatrix} \lambda_w & 0 & 0 \\ 0 & \lambda_h & 0 \end{bmatrix} \tag{10}$$

where T_m and T'_m are the width of the text area in the reference image and the image to be detected, T_n and T'_n are the height of the text area in the reference image and the image to be detected, λ_w and λ_h are the scaling factors for the width and height of the text area in the image to be detected, respectively.

Similarity Calculation. Based on the principle of Jaccard similarity [14], the set $\boldsymbol{R}$ of text region feature points of the image to be aligned under a specific rotation angle is extracted and compared with the text region of the reference image to calculate the Jaccard similarity between the two. Each similarity value corresponds to a specific rotation angle, reflecting the degree of overlap of the character regions under that angle. The Jaccard correlation coefficient as in Eq. (11).

$$J(\boldsymbol{Z}_1, \boldsymbol{Z}_2) = \frac{|\boldsymbol{Z}_1 \cap \boldsymbol{Z}_2|}{|\boldsymbol{Z}_1 \cup \boldsymbol{Z}_2|} \tag{11}$$

where, $|\boldsymbol{Z}_1 \cap \boldsymbol{Z}_2|$ denotes the number of common pixel points of the two images, $|\boldsymbol{Z}_1 \cup \boldsymbol{Z}_2|$ denotes the total number of pixel points of the two images, $J(\boldsymbol{Z}_1, \boldsymbol{Z}_2)$is the Jaccard similarity coefficient, which takes a value ranging from 0 to 1, with larger values indicating a higher degree of similarity between the images.

Determine the Tilt Angle. Starting from the original position of the image, clockwise and counterclockwise rotation operations are performed on the image, 1° at a time, until 90° is reached. Correspondingly, 180 Jaccard similarity values are generated and the maximum value is found. According to the definition of Jaccard similarity, when the similarity value reaches the maximum, the image to be aligned has the highest degree of match with the reference image under the corresponding rotation angle, and the character regions of the two images are most similar structurally, thus determining the tilt angle.

2.2 Minimum Outer Rectangle Analysis

The correction algorithm proposed in this paper achieves optimal alignment of images in terms of spatial location by maximizing the number of common pixel points. However, when the image to be aligned has an italicized feature, the distribution of pixels in space changes due to the tilt of the font, and the pixel correspondence between the base image and the image to be corrected is shifted, which affects the accurate calculation of the overlap degree. For this reason, this paper introduces an angle correction step in the alignment process, which effectively mitigates the error problem caused by the italicized font shape and improves the stability and accuracy of image registration.

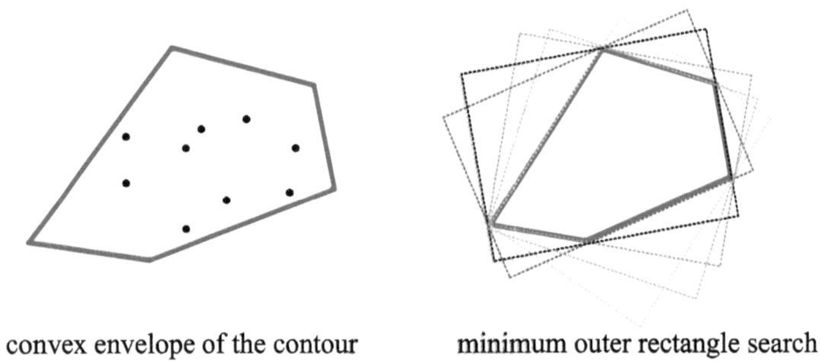

Fig. 3. Minimum outer rectangle searching process.

Minimum Outer Rectangle of the Merged Contour. Iterate over all the contours in the image and stack the point sets of each contour along the vertical direction, i.e., connect them sequentially in the row direction, to obtain a unified point set representing the pixel distribution of all the contours after merging, by which we integrate the contour information of each stroke of a Chinese character. Find the convex envelope [15] of the point set corresponding to the merged contours, which is the smallest convex polygon of this set of points that ensures that the rectangle can at least enclose these points. Find the smallest outer rectangle of the outline based on the method of rotating the rectangle around the box [16]. This is shown in Fig. 3.

Detecting the Angle of Correction. Minimum outer rectangle is a special kind of rectangle, which adapts to the shape of the contour of the enclosed object by rotating it to minimize the area while completely enclosing the object, so as to realize the precise framing of the target object, and then reflect the orientation information of the object through its rotation angle. The minimum outer rectangle of the merged contour is drawn, and the angle information of the minimum outer rectangle of the merged contour is obtained. In the field of computer vision, the x-axis of the image coordinate system is rotated clockwise, and the angle between it and the first side encountered is the angle of the smallest external rectangle, which is the angle of the smallest external rectangle. from the geometrical operation, it can be seen that, the font exhibits left tilt when $45^\circ < \text{angle} < 90^\circ$, and the tilted image needs to be corrected by rotating it 90-angle in the image coordinate system; $0^\circ < \text{angle} < 45^\circ$ when the font shows right tilt, the tilted image needs to be corrected by rotating the angle counterclockwise in the image coordinate system as shown in Fig. 4.

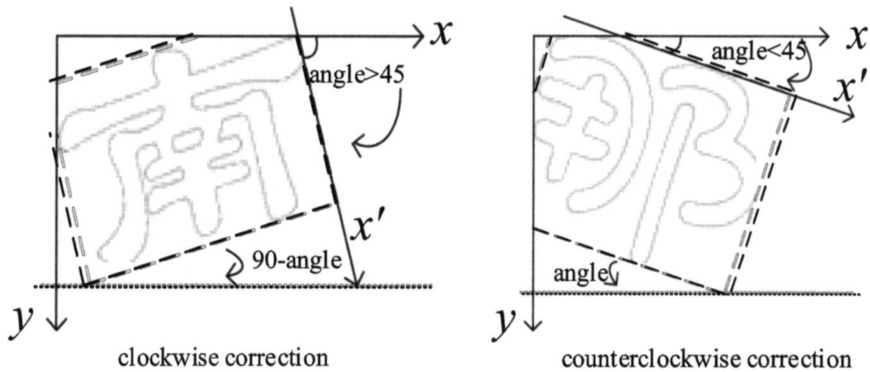

Fig. 4. Detection the Angle of Correction.

3 Experiments

The experiments in this paper are carried out on Windows 11 system, and the experimental techniques use OpenCV [17] computer vision learning library and Python 3.9. In order to test the effectiveness of the algorithm in image tilt correction, seven Chinese character images are selected as the dataset, and the experimental results are analyzed by using four kinds of evaluation indexes.

3.1 Dataset

In this paper, the fonts of the baseline image are SimHei, and the fonts to be corrected are fonts with serif [18] characteristics: FZDongRangChao, FZHuoLong, FZMaLi and fonts without serif characteristics: FZChaoZhi, FZDiaoBanSong, FZRuiSu. The baseline image and the target font image are shown in Fig. 5.

The 2312 commonly used Chinese characters in the List of Commonly Used Modern Chinese Characters are selected, and the font files are converted to font images using the ImageFont module [19], with an image size of $256 \times 256 \times 3$. There are a total of 16,184 font images of the 7 types of 2312 Chinese characters, and in addition, the 6 types of fonts to be corrected are tilted counterclockwise by 30°, 45°, 60° and clockwise by 30°, 45°, 60°, respectively, and each type of tilting is processed with a tilt of 30°, 45°, 60°. degree tilt processing, and 2312 images were generated at each tilt angle, totaling 83232 images. Finally, the whole dataset contains 99,416 images.

Fig. 5. Dataset glyphs.

3.2 Evaluation Indicators

In this paper, four evaluation metrics, OCR [20] character recognition accuracy, Chamfer Distance (CD) [21], Edge Similarity Dice Coefficient (ESDC) [22], and tilt angle (TA), are used to assess the tilt angle (TA) to evaluate the tilt correction effect of a single Chinese character image.

OCR character recognition technology that compares the recognition accuracy of pre-corrected and post-corrected images to effectively evaluate the effectiveness of character tilt correction. The accuracy formula is shown in Eq. (12).

$$\text{Accuracy} = \frac{C}{T} \tag{12}$$

where C denotes the number of correctly predicted characters and T denotes the total number of characters

Chamfer distance(CD) measures the similarity of two point sets by calculating, for each point, its minimum distance to the nearest point of the other point set, with smaller values indicating better corrections. The formula is shown in Eq. (13)

$$CD(\boldsymbol{A},\boldsymbol{B}) = \frac{1}{|\boldsymbol{A}|}\sum_{a\in \boldsymbol{A}} \min_{b\in \boldsymbol{B}} \|a-b\| + \frac{1}{|\boldsymbol{B}|}\sum_{b\in \boldsymbol{B}} \min_{a\in \boldsymbol{A}} \|b-a\| \tag{13}$$

where $\boldsymbol{A}$ and $\boldsymbol{B}$ denote two sets of points, $\|a-b\|$ is the Euclidean distance [23] between point a and point b, respectively. This formula calculates the shortest distance from each point to another set of points and then averages them.

Edge similarity dice coefficient(ESDC) is used to measure the degree of overlap between the edges of two images, the closer the value is to 1, the higher the overlap between the two edge sets and the greater the similarity, The formula is shown in Eq. (14)

$$\mathrm{ED}(\boldsymbol{M},\boldsymbol{N}) = \frac{2|\boldsymbol{M}\cap \boldsymbol{N}|}{|\boldsymbol{M}|+|\boldsymbol{N}|} \tag{14}$$

where $\boldsymbol{M}$ and $\boldsymbol{N}$ are two edge sets, $|\boldsymbol{M}|$ and $|\boldsymbol{N}|$ are the number of edge points in sets $\boldsymbol{M}$ and $\boldsymbol{N}$ respectively, and $|\boldsymbol{M}\cap \boldsymbol{N}|$ is the intersection of the two sets, which corresponds to the overlapping portion of the edge sets.

Tilt angle is the degree of inclination of some features or edges in the image relative to the horizontal line, which can directly indicate the correction effect. CCW30, CCW45, and CCW60 represent counterclockwise tilt states of 30°,45°, 60°, and CW30, CW45, and CW60 represent clockwise tilt states of 30°, 45°, 60°.

Table 1. Evaluation metrics for multiple glyph datasets.

Font categor	Image	*Accuracy* ↑	*CD* ↓	*ED* ↑
FZDongRangChao	Tilt image	0.2251	5.4615	0.3227
	Result image	**0.9450**	**1.6997**	**0.7410**
FZHuoLong	Tilt image	0.2996	7.7823	0.4565
	Result image	**0.9497**	**2.2769**	**0.8498**
FZMaLi	Tilt image	0.1982	6.6200	0.3252
	Result image	**0.9524**	**1.9635**	**0.7383**
FZChaoZhi	Tilt image	0.2628	8.5184	0.2077
	Result image	**0.9408**	**1.7769**	**0.7023**
FZDiaoBanSong	Tilt image	0.2046	5.8158	0.4384
	Result image	**0.9701**	**1.3188**	**0.8940**
FZRuiSu	Tilt image	0.2117	8.1529	0.4614
	Result image	**0.8881**	**2.5968**	**0.8470**

3.3 Experimental Results and Analysis

The algorithm in this paper makes correction on the images of Chinese characters with 6 different fonts, and the results of evaluation indexes are shown in Table 1, the recognition effect of the corrected image is obviously improved, and the similarity between corrected image and the target image is improved by 30%–40% compared to that between tilted image and the target image.

Table 2. Minimum outer rectangular analysis module ablation experiment.

Font categor	Image	$CCW30$	$CCW45$	$CCW60$	$CW30$	$CW45$	$CW60$
FZDongRangChao	Result1	6.52	6.66	6.32	6.87	7.10	6.79
	Result2	**5.79**	**5.69**	**5.61**	**6.03**	**6.05**	**6.13**
FZHuoLong	Result1	4.29	4.33	4.45	3.99	4.35	4.60
	Result2	**3.40**	**3.47**	**3.41**	**3.71**	**3.74**	**3.69**
FZMaLi	Result1	6.22	5.98	6.10	6.29	6.14	5.89
	Result2	**5.16**	**5.01**	**5.02**	**4.98**	**5.10**	**5.00**
FZChaoZhi	Result1	4.26	4.33	4.95	4.50	5.11	4.99
	Result2	**3.74**	**3.69**	**4.02**	**3.79**	**4.01**	**4.11**
FZDiaoBanSong	Result1	2.59	2.96	2.63	2.50	2.73	3.04
	Result2	**2.07**	**1.99**	**2.10**	**1.89**	**2.10**	**2.20**
FZRuiSu	Result1	5.88	5.78	5.92	5.85	5.94	5.79
	Result2	**4.86**	**4.89**	**4.88**	**5.16**	**5.19**	**5.15**

Fig. 6. Corrective effect of different glyphs.

By testing the counterclockwise and clockwise 30°, 45°, and 60° tilt images, the Table 2 shows that compared to the original image result1 after image registration, the tilt angle of the image result2 is reduced by an average of 1-degree after the addition of the Minimum Outer Rectangle Analysis Module. It proves that this module can effectively improve the tilt correction effect of italic style images. The actual correction effect of 6 fonts is shown in Fig. 6.

4 Conclusion

In this paper, a tilt correction algorithm for a single Chinese character image is proposed to solve the problem of tilted text region in Chinese character images. The text region of a Chinese character image is first extracted as the feature space. Then, the pixel point similarity and width-to-height ratio of the feature space under different rotation angles are calculated, and then the optimal feature space is confirmed by using the double detection of pixel point similarity and image rotation angle, and the parameters of the transformation model are estimated on the basis of which, and finally the character correction is realized by the model transformation. In order to reduce the correction error caused by italic style, the minimum outer rectangle analysis algorithm is introduced to calculate the tilt angle and correct it on the basis of image registration. The algorithm in this paper is experimented on six Chinese character image datasets with large differences in font types, and it can effectively improve the correction effect of tilted images in terms of four evaluation indexes.

Despite the positive progress made in this study in the task of multi-font tilt correction, there are still several issues that deserve to be explored in depth. Future research can focus on handwritten input in real-world scenarios, and due to the high uncertainty and deformation of character structures, more robust structural modeling techniques can be subsequently introduced to enhance the adaptability of the method under handwriting tilt and diverse deformations.

References

1. Biswas, B., Bhattacharya, U., Chaudhuri, B.B.: Document image skew detection and correction: a survey. Int. J. Innov. Res. Technol. (IJIRT) **9**(10) (2023)
2. Zhao, K., Han, Q., Zhang, C., Xu, J., Cheng, M.: Deep Hough transform for semantic line detection. IEEE Trans. Pattern Anal. Mach. Intell. **44**(9), 4793–4806 (2021)
3. Zitová, B., Flusser, J.: Image registration methods: a survey. Image Vis. Comput. **21**(11), 977–1000 (2003)
4. Li, Y., Li, Y., Sui, X.: Review of infrared and visible image registration. Infrared Technology **44**(7), 641–651 (2022)
5. Bharati, S., Mondal, M., Podder, P., et al.: Deep learning for medical image registration: a comprehensive review, arXiv preprint arXiv:2204.11341 (2022)
6. Zhu, K., et al.: AAformer: auto-aligned transformer for person re-identification. IEEE Trans. Neural Netw. Learn. Syst. **35**(12), 17307–17317 (2024)
7. Liu, C., Xu, J., Wang, F.: A review of keypoints' detection and feature description in image registration, Sci. Program. **2021**(1), 8509164 (2021)

8. Cai, H., Mao, Y., Zhu, X., Fu, Y., Zhou, R.: Comparison of the minimum bounding rectangle and minimum circumscribed ellipse of rain cells from TRMM. Adv. Atmos. Sci. **41**(3), 391–406 (2024)
9. Distante, A., Distante, C., Distante, A., Distante, C.: Geometric Transformations. In: Handbook of Image Processing and Computer Vision: Volume 2: From Image to Pattern, pp. 149–208. Springer, Cham (2020). https://doi.org/10.1007/978-3-030-42374-2_3
10. Guo, Z., Zhang, J., Guo, X., Mei, L.: Seam Carving image scaling method with visual significant graph. J. Yunnan Univ.: Natl. Sci. Edit. **40**(2), 222–227 (2018)
11. Yang, Z., Zuo, S., Zhou, Y., He, J., Shi, J.: A review of document binarization: main techniques, new challenges, and trends. Electronics **13**(7), 1394 (2024)
12. Li, Z., Guo, B., Meng, F.: Fast shape recognition method using feature richness based on the walking minimum bounding rectangle over an occluded remote sensing target. Remote Sensing **14**(22), 5845 (2022)
13. Sun, F., Li, H., Liu, Z., Li, X., Wu, Z.: Arbitrary-angle bounding box based location for object detection in remote sensing image. Eur. J. Remote Sens. **54**(1), 102–116 (2021)
14. Lomov, L., Seredin, O., Liakhov, D., Kushnir, O.: Constraints for Jaccard index-based rotational symmetry focus position in binary images. Kompyuternaya optika **47**(6), 948–957 (2023)
15. Liang, H., et al.: Deephullnet: a deep learning approach for solving the convex hull and concave hull problems with transformer. Int. J. Digit. Earth **17**(1), 2358843 (2024)
16. Yang, D., Peng, B., Al-Huda, Z., Malik, A., Zhai, D.: An overview of edge and object contour detection. Neurocomputing **488**, 470–493 (2022)
17. Rakesh, V., Chilukuri, P., Vaishnavi, P., Sreekaran, P., Sujala, P., Yadav, D.R.: Real time object recognition using opencv and numpy in python. In: Proceedings of the 2023 International Conference on Innovative Data Communication Technologies and Application (ICIDCA), IEEE, pp. 421–426 (2023)
18. Jin, X., Jin, X.: Analysis of the development status of Chinese font design for mobile phones. Design **8**(2), 679–684 (2023)
19. ImageFont Module. https://pillow.readthedocs.io/en/latest/reference/ImageFont.html. Accessed 8 July 2025
20. Sarkar, O., Sinha, S., Jena, A.K., Parida, A., Parida, N., Parida, R.K.: Automatic number plate character recognition using paddle-OCR. In: 2024 International Conference on Innovations and Challenges in Emerging Technologies (ICICET), IEEE, pp. 1–7 (2024)
21. Al-Shurbaji, T.A., AlKaabneh, K.A., Alhadid, I., Masa'deh, R.: An optimized scale-invariant feature transform using chamfer distance in image matching. Intell. Autom. Soft Comput. **71**, 971–985 (2022)
22. Zhao, S.: Research on brain tumor image segmentation methods based on deep learning. Model. Simul. **14**(4), 555–565 (2025)
23. Wang, L., Zhang, Y., Feng, J.: On the Euclidean distance of images. IEEE Trans. Pattern Anal. Mach. Intell. **27**(8), 1334–1339 (2005)

SRBDMA Workshop

DITS: Double Imputation for Enhanced Irregular Multivariate Time Series Classification

Shilong Deng[1], Hongting Qu[1], Zhenkun Wen[1(✉)], Shuyan Jin[2], and Ping Dai[3]

[1] Shenzhen University, Shenzhen, China
wenzk@szu.edu.cn
[2] Shenzhen Maternity and Child Healthcare Hospital, Shenzhen, China
[3] Shenzhen University General Hospital, Shenzhen, China

Abstract. Irregularly sampled multivariate time series classification is challenging and widely used in various domains, including health care, biology, and space science. Imputation methods treat irregularly sampled multivariate time series as having missing values and achieve superior performance. However, most of the existing methods impute time series by the correlation within data streams ignoring the correlation across data streams (both correlations are important information of multivariate time series). Few methods exploit both two correlations in a sequential way which will cause sub-optimal results. To address this problem, we propose a novel imputation-reconstruction-prediction model with double imputation. The double imputation network exploits two kinds of correlations in a currency way and achieves excellent performance without architecture changes. We apply a joint-optimization training approach with three learning tasks to fit our imputation-reconstruction-prediction architecture and the model is end-to-end trainable. Experiments on public real-world datasets demonstrate the effectiveness of our method.

Keywords: Time series classification · Multivariate time series · Irregular time series

1 Introduction

Multivariate time series data is abundant in many application domains including health care, financial forecasting and cyber security. The majority of the models assume that the time series is fully observed and fixed-size [15]. However, in the field of practical applications, multivariate time series is always irregularly sampled. Irregularly sampled time series poses considerable challenges.

Existing deep learning methods have different understandings of the characteristics of irregularly sampled time series [15]. One perspective focuses on the structure of the raw data itself. The other perspective treats irregular time series as having missing values. Neither perspective can defeat the other. Our work follows the second perspective. As shown in Fig. 1, the time series treatment of the

J. Li et al. (Eds.): APWeb-WAIM 2025, CCIS 2660, pp. 213–224, 2026.
https://doi.org/10.1007/978-981-95-7996-9_19

second perspective causes a large number of missing values. The advantage of the second perspective is once we obtain the complete and fixed-size time series by deletion or imputation, many excellent methods can be applied, such as TARnet [4]. How to obtain the complete time series to assist downstream predictive tasks is crucial.

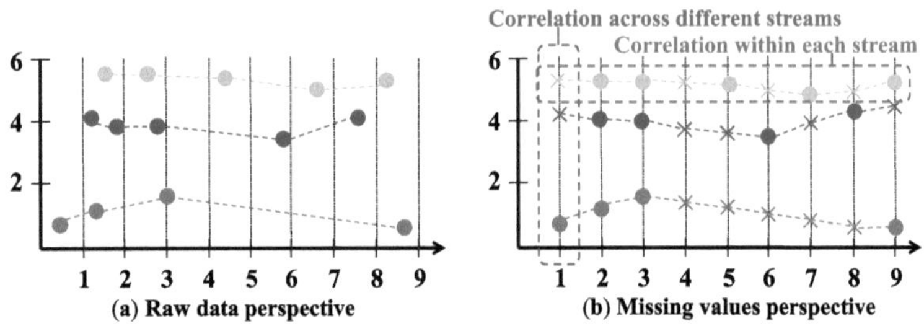

Fig. 1. Example of two perspectives on irregularly sampled multivariate time series and two kinds of multivariate time series correlations.

Much prior work proposed to impute the missing values of multivariate time series. However, most of them do not explore both the correlation at different time steps within each stream (step-wise correlation) and the correlation at the same time step across different streams (channel-wise correlation). Figure 1 (b) shows two kinds of correlations. M-RNN [22] and BRITS [1] exploit both correlations within streams and across streams to estimate missing values. However, both M-RNN and BRITS utilize the two correlations in a sequential way. They first estimate missing values through the step-wise correlation. Then, they impute missing values through the channel-wise correlation. The order of two kinds of imputations affects the performance of the model. The best order may change when taking different datasets. M-RNN and BRITS do not change the order and obtain sub-optimal results on some datasets. However, such a change of order lead to the alteration of network architecture.

To address this problem, we propose a novel model with double imputation that exploits the two correlations in a concurrency way and executes irregular time series classification, termed DITS. DITS has two imputation modules, a reconstruction module, and a prediction module. We apply the self-attention mechanism to construct the channel-wise imputation module that exploits the correlation across streams. The step-wise imputation module that exploits the correlation within each stream is based on the bidirectional recurrent network. DITS imputes the time series simultaneously by two kinds of correlations. Two imputed data from different imputation modules are concatenated in the channel direction and reconstructed by the reconstruction module. The prediction module takes the estimated data from the reconstruction module and outputs the prediction. DITS is constructed to be end-to-end trainable. We apply a joint-optimization training approach to train our model. The combination of imputation, reconstruction, and classification tasks allows the model to understand the time series better. Our contributions are summarized as follows:

- We propose DITS, a novel end-to-end Imputation-Reconstruction-Prediction model trained by the joint-optimization training approach for irregularly sampled multivariate time series classification.
- We treat irregular series as having missing data, and design a new double imputation structure to impute time series by exploiting two kinds of multivariate time series correlations simultaneously.
- We conduct extensive experiments on numerous real-world datasets to validate the effectiveness of our proposed method.

2 Related Work

The raw data-based methods focus on the structure of the irregular time series itself. Such methods directly model irregularly sampled multivariate time series using irregular time information. A novel deep learning architecture based on semi-parametric interpolation network and prediction network is proposed to address the supervised learning with irregular multivariate time series [13]. SeFT transforms irregularly sampled time series into a set of observations using differentiable set functions [9]. Tan et al. propose a new end-to-end Dual-Attention Time-Aware Gated Recurrent Unit, named DATA-GRU [16]. The novel time-aware mechanism in DATA-GRU can handle irregular time intervals and preserve useful information. mTAND introduces a multi-time attention network, which learns temporal similarity from irregularly collected observations and generates continuous time embeddings [14]. Raindrop, based on the graph neural network, models irregularly sampled multivariate time series as graphs [24]. PrimeNet introduces a self-supervised representation learning approach to model irregular multivariate time-series [3].

The missing data-based methods treat irregular time series as having missing data. Most of them focus on deep learning time series imputation. M-RNN exploits correlation within data streams and correlation across data streams to impute multivariate time series with missing values [22]. Similar to M-RNN, BRITS imputes multivariate time series through two kinds of correlations, but it treats missing values as variables of the RNN graph [1]. GRU-D introduces a temporal decay mechanism and utilized gated recurrent unit (GRU) to impute multivariate time series [2]. LGnet jointly models local and global temporal dynamics of multivariate time series with missing values by using a memory module and adversarial training [17]. To impute missing values in multivariate time series, a new semi-supervised generative adversarial network model (SSGAN) is proposed [11]. SSGAN consists of a generator, a discriminator, and a classifier. For multivariate time series imputation, CSDI introduces the Conditional Score-based Diffusion model that can exploit correlations between observed values [18]. Du et al. propose a self-attention-based method for multivariate time series imputation, named SAITS [5]. SAITS trained by the joint-optimization training approach, learns missing values through a weighted combination of two diagonally-masked self-attention blocks.

3 DITS

Let an irregular time series dataset $\mathbf{A} = \{(S_z, y_z)|z = 1, ..., Z\}$ with Z labeled samples. Each individual data case S_z is an irregular multivariate time series with a label y_z. We discretize the irregular multivariate time series in $\mathbf{A}$ into multivariate time series with missing values to obtain $A = \{(X_z, y_z)|z = 1, ..., Z\}$. Each sample $X_z = \{\mathbf{x}_{z,1}, \mathbf{x}_{z,2}, ..., \mathbf{x}_{z,T}\} \in \mathbb{R}^{T \times D}$ is a multivariate time series with T time steps and D dimensions, where the t-th step $\mathbf{x}_{z,t} = \left\{x^1_{z,t}, x^2_{z,t}, ..., x^D_{z,t}\right\} \in \mathbb{R}^{1 \times D}$ and every value in it could be missing. The goal is to learn a function f that maps each X to the final classification score $\widehat{y}$ that close to label y.

Figure 2 illustrates the general architecture of DITS. The step-wise imputation module imputes time series by exploiting the correlation among measurements at different times within each stream. Our channel-wise imputation module can exploit the correlation among measurements at the same time across different streams (univariate time series) to impute time series. With such two imputation modules, DITS imputes the time series simultaneously by two kinds of correlations. The reconstruction module reconstructs the concatenated time series from two imputation networks. The prediction module takes the reconstructed time series and outputs the prediction.

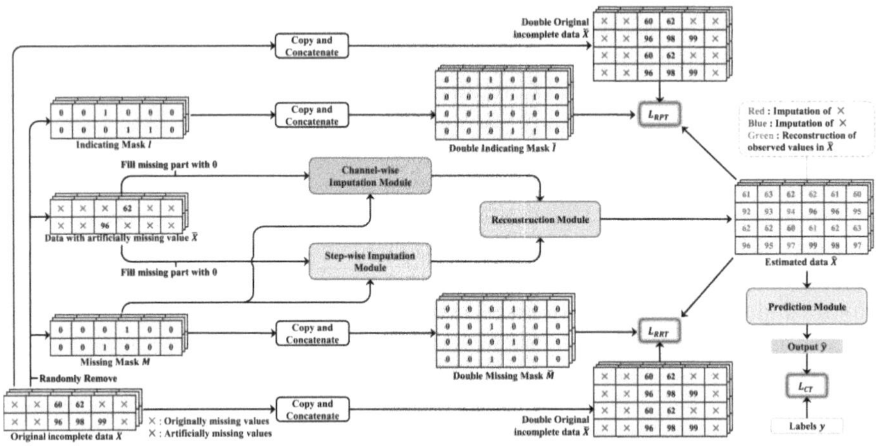

Fig. 2. Overview of DITS.

3.1 Training Approach

For the end-to-end Imputation-Reconstruction-Prediction architecture, we design the joint-optimization training approach consisting of an imputation task, a reconstruction task, and a prediction task. We assume that the outermost data in Fig. 2 is the sample z.

The Removed Imputation Task (RIT) is an imputation task that makes the model impute the artificially-removed values. As shown in Fig. 2, we randomly remove some observed values of the multivariate time series X to obtain the time series $\bar{X}$, the blue crosses indicate the artificially-removed values. The indicating mask I indicates which values are removed artificially and the removed values are the ground truth of the Removed Imputation Task. If the value of $x_{z,t}^d$ is artificially removed, $i_{z,t}^d = 1$, otherwise $i_{z,t}^d = 0$. The indicating mask is only used to calculate the loss of the Removed Imputation Task. The imputation model does not take the indicating mask and the removed values are not visible to the model. For our double imputation, we copy the indicating mask I and concatenate them in the channel direction. We perform the same operation for time series X. The double indicating mask $\check{I}$, the double incomplete time series $\check{X}$, and the estimated time series $\widehat{X}$ are used to obtain the loss. With The Removed Imputation Task, the model is able to impute missing values accurately.

The Remained Reconstruction Task (RRT) is a task that forces the model to converge to the distribution of observed data. The missing mask vector $\bar{M}$ indicated which values are observed in $\bar{X}$ (with artificially-missing values), if $\bar{x}_{z,t}^d$ is observed, $m_{z,t}^d = 1$, otherwise, $m_{z,t}^d = 0$. The missing mask is input to the imputation model. We copy the missing mask M and concatenate them in the channel direction, as shown in Fig. 2. The same operation is performed for time series X. The double missing mask $\check{M}$, the double incomplete time series $\check{X}$, and the estimated time series $\widehat{X}$ are used to obtain the loss. The observed values in $\bar{X}$ are the target values of the Remained Reconstruction Task.

The Classification Task (CT) ensures that the model predicts the classification prediction $\widehat{y}$ as close as possible to labels y. The combination of the three tasks can help the model better understand the time series, and achieve excellent imputation and classification performance.

3.2 Channel-Wise Imputation Module

To exploit the correlation among measurements at the same time across different streams, the self-attention mechanism is utilized to impute missing values within data streams (channel-wise imputation):

$$SelfAttention(Q, K, V) = Softmax\left(\frac{QK^{\mathsf{T}}}{\sqrt{d_k}}\right)V \tag{1}$$

Multi-Head Attention (MHA) is constructed as follows:

$$MHA(x) = Concat(head_1, ..., head_i, ..., head_h)W^O \tag{2}$$

$$head_i = SelfAttention(xW_i^Q, xW_i^K, xW_i^V) \tag{3}$$

where $W_i^Q \in \mathbb{R}^{d_{model} \times d_k}$, $W_i^K \in \mathbb{R}^{d_{model} \times d_k}$, $W_i^V \in \mathbb{R}^{d_{model} \times d_v}$ are parameters of the linear layers projecting input x to Q, K, and V separately. $W^O \in \mathbb{R}^{hd_v \times d_{model}}$ is the parameter of the output layer in MHA.

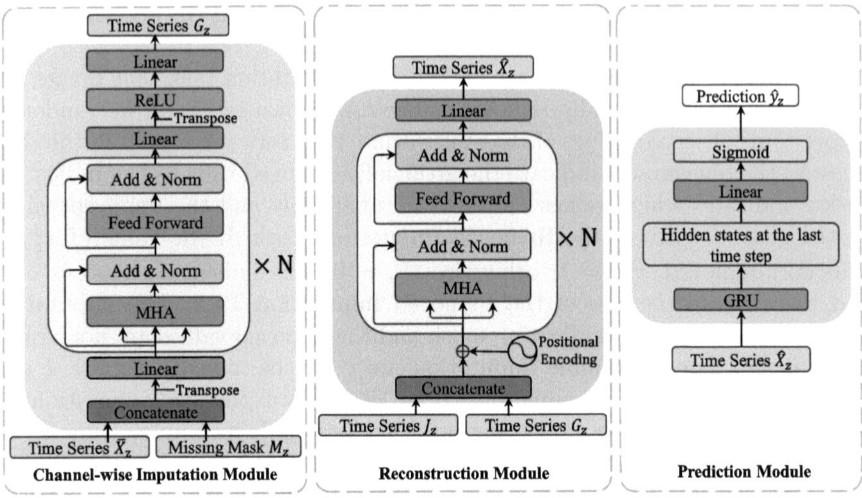

Fig. 3. Details of three modules.

As Fig. 3 shows, the Module concatenates a given time series sample $\bar{X}_z$ and its missing mask $\bar{M}_z$. $\bar{X}_z$ and $\bar{M}_z$ are concatenated in channel direction. The concatenated data is transposed and projected to obtain C_z:

$$C_z = [Concat(\bar{X}_z, M_z)]^{\top} W_\alpha + b_\alpha \tag{4}$$

The Multi-Head Attention and the feed-forward network with N stacked layer are utilized to transfer C_z to E_z:

$$E_z = \{FFN[MHA(C_z)]\}^N = \{ReLU[MHA(C_z)W_\beta + b_\beta]W_\gamma + b_\gamma\}^N \tag{5}$$

where FFN represents the feed-forward network, $W_\beta \in \mathbb{R}^{d_{model} \times d_{ffn}}$, $W_\gamma \in \mathbb{R}^{d_{ffn} \times d_{model}}$, $b_\beta \in \mathbb{R}^{d_{ffn}}$, $b_\gamma \in \mathbb{R}^{d_{model}}$. Since the position of the channel in the multivariate time series has no relative or absolute correlation, positional encoding is not applied. Input E_z into the linear layer and transpose it to be activated by the ReLU activation function, then pass it through another linear layer to get the final result:

$$G_z = ReLU[(E_z W_\delta + b_\delta)^{\top}]W_\omega + b_\omega \tag{6}$$

where $W_\delta \in \mathbb{R}^{d_{model} \times T}$, $W_\omega \in \mathbb{R}^{2D \times D}$, $b_\delta \in \mathbb{R}^T$, $b_\omega \in \mathbb{R}^D$.

3.3 Step-Wise Imputation Module

As shown in Fig. 4, the revised Bidirectional Recurrent Imputation Network is used to exploit the correlation among measurements at different times within each stream (univariate time series). The imputation function ψ_k imputes missing values of time series $\bar{x}^k$ (time step-wise imputation). Concatenating the time series X_z and its missing mask M in the channel direction to obtain $XX_z \in \mathbb{R}^{T\times 2D}$. In the forward GRU direction, the timing of the time series input to the hidden layer is lagged. The hidden layer at time step t takes the feature vector at time step $t-1$. In the backward GRU direction, the input to the hidden layer is advanced and the hidden layer at time step t receives features at time step $t+1$. This approach enables the imputation results of the model at t time step to be independent of the observed values at t time step.

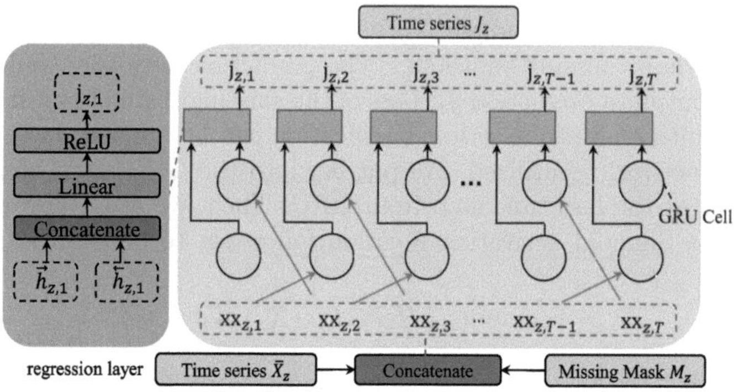

Fig. 4. Step-wise imputation module.

3.4 Reconstruction Module

The self-attention imputation network operates channel-wise imputation and obtains time series G_z. The revised bidirectional recurrent network performs step-wise imputation and obtains time series J_z. To utilize both G_z and J_z, a simple way is to concatenate them. We can concatenate two multivariate time series in time step and channel direction. However, concatenation in the time step direction may lead to mutation. Such mutation is detrimental to the prediction of the classification network. Thus, we concatenate G_z and J_z in the channel direction and add up with the positional encoding p:

$$O_z = Concat(G_z, J_z) + p \tag{7}$$

where p is the positional encoding in Transformer [19]. As shown in Fig. 3, the self-attention-based network reconstruct the concatenated time series:

$$Q_z = \{FFN[MHA(O_z)]\}^N = \{ReLU[MHA(O_z)W_\varsigma + b_\varsigma]W_\eta + b_\eta\}^N \tag{8}$$

where *FFN* represents the feed-forward network, *MHA* indicates the multi-head attention. $W_\varsigma \in \mathbb{R}^{d_{model} \times d_{ffn}}$, $W_\eta \in \mathbb{R}^{d_{ffn} \times d_{model}}$, $b_\varsigma \in \mathbb{R}^{d_{ffn}}$, $b_\eta \in \mathbb{R}^{d_{model}}$. The linear layer takes Q_z to obtain the reconstruction result $\widehat{X}_z$:

$$\widehat{X}_z = (Q_z W_\epsilon + b_\epsilon) \tag{9}$$

where $W_\epsilon \in \mathbb{R}^{d_{model} \times 2D}$, $b_\epsilon \in \mathbb{R}^{2D}$.

3.5 Prediction Module

Following the application of the imputation modules and reconstruction module, the given time series sample $\bar{X}_z$ has been imputed and reconstructed as time series without missing values $\widehat{X}_z$. The prediction network should take $\widehat{X}_z$ as input and output a prediction $\widehat{y}_z$. Lots of the existing state-of-the-art methods for regular multivariate time series classification can be applied. In this work, a simple GRU network is utilized. We put $\widehat{X}_z$ into the GRU layer and take the hidden states of the last time step from GRU. Then, we use a linear transformation with a Sigmoid activation function following, to obtain the prediction $\widehat{y}_z$:

$$\widehat{y}_z = Sigmoid\left\{LHS[GRU(\widehat{X}_z)]W_\theta + b_\theta\right\} \tag{10}$$

where $W_\theta \in \mathbb{R}^{d_{hs} \times 1}$, $b_\theta \in \mathbb{R}^1$. $\widehat{y}_z$ is the classification result of sample z.

3.6 Loss Functions

Our total loss function consists of three components: the imputation loss of Removed Imputation Task (L_{RIT}), the reconstruction loss of Remained Reconstruction Task (L_{RRT}), and the classification loss of Classification Task (L_{CT}). Both L_{RIT} and L_{RRT} are calculated by Mean Absolute Error (MAE):

$$L_{MAE}(v, a, m) = \frac{\sum_d^D \sum_t^T |(v-a) \odot m|_t^d}{\sum_d^D \sum_t^T m_t^d} \tag{11}$$

where v, a, m denotes valuation, actual, and mask.

$$L_{RIT} = \sum\nolimits_{z=1}^{Z} L_{MAE}(\widehat{X}_z, \check{X}_z, \check{I}_z), L_{RRT} = \sum\nolimits_{z=1}^{Z} L_{MAE}(\widehat{X}_z, \check{X}_z, \check{M}_z) \tag{12}$$

where $\check{M}$ indicate which values are observed in $\bar{X}$. $\check{I}$ indicates which values are artificially removed, and the removed values can be found in the original time series X. $\widehat{X}$ is the estimated data from the reconstruction network.

We apply the binary cross-entropy to produce the loss of classification task:

$$L_{CT} = -\frac{1}{Z}\sum_{z=1}^{Z} y_z \cdot log(\widehat{y}_z) + (1-y_z) \cdot log(1-\widehat{y}_z) \tag{13}$$

where Z denotes the number of samples in the datasets, y_z refers to the label of the z-th sample. The total loss L is defined as:

$$L = L_{RRT} + L_{RIT} + \psi L_{CT} \tag{14}$$

where ψ indicates the hyperparameter which is set to 2 in our method. Our model is updated by minimizing the total loss L.

4 Experiments

4.1 Experimental Setup

To benchmark our method, we choose three real-world irregularly-sampled multivariate time series datasets. We predict patient mortality on two datasets and the onset of sepsis on another dataset. Including the Physionet 2012 [8], Physionet 2019 [12] and MIMIC-III [10]. For irregular time series classification, we compare our model with some baselines: Transformer [19], GRU-D [2], IP-Net [13], SeFT [9], MTGNN [21], mTAND [14], DGM^2-O [20], Raindrop [24] and PrimeNet [3]. We use three metrics to evaluate the classification performance of methods: AUROC (Area Under the Receiver Operating Characteristic), AUPRC (Area Under the PrecisionRecall Curve), and F1-score to measure the classification performance of models.

Table 1. Overall classification performance of methods on P12, P19 and MIMIC-III. The reported values are means ± standard deviations.

Methods	P12			P19			MIMIC-III		
	AUROC	AUPRC	F1-score	AUROC	AUPRC	F1-score	AUROC	AUPRC	F1-score
GRU-D	84.0 ± 1.2	47.5 ± 1.0	39.8 ± 1.4	83.9 ± 1.7	46.9 ± 2.1	39.5 ± 1.1	**85.6 ± 0.2**	52.0 ± 0.8	40.4 ± 1.5
IP-Net	83.8 ± 2.1	46.5 ± 4.3	37.8 ± 1.9	84.6 ± 1.3	38.1 ± 3.7	38.2 ± 1.0	83.1 ± 0.7	48.3 ± 0.4	38.2 ± 1.0
Transformer	80.7 ± 4.9	46.5 ± 1.3	36.4 ± 2.3	83.2 ± 1.3	47.6 ± 2.3	37.0 ± 1.6	82.3 ± 0.9	42.6 ± 1.3	36.1 ± 1.6
SeFT	74.6 ± 1.4	30.6 ± 1.4	31.3 ± 2.1	78.8 ± 2.5	31.1 ± 2.9	32.0 ± 1.8	83.9 ± 0.6	46.5 ± 0.5	38.4 ± 1.0
mTAND	82.4 ± 1.3	33.6 ± 1.8	37.1 ± 2.1	80.4 ± 1.3	32.4 ± 1.9	36.9 ± 2.0	82.9 ± 1.6	45.7 ± 2.3	38.3 ± 1.4
DGM^2-O	84.1 ± 1.5	47.8 ± 1.2	39.4 ± 2.0	86.7 ± 1.3	44.7 ± 1.2	41.5 ± 0.8	84.2 ± 2.0	49.7 ± 1.2	40.5 ± 1.7
MTGNN	75.1 ± 6.4	35.9 ± 5.5	31.6 ± 2.3	81.9 ± 6.2	39.9 ± 8.9	37.3 ± 2.1	80.1 ± 5.9	44.8 ± 7.3	35.6 ± 2.7
RAINDROP	82.1 ± 1.3	45.2 ± 1.4	36.8 ± 1.9	87.0 ± 1.9	51.8 ± 2.0	41.7 ± 1.3	83.7 ± 1.6	50.0 ± 2.3	38.7 ± 1.1
PrimeNet	84.3 ± 0.6	48.1 ± 1.0	39.9 ± 1.3	88.6 ± 1.7	53.1 ± 1.4	43.3 ± 1.2	84.6 ± 1.2	49.8 ± 1.6	40.4 ± 1.0
DITS	**85.6 ± 1.7**	**51.8 ± 1.3**	**41.0 ± 1.8**	**89.5 ± 1.9**	**56.2 ± 1.7**	**45.1 ± 1.1**	84.5 ± 1.4	**52.3 ± 2.0**	**42.2 ± 1.6**

4.2 Comparison Study

As shown in Table 1, the first set of experiments explores the performance of different methods on the classification task. On all the benchmark datasets, DITS obtains competitive performance. In the binary classification task of P12, DITS outperforms all baselines. Our method performs 1.3, 3.7, and 1.1 higher than the second-best PrimeNet in terms of AUROC, AUPRC, and F1-score points (%). On P19, DITS exceeds the best-performing state-of-the-art baseline (PrimeNet)

Table 2. Performance comparison between methods on four datasets. Metrics are reported in the order of MAE/RMSE/MRE.

Model	PhysioNet-2012	Air-Quality	Electricity	ETT
M-RNN	0.533/0.776/76.0%	0.294/0.643/41.4%	1.244/1.867/66.6%	0.376/0.428/31.6%
GP-VAE	0.398/0.630/56.7%	0.268/0.614/37.7%	1.094/1.565/58.6%	0.274/0.307/15.5%
BRITS	0.256/0.767/36.5%	0.153/0.525/21.6%	0.847/1.322/45.3%	0.130/0.259/12.5%
SAITS	0.186/0.431/26.6%	0.137/0.518/19.3%	0.735/1.162/39.4%	**0.092/0.139/8.8%**
DITS-I	**0.126/0.417/17.8%**	**0.126/0.417/17.8%**	**0.553/0.866/29.6%**	**0.102/0.150/11.8%**

by 1.1, 3.1, and 2.2 in AUROC, AUPRC, and F1-score points. As for the MIMIC-III dataset, DITS outperforms the GRU-D in terms of AUPRC and F1-score, but the AUROC of our method is lower than GRU-D. DGM^2-O, which follows the end-to-end imputation-prediction architecture and is designed for time series forecasting, exhibits competitive performance. The performance of DGM^2-O also shows the effectiveness of the end-to-end imputation and prediction architecture.

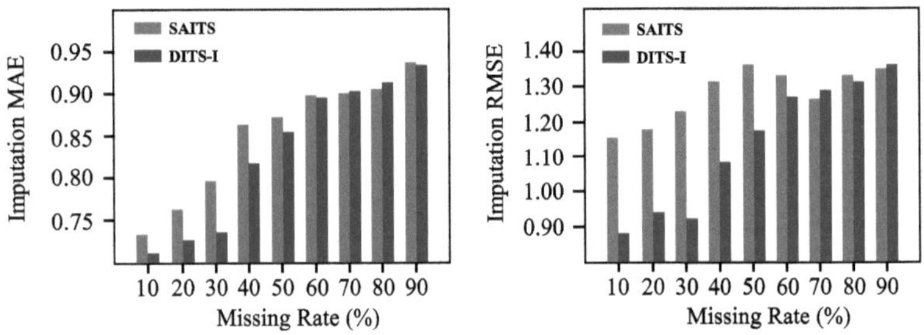

Fig. 5. Comparison at different missing rates.

4.3 Time Series Imputation

To study the pure time series imputation performance of the double imputation network, we compare our method with some state-of-the-art baselines on four public real-world datasets: PhysioNet 2012 Mortality Prediction Challenge, Beijing Multi-Site Air-Quality [23], Electricity Load Diagrams [6], and Electricity Transformer Temperature [25]. The baselines are: M-RNN [22], BRITS [1], GP-VAE [7], SAITS [5]. We randomly removed 10% values of the datasets as the target values. Three metrics are utilized to evaluate the imputation performance of methods: MAE (Mean Absolute Error), RMSE (Root Mean Square Error), and MRE (Mean Relative Error). While M-RNN and BRITS utilize the two kinds of time series correlations in a sequential way, DITS exploits the two correlations in a concurrency way. We remove the prediction module of DITS and set ψ as 0 to obtain DITS-I. As shown in Table 2, DITS-I outperforms M-RNN and BRITS. Experimental results demonstrate the effectiveness of double imputation. To further study the performance of the methods on different missing rates, the artificial missing rate of the Electricity dataset is set from 10% to 90%. As shown in Fig. 5, DITS is superior to SAITS in most cases.

5 Conclusion

This paper proposes a novel DITS for irregularly sampled multivariate time series classification. The joint-optimization training approach is introduced to fit our end-to-end Imputation-Reconstruction-Prediction architecture. We design the double imputation network to exploit two kinds of multivariate time series correlations in a concurrency way. The extensive experimental results on a variety of real-world datasets demonstrated the effectiveness of our proposed method.

Acknowledgments. This work was supported by the Marshall Laboratory of Biomedical Engineering, Shenzhen University and Shenzhen Institute of Artificial Intelligence and Robotics for Society.

Disclosure of Interests. The authors have no competing interests to declare that are relevant to the content of this article.

References

1. Cao, W., Wang, D., Li, J., Zhou, H., Li, L., Li, Y.: BRITS: bidirectional recurrent imputation for time series. Adv. Neural Inf. Process. Syst. **31** (2018)
2. Che, Z., Purushotham, S., Cho, K., Sontag, D., Liu, Y.: Recurrent neural networks for multivariate time series with missing values. Sci. Rep. **8**(1), 6085 (2018)
3. Chowdhury, R.R., Li, J., Zhang, X., Hong, D., Gupta, R.K., Shang, J.: PrimeNet: pre-training for irregular multivariate time series. In: Proceedings of the AAAI Conference on Artificial Intelligence (2023)
4. Chowdhury, R.R., Zhang, X., Shang, J., Gupta, R.K., Hong, D.: TARNet: task-aware reconstruction for time-series transformer. In: Proceedings of the 28th ACM SIGKDD Conference on Knowledge Discovery and Data Mining, Washington, DC, USA, pp. 14–18 (2022)
5. Du, W., Côté, D., Liu, Y.: SAITS: self-attention-based imputation for time series. Expert Syst. Appl. **219**, 119619 (2023)
6. Dua, D., Graff, C., et al.: UCI machine learning repository (2017)
7. Fortuin, V., Baranchuk, D., Rätsch, G., Mandt, S.: GP-VAE: deep probabilistic time series imputation. In: International Conference on Artificial Intelligence and Statistics, pp. 1651–1661. PMLR (2020)
8. Goldberger, A.L., et al.: PhysioBank, PhysioToolkit, and PhysioNet: components of a new research resource for complex physiologic signals. circulation **101**(23), e215–e220 (2000)
9. Horn, M., Moor, M., Bock, C., Rieck, B., Borgwardt, K.: Set functions for time series. In: International Conference on Machine Learning, pp. 4353–4363. PMLR (2020)
10. Johnson, A.E., et al.: MIMIC-III, a freely accessible critical care database. Sci. Data **3**(1), 1–9 (2016)

11. Miao, X., Wu, Y., Wang, J., Gao, Y., Mao, X., Yin, J.: Generative semi-supervised learning for multivariate time series imputation. In: Proceedings of the AAAI Conference on Artificial Intelligence, vol. 35, pp. 8983–8991 (2021)
12. Reyna, M.A., et al.: Early prediction of sepsis from clinical data: the PhysioNet/computing in cardiology challenge 2019. In: 2019 Computing in Cardiology (CinC), pp. Page–1. IEEE (2019)
13. Shukla, S.N., Marlin, B.: Interpolation-prediction networks for irregularly sampled time series. In: International Conference on Learning Representations (2019)
14. Shukla, S.N., Marlin, B.M.: Multi-time attention networks for irregularly sampled time series. In: International Conference on Learning Representations (2021)
15. Sun, C., Hong, S., Song, M., Li, H.: A review of deep learning methods for irregularly sampled medical time series data. arXiv preprint arXiv:2010.12493 (2020)
16. Tan, Q., et al.: Data-GRU: dual-attention time-aware gated recurrent unit for irregular multivariate time series. In: Proceedings of the AAAI Conference on Artificial Intelligence, vol. 34, pp. 930–937 (2020)
17. Tang, X., Yao, H., Sun, Y., Aggarwal, C., Mitra, P., Wang, S.: Joint modeling of local and global temporal dynamics for multivariate time series forecasting with missing values. In: Proceedings of the AAAI Conference on Artificial Intelligence, vol. 34, pp. 5956–5963 (2020)
18. Tashiro, Y., Song, J., Song, Y., Ermon, S.: CSDI: conditional score-based diffusion models for probabilistic time series imputation. Adv. Neural. Inf. Process. Syst. **34**, 24804–24816 (2021)
19. Vaswani, A., et al.: Attention is all you need. Adv. Neural. Inf. Process. Syst. **30** (2017)
20. Wu, Y., et al.: Dynamic gaussian mixture based deep generative model for robust forecasting on sparse multivariate time series. In: Proceedings of the AAAI Conference on Artificial Intelligence, vol. 35, pp. 651–659 (2021)
21. Wu, Z., Pan, S., Long, G., Jiang, J., Chang, X., Zhang, C.: Connecting the dots: multivariate time series forecasting with graph neural networks. In: Proceedings of the 26th ACM SIGKDD International Conference on Knowledge Discovery & Data Mining, pp. 753–763 (2020)
22. Yoon, J., Zame, W.R., van der Schaar, M.: Estimating missing data in temporal data streams using multi-directional recurrent neural networks. IEEE Trans. Biomed. Eng. **66**(5), 1477–1490 (2018)
23. Zhang, S., Guo, B., Dong, A., He, J., Xu, Z., Chen, S.X.: Cautionary tales on air-quality improvement in Beijing. Proc. Roy. Soc. A: Math. Phys. Eng. Sci. **473**(2205), 20170457 (2017)
24. Zhang, X., Zeman, M., Tsiligkaridis, T., Zitnik, M.: Graph-guided network for irregularly sampled multivariate time series. In: International Conference on Learning Representations (2022)
25. Zhou, H., et al.: Informer: beyond efficient transformer for long sequence time-series forecasting. In: Proceedings of the AAAI Conference on Artificial Intelligence, vol. 35, pp. 11106–11115 (2021)

Transmission Tower Inspection Drone Hub Deployment Optimization System Based on DeepSeek Large Language Model

Cheng Su[1,2], Yang Yang[3], Shijie Li[4], Zixuan Zhao[1], Shaohua Wang[1,2](✉), Xiaohan Jiang[5], and Chang Liu[1,2]

[1] State Key Laboratory of Remote Sensing and Digital Earth, Aerospace Information Research Institute, Chinese Academy of Sciences, Beijing 100094, China
wangshaohua@aircas.ac.cn

[2] University of Chinese Academy of Sciences, Beijing 100049, China

[3] China Southern Power Grid Digital Platform Technology Company, Guangzhou 510700, China

[4] China Southern Power Grid Company Limited, Guangzhou, China

[5] School of Architecture and Urban Planning, Lanzhou Jiaotong University, Lanzhou 730070, China

Abstract. Ensuring the safety of high-voltage transmission towers is critical for urban power grid resilience, particularly with increasing urbanization and extreme weather. Traditional manual inspections are inefficient, costly, and hazardous for large-scale networks. This study proposes an intelligent decision support system (DSS) driven by a Large Language Model (LLM) to optimize drone hub deployment for transmission tower inspection. By integrating the DeepSeek LLM with the HiSpot optimization toolkit, our system formulates this task as a p-Hub median problem and automates the end-to-end solution process. The system enables users to describe objectives in natural language, then autonomously identifies the problem, extracts parameters, and invokes solution algorithms. A case study in Zengcheng District, Guangzhou, utilizing OpenStreetMap data for thousands of towers, validates the system's effectiveness on large-scale p-Hub problems. Results show the system lowers technical barriers for geospatial optimization, offers a novel approach for intelligent infrastructure management, and significantly enhances urban power system resilience.

Keywords: Large Language Model · Spatial Optimization · p-Hub Median Problem · Intelligent Decision Support System · DeepSeek · HiSpot

1 Introduction

The rapid evolution and increasing complexity of modern power grid systems have placed unprecedented demands on inspection and maintenance operations. With the continuous expansion of grid scale and the growing intricacy of operating environments, traditional manual inspection methods—characterized by low efficiency and high

C. Su and Z. Zhao---These authors contributed equally to this work.

J. Li et al. (Eds.): APWeb-WAIM 2025, CCIS 2660, pp. 225–234, 2026.
https://doi.org/10.1007/978-981-95-7996-9_20

risk—are increasingly inadequate for meeting the requirements of safe, efficient, and continuous maintenance. As a significant complementary approach, unmanned aerial vehicles (UAVs) have been widely adopted in power grid inspection and maintenance due to their flexibility, maneuverability, and controllable risk profiles [1, 2]. However, current UAV scheduling strategies are predominantly rule-based and heavily reliant on conventional image recognition techniques [3]. These approaches tend to exhibit delayed responsiveness and limited task reorganization capabilities in complex scenarios involving concurrent tasks, abrupt environmental changes, or unexpected failures [4, 5]. Consequently, there is a pressing need for intelligent systems capable of enhanced task comprehension and dynamic planning.

In recent years, large language models (LLMs) have demonstrated remarkable capabilities in knowledge reasoning, decision-making, and tool integration [6]. Toolformer, proposed by Schick et al. [7], employs a self-supervised learning framework to automatically determine the timing and method of tool invocation, thereby improving task performance under zero-shot conditions. The ReAct framework [8] integrates reasoning traces with actionable operations, enabling dynamic interaction and execution in complex tasks. ToolLLM [9] further advances this line of work by constructing the large-scale ToolBench dataset and introducing a dependency-tree-based planning method, significantly enhancing the generalizability and depth of multi-tool scheduling. These innovations collectively establish a methodological foundation for LLM-based centralized control in complex task environments.

Initial progress has also been made in integrating LLMs into robotics and unmanned systems. The SMART-LLM framework proposed by Kannan et al. [10] utilizes few-shot prompting for instruction decomposition and multi-robot collaborative planning, with its feasibility demonstrated in both simulation and real-world platforms. Wang et al. subsequently introduced DART-LLM, which incorporates task dependency mod-eling and multi-stage execution to improve temporal logic and scheduling efficiency. While these studies showcase the potential of LLMs in autonomous systems, most applications remain confined to structured environments and have not yet addressed real-world, mission-critical scenarios such as power grid inspection [11].

Building on these developments, the present study proposes a novel UAV scheduling and deployment framework designed for complex operational contexts. The system leverages DeepSeek as a language-centric control module, integrating path planning, state perception, and task execution components. This architecture is intended to enhance the adaptability and intelligence of UAV-based power grid inspection systems, with the ultimate goal of supporting more robust and autonomous maintenance workflows in modern power grids.

2 System Architecture

2.1 System Overall Architecture

This study designs and implements an intelligent decision support system that integrates a large language model with geospatial optimization algorithms, aiming to bridge the gap between complex optimization tasks and user requirements expressed in natural language. The system architecture adheres to the principles of modularity and scalability, ensuring the independence of functional components, standardization of interfaces, and ease of future maintenance and upgrades. The overall architecture (Fig. 1) is composed of four highly synergistic core layers: the User Interaction and Task Input Layer, the Language Model Understanding and Translation Layer, the Geospatial Optimization Modeling and Solving Layer, and the Data Management and Results Visualization Layer.

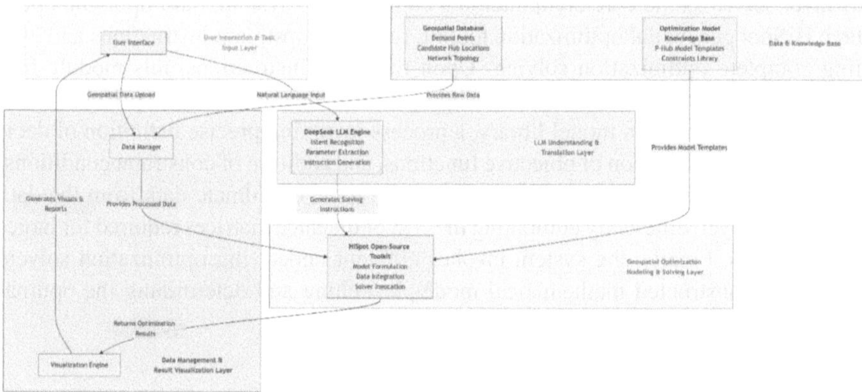

Fig. 1. System Overall Architecture

2.1.1 User Interaction and Task Input Layer

This layer serves as the human-computer interface frontend of the system, providing users with a unified entry point for task definition and data submission. It not only supports users in providing high-level semantic descriptions of optimization tasks through natural language, such as directly articulating the number of hubs to be selected, spatial distribution characteristics of demand points, and optimization objectives, thereby eliminating the need to construct complex mathematical models; simultaneously, to support large-scale spatial optimization problems, this layer integrates robust data upload functionality capable of processing geographic datasets containing hundreds or even thousands of spatial entities, and supports direct import of industry-standard formats such as Shapefile.

2.1.2 Language Model Understanding and Translation Layer

This layer constitutes the intelligent core of the system, supported by the DeepSeek large language model technology, responsible for translating unstructured natural language input from users into structured, formalized instructions executable by downstream modules. Its workflow follows a clear logical sequence: first, the DeepSeek model performs intent recognition on user descriptions through deep semantic analysis, identifying the problem as belonging to the p-Hub median problem or its variants by capturing domain-specific terminology such as "hub" and "transfer"; subsequently, the model automatically extracts key parameters required for mathematical model construction within the identified framework, such as the number of hubs P and related business constraints; finally, this layer integrates the parsed model type and parameters into a standardized solving instruction, providing clear, unambiguous computational task specifications for downstream optimization modules.

2.1.3 Geospatial Optimization Modeling and Solving Layer

This layer serves as the core computational engine of the system, built upon the open-source HiSpot geospatial optimization toolkit, receiving upstream instructions and executing complete optimization solving. Upon receiving instructions, this module first dynamically instantiates the mathematical programming paradigm of the p-Hub median problem from the built-in model library, a process involving precise definition of decision variables, construction of objective functions, and addition of constraint conditions. Subsequently, the module retrieves massive geographic coordinate data from the data management layer, efficiently computing the cost or distance matrices required for large-scale problems. Finally, the system invokes efficient underlying optimization solvers to solve the constructed mathematical model, searching and determining the optimal solution within the vast solution space.

2.1.4 Data Management and Results Visualization Layer

This layer is responsible for the input and preprocessing of all system data, as well as the output and presentation of optimization results. Its data management functionality supports efficient parsing and standardized preprocessing of various geographic data formats uploaded by users (particularly Shapefile), including data format validation, coordinate system transformation, and data cleaning, aimed at providing high-quality input for optimization analysis. Upon obtaining optimization results, its visualization functionality renders the solving solutions, generating relevant geographic information result maps for presentation to users, clearly indicating optimal hub locations and service allocation relationships for further analysis.

2.2 HiSpot

HiSpot (High Intelligence Spatial Optimization) is an open-source, high-intelligence geospatial optimization tool designed specifically for addressing complex spatial location and routing problems. The system integrates a variety of classical spatial optimization models, including the capacitated Location Routing Problem (LRP) and the p-Hub Median Problem, and incorporates efficient mathematical programming solvers, approximation algorithms, and heuristic methods. This enables fast modeling and efficient solving of large-scale, multi-constraint geospatial optimization tasks. In the context of UAV applications for intelligent power grid inspection, HiSpot can leverage geographic information and task requirements to optimize the siting of inspection facilities and the allocation of inspection tasks.

3 Real-World Case

3.1 p-Hub Median Problem

The p-Hub median problem is a classic spatial optimization model widely applied in aviation, logistics, emergency dispatch, and other fields [12, 13]. This model aims to select p hub nodes from a set of candidate locations and allocate transit routes for all non-hub nodes (spokes) to minimize overall transportation costs. Its core feature lies in leveraging the centralized transfer mechanism of hub nodes to simplify network structure while controlling total route costs.

This study is based on the spatial modeling scheme provided by the Hispot project, abstracting the UAV power grid inspection task as a p-Hub problem: power facilities are abstracted as a set of demand nodes N, from which P hubs are selected as UAV deployment points. The task requires transiting through hubs to reach target points to minimize the total system cost. Considering that the inspection task originates from the starting node i, transits through hubs k and h, and arrives at the destination node j, the total system cost can be divided into three parts: collection cost (from i to k), transfer cost (from k to h), and distribution cost (from h to j).

The model is defined as follows:

$$min\sum_{ik} PC \cdot C_{ik} \cdot O_i \cdot x_{ik} + \sum_{ikh} PT \cdot C_{kh} \cdot y_{ikh} + \sum_{ik} PD \cdot C_{ki} \cdot D_i \cdot x_{ik} \tag{1}$$

$$\text{s.t.} \sum_{k} x_{ik} = 1, \forall i \in N \tag{2}$$

$$x_{ik} \le x_{kk}, \forall i, k \in N \tag{3}$$

$$\sum_{k} x_{kk} = P \tag{4}$$

$$\sum_{h} y_{ikh} - \sum_{h} y_{ihk} = O_i x_{ik} - \sum_{j} W_{ij} x_{jk}, \forall i, k \in N \tag{5}$$

$$\sum_{h \neq k} y_{ikh} \le O_i x_{ik}, \forall i, k \in N \tag{6}$$

$$x_{ik} \in \{0,1\}, y_{ikh} \ge 0 \tag{7}$$

Where:

C_{ij}: Distance or cost between nodes i and j;

W_{ij}: Task demand volume from node i to node j;

$O_i = \sum_j W_{ij}, D_i = \sum_j W_{ji}$;

x_{ik}: Whether node i is assigned to hub k;

y_{ikh}: The flow of node i passes through $k \to h$;

PC, PT, PD: represent the cost coefficients for collection, transportation, and distribution respectively;

P: Maximum number of selected hubs.

By solving this model, the hub deployment scheme can be optimized under the constraints of UAV range and response time limits, thereby improving the global efficiency of inspection task scheduling and regional coverage capability.

3.2 Case Study: Transmission Towers in Zengcheng District, Guangzhou

Power infrastructure serves as a crucial pillar supporting the normal functioning of modern socioeconomic systems, with transmission towers acting as key nodes in power grids whose operational safety directly impacts regional energy supply stability. This study selects Zengcheng District in Guangzhou City, Guangdong Province, China as the empirical analysis area. As a subtropical megacity with over 10 million residents and a highly developed economy, Guangzhou's power system faces complex and severe challenges regarding safety and resilience.

Zengcheng District is frequently affected by extreme meteorological disasters, experiencing relatively high annual typhoon frequency and over 80 thunderstorm days per year. These natural hazards can easily cause physical damage to exposed transmission tower facilities, subsequently triggering power outages. Meanwhile, Zengcheng belongs to a high-density urban built-up area where 3,057 transmission towers spatially intertwine with dense road networks and building clusters. This not only increases routine maintenance complexity but also raises secondary risks during disasters. Should transmission towers sustain damage, consequences extend beyond industrial, commercial, and residential power interruptions—they may additionally obstruct emergency access routes and delay equipment repairs.

Therefore, how to efficiently and promptly conduct routine inspections and post-disaster emergency repairs for the numerous transmission towers in the Zengcheng District has become a critical issue in enhancing the resilience of the urban power system. The adoption of drone-based automated inspections represents a relatively advanced solution currently available, yet it introduces the challenge of geospatial optimization: how should the hub base stations for drone inspections be rationally deployed to cover the entire transmission tower network at the lowest overall cost? This study models this problem as a p-Hub median problem, aiming to plan an optimal drone inspection hub network for the Zengcheng District.

The core dataset used in this study consists of spatial distribution information of transmission towers, sourced from the global open-source geographic information collaboration project OpenStreetMap (OSM). As an open database maintained by global volunteers, OSM stores rich and structured geospatial entity data. After acquisition, the data was converted into Shapefile format and underwent coordinate system calibration and data cleaning to ensure data quality (Fig. 2).

In the case of Zengcheng District, the multiple hub location problem (p-Hub) is specifically defined as follows:

Demand Points: Refer to all transmission tower locations obtained through OpenStreetMap (OSM) within the study area, representing the target mission points for drone inspections.

Candidate Hub Points: Theoretically, any transmission tower location can serve as a candidate site for a drone and its operations team's hub base station.

Number of Hubs (P): The predetermined total number of hubs to be constructed, based on the decision-maker's budget constraints, resource conditions, and inspection response time requirements.

Optimization Objective: Under the constraint of meeting the specified number of hubs, select the optimal geographical locations for P hubs and assign each transmission

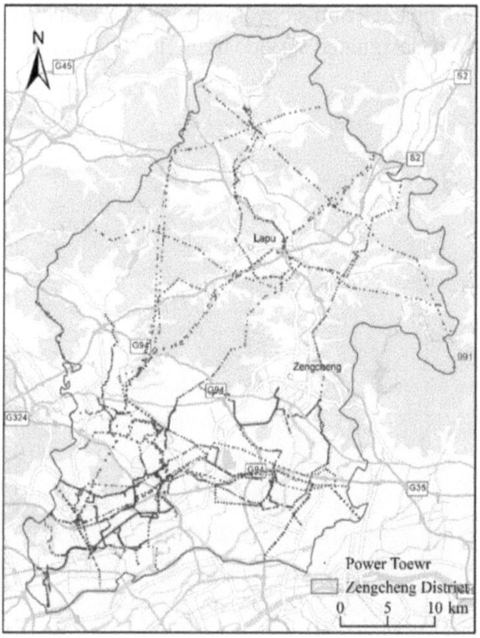

Fig. 2. Distribution of Transmission Tower Locations in Zengcheng District

tower to a service hub, with the goal of minimizing the total flight distance or time for drones to depart from their assigned hubs, complete all allocated tasks, and return. The model also allows drones to transfer between different hubs, optimizing cross-regional inspection routes to further reduce scheduling costs.

By solving this multi-hub location model, this study aims to provide the power management department of Zengcheng District, Guangzhou City, with a scientific and quantifiable decision-making support solution, enhancing the efficiency and resilience of urban power grid inspections to address increasingly complex operational environments and safety challenges.

3.3 Optimization Results

This study examined 3,057 transmission towers in Zengcheng District, Guangzhou, within a geographic area spanning 113.6°–114.0°E longitude and 23.1°–23.5°N latitude. To evaluate the impact of varying hub quantities on network optimization performance, p-Hub location optimization was conducted for two scenarios: $p = 50$ and $p = 75$. Results demonstrate that with 50 hubs, the optimization model successfully determined the optimal hub location configuration for the transmission tower network in Zengcheng District. The 50 hubs exhibited a scientifically sound distribution pattern throughout the study area, effectively covering the primary electrical load centers of Zengcheng District. When the hub count increased to 75, the network displayed more refined characteristics. The rational increase in hub density enabled further subdivision of service areas, optimizing each hub's coverage zone and reducing the total connection costs from towers to their designated service hubs (Figs. 3 and 4).

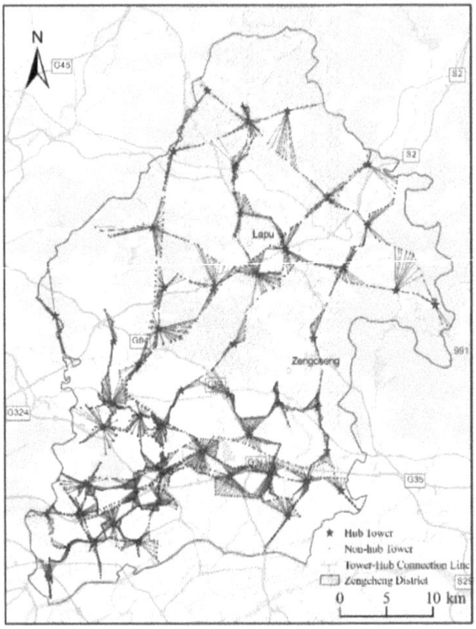

Fig. 3. Drone Hub Optimization Results (p = 50)

Comparative analysis of the optimization results for the two different p-values reveals that the p = 50 scenario assigns larger service areas to each hub, achieving favorable spatial optimization while maintaining reasonable management complexity. In contrast, the p = 75 scenario produces more compact service regions, where the increased hub count further enhances network spatial efficiency and service quality. Both configurations achieved complete coverage of the study area, with hubs serving as regional centers effectively connecting surrounding towers. Experimental analysis indicates that both scenarios successfully minimized service distances, with the p = 75 configuration demonstrating shorter average service distances and more balanced load distribution among hubs, thereby avoiding potential overload issues at individual hubs. The optimized network structure exhibits robust characteristics, where failure of a single hub would not severely compromise overall network performance, ensuring power system reliability.

Results demonstrate that the p-Hub optimization model effectively addresses location optimization challenges in transmission tower networks, with different p-value settings significantly influencing optimization outcomes, necessitating rational selection based on practical application requirements. The p = 50 configuration is more suitable for scenarios requiring balanced optimization effectiveness and operational costs, while the p = 75 configuration is appropriate for applications demanding higher network service quality. The optimization results provide scientific foundations for power system

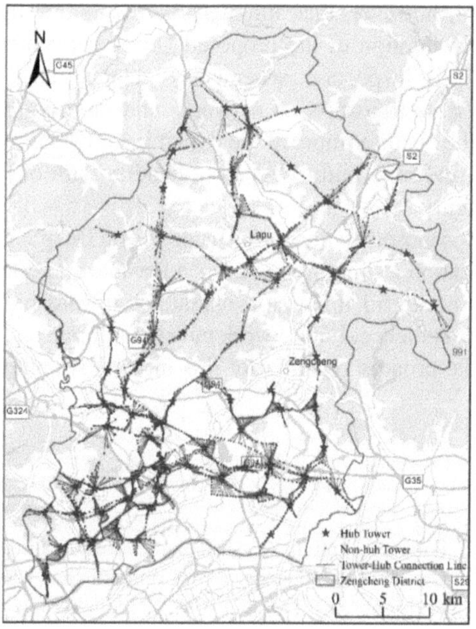

Fig. 4. Drone Hub Optimization Results (p = 75)

operation and maintenance management, fault response, and resource allocation, offering effective solutions and methodological references for similar power infrastructure network optimization problems.

4 Conclusion and Discussion

With the continuous expansion of power grid scale and the increasing complexity of operational environments, traditional UAV-based inspection scheduling and path planning approaches have become insufficient to meet the growing demands of smart grids for efficiency, autonomy, and emergency responsiveness. To address the challenges posed by complex and dynamic inspection scenarios, this study proposes a unified UAV scheduling and deployment mechanism with the large language model DeepSeek serving as the intelligent control center. The objective is to significantly enhance the adaptability and intelligence of power grid inspection systems.

To evaluate the feasibility and effectiveness of the proposed mechanism, a series of experiments were conducted using the open-source geospatial optimization toolkit HiSpot, based on real-world data from Zengcheng District, Guangzhou. The results demonstrate the applicability of the p-hub model in optimizing the aggregation of UAV inspection nodes, task relay, and interregional coordination. The deep integration of DeepSeek with functional modules further reveals that large language models can effectively support and enhance UAV inspection scheduling through knowledge reasoning, task comprehension, and dynamic instruction generation.

It should be noted, however, that although the case study in Zengcheng District provides preliminary validation of the proposed DeepSeek-centered UAV scheduling framework in power grid inspection, several key technical challenges remain for its application in larger-scale transmission networks and more volatile operational contexts. These include the need to improve the dynamic knowledge updating capabilities of language models, real-time multi-UAV coordination with obstacle avoidance, and communication reliability.

In conclusion, the proposed architecture—centered on DeepSeek and integrated with the geospatial optimization tool HiSpot—offers a feasible and effective technical pathway for advancing power grid inspection operations toward higher levels of intelligence, autonomy, and resilience. Future work may incorporate real-time task prediction and further explore the broader potential of this mechanism within intelligent energy systems.

References

1. Chehri, A., Jeon, G., Fofana, I., Imran, A., Saadane, R.: Accelerating power grid monitoring with flying robots and artificial intelligence. IEEE Commun. Stand. Mag. **5**(4), 48–54 (2021)
2. Tang, N., Landas, N., Rodrigues, Y.R., Monteiro, M.R.: Industry state-of-art and opportunities for the use of drones in smart grids inspections. In: Proceedings of SPEEDAM 2024, pp. 327–331. IEEE, Amalfi (2024)
3. Jing, L., Rahman, A.: Fault diagnosis in power grids with large language model. arXiv:2407.08836 (2024)
4. Ding, Z., Yüksel, S., Dinçer, H.: An integrated Pythagorean fuzzy soft computing approach to environmental management systems for sustainable energy pricing. Energy Rep. **7**, 5575–5588 (2021)
5. Liu, Y.T., Ni, Y.J.: A new method of location detection with knowledge transfer learning. J. Phys.: Conf. Ser. **1087**(6), 062017 (2018)
6. Wang, J., et al.: Power-Llava: Large language and vision assistant for power transmission line inspection. In: Proceedings of ICIP 2024, pp. 963–969. IEEE, Singapore (2024)
7. Schick, T., et al.: Toolformer: language models can teach themselves to use tools. Adv. Neural. Inf. Process. Syst. **36**, 68539–68551 (2023)
8. Yao, S., et al.: ReAct: synergizing reasoning and acting in language models. In: Proceedings of ICLR (2023)
9. Qin, Y., et al.: ToolLLM: facilitating large language models to master 16000+ real-world APIs. arXiv:2307.16789 (2023)
10. Kannan, S.S., Venkatesh, V.L.N., Min, B.-C.: SMART-LLM: smart multi-agent robot task planning using large language models. arXiv:2309.10062 (2024)
11. Wang, Y., et al.: DART-LLM: dependency-aware multi-robot task decomposition and execution using large language models. arXiv:2411.09022 (2025)
12. Campbell, J.F.: Hub location and the p-hub median problem. Oper. Res. **44**(6), 923–935 (1996)
13. Alumur, S., Kara, B.Y.: Network hub location problems: The state of the art. Eur. J. Oper. Res. **190**(1), 1–21 (2008)

SGDA: Class Augmentation for Imbalanced Graph Learning at Scale

Zeshun Peng, Chuhan Zhang, Yanfeng Zhang(✉), and Ge Yu

Northeastern University, Shenyang 110819, China
{pengzeshun,2101846}@stumail.neu.edu.cn, {zhangyf,yuge}@mail.neu.edu.cn

Abstract. Graph Neural Networks (GNNs) are widely applied in node classification tasks across social networks, bioinformatics, and healthcare. However, datasets in these applications typically exhibit class imbalance, where the majority classes contain abundant labeled samples while the minority classes have limited annotations. During the training process, nodes from the majority classes are more likely to be labeled. In contrast, nodes from minority classes are more difficult to annotate. This labeling bias causes GNNs to favor majority classes while under-representing minority classes, leading to poor classification performance.

This paper proposes Self-supervised Graph Data Augmentation (SGDA) to improve accuracy in imbalanced learning. First, we utilize graph partitioning to identify nodes that require augmentation and generate new nodes with edges based on their neighborhoods, ensuring topological balance. Second, to address the processing of large-scale graphs, we propose mini-batch training where nodes in a batch are selected based on their label distribution. Experimental results show that SGDA achieves up to 0.38 points improvement in balanced accuracy compared to GraphSMOTE and ReNode. The mini-batch training achieves 1.5× speedup over GCN.

Keywords: Imbalanced Graph Data · Graph Neural Networks · Deep Learning · Mini-batch Training · Semi-supervised Learning

1 Introduction

Graph Neural Networks (GNNs) are widely applied in node classification tasks across various domains. For example, in social networks, each node represents a user, and each edge represents a user relationship. We can predict user interests and categorize them into groups by analyzing node features within the graph structure. To handle large-scale graph data, GNNs are typically trained in a semi-supervised manner, using a small number of nodes with ground truth labels to guide the learning of many unlabeled nodes [6]. Although labeled nodes constitute only a small proportion of all nodes, we can infer the classes of unlabeled nodes by analyzing their features and connectivity.

Existing semi-supervised learning methods [3,7] typically assume that the number of nodes in each class within the training set is balanced. However, real-world datasets usually exhibit class imbalance. As shown in Fig. 1a, solid nodes

J. Li et al. (Eds.): APWeb-WAIM 2025, CCIS 2660, pp. 235–247, 2026.
https://doi.org/10.1007/978-981-95-7996-9_21

represent labeled nodes during training, while hollow nodes represent unlabeled nodes. In a sports social network, only a small portion of nodes are labeled, and the number of nodes of popular sports like basketball players is significantly higher than that of niche sports like squash players. This causes classes with more labels to dominate the learning process, making GNN models focus more on the node features of classes with abundant labels while neglecting the node features of classes with few labels, resulting in lower classification accuracy for minority class nodes. In the Cora [10] dataset, the class imbalance is also significant. As shown in Fig. 1b, this dataset contains seven classes (L0-L6), with highly uneven node distribution across classes. Specifically, the L3 class accounts for $\approx 30\%$ of the nodes, while the L1 class only comprises $\approx 8\%$ of the nodes.

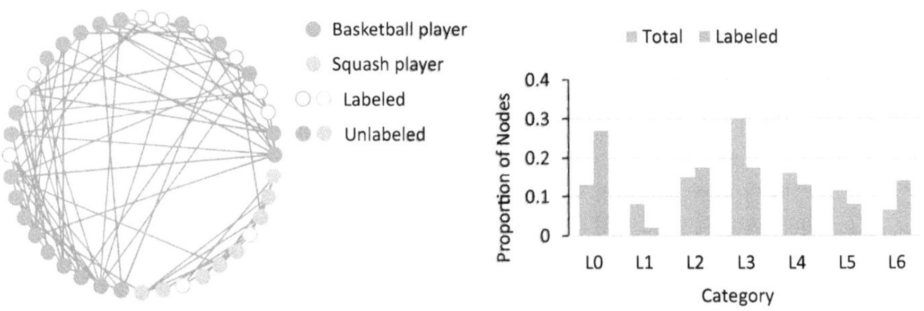

(a) An example of sport social network **(b) Category distribution of nodes in Cora dataset**

Fig. 1. Examples of Class Imbalance in Graph Data.

Furthermore, many minority class nodes are located in unfavorable structural positions within the graph topology, exacerbating the difficulty of learning minority class patterns. GNN models also exhibit bias toward densely connected regions of the graph. When minority class nodes are positioned in unfavorable structural positions, such as peripheral areas with low connectivity, this leads to insufficient feature learning and poor recognition performance for these under-represented nodes. Traditional rebalancing methods (such as SMOTE [2]) primarily focus on feature space similarity and cannot effectively address such structured imbalance problems inherent in graph data.

Moreover, existing approaches typically adopt mini-batch training strategies when graph data scales to millions of nodes due to memory constraints. However, traditional mini-batch training randomly selects nodes in a batch. When the number of labels of different classes is already imbalanced, this further leads to sampling bias, causing minority class nodes to be further marginalized during the training process, thereby reducing the classification accuracy.

To address these challenges, we propose Self-supervised Graph Data Augmentation (SGDA) to improve the accuracy and training performance of large-scale GNNs in semi-supervised classification tasks.

First, we augment the labeled data for minority classes to reduce learning bias. Specifically, we employ a self-supervised learning model to learn patterns from the majority of unlabeled nodes in the graph and predict their potential class labels (i.e., pseudo-labels). Based on these predicted labels, we identify which classes and topology positions require additional samples. We then generate new nodes and their connections to existing nodes by analyzing neighborhood relationships and feature patterns, ensuring that the generated data not only balances all classes numerically but also maintains reasonable topological structure in the graph. Meanwhile, to further mitigate the negative impact of class imbalance on model training, we improve the loss function of GNNs. By assigning different weights to samples of different classes, the model can focus more fairly on all classes during training rather than being biased toward the majority of classes with larger sample sizes.

Second, we propose data-enhanced mini-batch training. The approach ensures that nodes of different classes in each batch have relatively balanced representations while preserving the topological connectivity information. We also pre-identify and cache embeddings of nodes shared across batches to avoid redundant forward propagation computations, reducing computational overhead.

In summary, the contributions of this paper are as follows:

1. We propose SGDA to augment the labeled data for minority classes and unfavorable positions, reducing learning bias in imbalanced graph datasets.
2. We introduce data-enhanced mini-batch training to alleviate the reduction of minority class samples caused by traditional random sampling.
3. Experimental results show that SGDA achieves up to 0.38 points improvement in balanced accuracy compared to GraphSMOTE [20] and ReNode [3]. The data-enhanced mini-batch training achieves a 1.5× speedup over GCN.

2 Background and Motivation

2.1 Learning on Imbalanced Graphs

GNNs are used to learn representations from graph-structured data to perform various tasks, such as node classification and link prediction. Most real-world graph datasets exhibit class imbalance [8], where some classes contain many nodes while others have only a few samples. This uneven distribution causes GNNs to perform well on common classes but poorly on rare classes.

During training, many nodes in minority classes may suffer from topological imbalance [3], resulting in poor representation learning by GNNs, which further exacerbates the class imbalance problem. Topological imbalance refers to the uneven information propagation among nodes due to structural position differences. For example, nodes at the center of the graph can easily receive and share information, while nodes at the edges or isolated areas struggle to get enough information. This leads to poor performance for nodes in unfavorable positions. When classes are unevenly distributed, rare class nodes tend to be more isolated with fewer connections, making it even harder for them to learn effectively.

Existing approaches address imbalanced graphs through data optimizations or algorithm optimizations.

Data optimization includes oversampling minority classes and undersampling majority classes. GraphSMOTE [20] and ImGAGN [12] augment minority class samples by generating new nodes and edges. Mixup [19], GraphMixup [16], and GraphENS [11] create new nodes by blending semantic features with contextual edges, aiming to generate representative minority class samples to balance the training set. ReNode [3] and TAM [13] consider topological information and adaptively modify boundaries based on each node's connectivity patterns and neighbor features. G2GNN [15] leverages global and local properties of the graph structure to enhance supervision signals.

Algorithm optimization addresses class imbalance by improving learning algorithms. [1] modifies loss functions to assign higher weights to minority class nodes, enhancing model sensitivity to these nodes. [4] adjusts decision boundaries to reduce classification errors between minority and majority classes. RAHNet [9] jointly learns a balanced feature extractor and an unbiased classifier with decoupled training, addressing class imbalance by regularizing classifier weights while keeping the feature extractor fixed. CoMe [17] integrates multiple expert classifiers that learn different features on the graph and collaboratively optimize overall model performance through mutual information exchange and guidance.

2.2 Limitation of Existing Data Augmentation Approaches

Existing approaches (e.g., GraphENS [11] and GraphSMOTE [20]) only generate synthetic nodes or edge structures based on node relationships in the embedding space. Specifically, they apply the KNN algorithm to graph data. KNN maps data to an embedding space and evaluates node similarity through the Euclidean distance between two nodes. Therefore, these approaches assume that nodes close in the embedding space are also adjacent in the topological space. For example, in Fig. 2a, node v_1's nearest neighbor in the embedding space is node v_2, which also corresponds to a close distance in the topological space, as shown in Fig. 2b. However, although nodes v_1 and v_3 are adjacent in the embedding space, they have a large topological distance and inconsistent topological structures. In such cases, existing methods may generate inappropriate edges, leading to incorrect information propagation in the graph and affecting classification correctness.

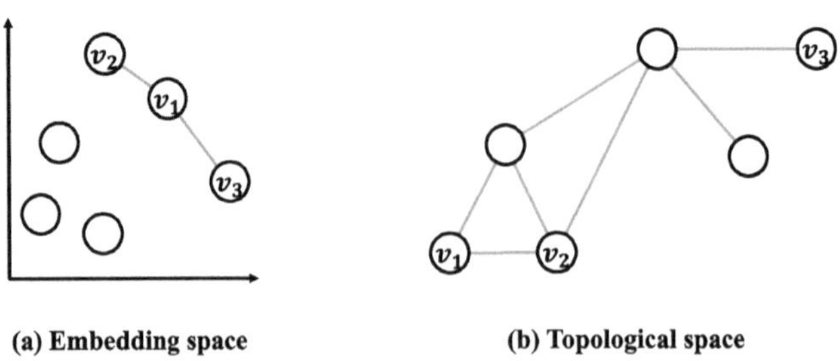

Fig. 2. Node relationships in embedding and topological spaces.

We also perform an experiment on the Cora dataset [10] to better understand the differences in relationships between embedding and topological spaces.

For neighbors in the embedding space, we first compute the embedding vector of each node using GraphSAGE, then calculate the relative proximity of each node through cosine similarity. Each node selects K nearest neighbors, where we set $k = 5$ [2]. For neighbors in the topological space, we directly select K nearest neighbors on the graph. If a node has fewer than K one-hop neighbors, we randomly repeat the selection of existing neighbors until reaching K neighbors.

We use the Common Neighbors (CN) metric to compare the overlap between node neighbors in the embedding and topological spaces. As shown in Eq. 1, $N_e(u)$ and $N_t(u)$ represent the sets of K nearest neighbor nodes of node u in the embedding space and topological space, respectively.

$$CN(u) = |N_e(u) \cap N_t(u)| \tag{1}$$

The Average Shortest Path Length (SL) measures the reachability of neighbors identified in the embedding space within the graph's topological structure. It represents the average shortest topological distance of a node's K nearest neighbors in the embedding space on the graph structure, as shown in Eq. 2, where $d(u, v)$ denotes the shortest path distance between nodes u and v in the graph topology.

$$SL(u) = \frac{1}{K} \sum_{v \in N_e(u)} d(u, v) \tag{2}$$

As shown in Fig. 3, the nearest neighbors of nodes in the embedding space often have topological distances greater than 1-hop from the target nodes in the topological space (i.e., they are not topological neighbors). Specifically, nodes in L1 and L2 exhibit clustering behavior in the topological space, with a larger number of common neighbors between the embedding space and the topological space. For categories that are relatively dispersed in the topological space, such as nodes in L3 and L5, the overlap between one-hop neighbors in the graph structure and nearest neighbors in the embedding space is low.

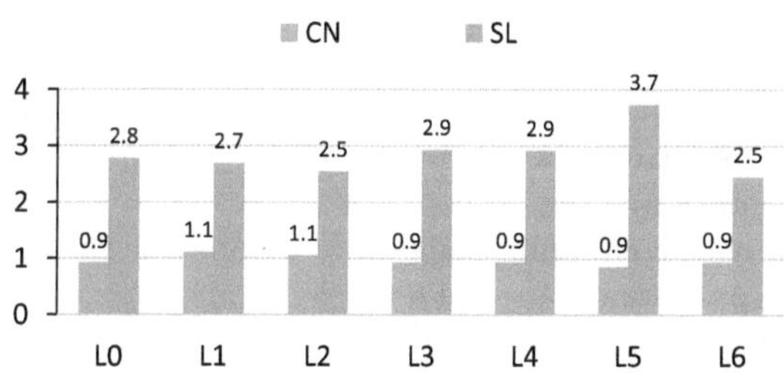

Fig. 3. Attributes SL and CN across Cora dataset categories.

3 Self-supervised Graph Data Augmentation

SGDA enhances model performance on imbalanced graph data by analyzing neighborhood features and structures to generate synthetic nodes and edges, thereby optimizing the graph data for training. As illustrated in Fig. 4, the SGDA workflow consists of four sequential stages:

1. **Identify Representative Nodes:** Identify representative nodes (v_r) from classes requiring augmentation with graph partitioning.
2. **Generate Synthetic Nodes:** Generate labeled synthetic nodes (v_s) that are similar to v_r nodes. These v_s nodes inherit the same labels as their corresponding v_r nodes and are added to the training set.
3. **Generate Synthetic Edges:** Generate synthetic edges between v_s nodes and existing nodes.
4. **Training:** Train on the augmented data using a weighted GNN model $\psi(\cdot)$.

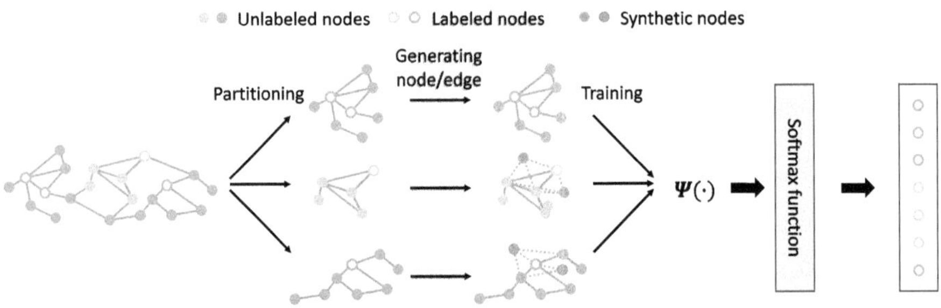

Fig. 4. The workflow of SGDA.

3.1 Partition-Based Node Selection

To simultaneously consider node relationships in topological space, we adopt the Louvain community detection algorithm to partition the graph and then perform data augmentation separately for nodes within each partition. Nodes within each community have tighter topological connections. Since communities with fewer nodes and sparse connections tend to contain more minority class nodes, labeled nodes located in sparse community subgraphs, serve as target samples for augmentation.

Specifically, for each community subgraph, we analyze the labeled nodes contained within it. Suppose the degree of a labeled node is greater than the density ratio dr. In that case, we consider that this labeled node has a sparse influence on its community, and data augmentation is needed to generate new nodes to increase the number of labeled node samples in this subgraph.

3.2 Self-supervised Synthetic Node Generation

SGDA derives node labels in community subgraphs through self-supervised learning and assigns different weights to these inferred pseudo-labels based on their consistency with the categories of representative nodes v_r to aggregate and generate features for synthetic nodes v_s. Since the neighbors used for aggregating synthetic node embeddings belong to the same subgraph, this avoids potential noise interference from nodes in other subgraphs.

Pre-training Phase. Self-supervised learning consists of two stages: pre-training and fine-tuning. In the pre-training phase, a teacher network is pre-trained using confirmed node labels to achieve accurate classification performance. Specifically, node embedding is obtained through a single-layer graph convolutional network $\psi(x)$, which is input to a Bayesian neural network. The posterior distribution of model parameters is learned based on accurate node labels, as shown in Eq. 3:

$$p(\zeta|\psi(x)) \propto p(\psi(x)|\zeta) \cdot p(\zeta) \tag{3}$$

where ζ represents the parameters in the Bayesian neural network, $p(\zeta)$ denotes the prior distribution of ζ, and $p(\psi(x)|\zeta)$ represents the probability of observing node embeddings given parameters ζ. The posterior distribution $p(\zeta|\psi(x))$ obtained through the Bayesian theorem is used to update the parameter distribution of the Bayesian neural network.

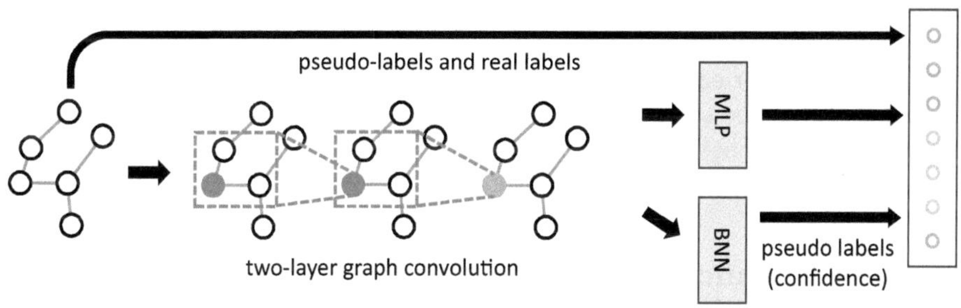

Fig. 5. Self-supervised training for reasoning node categories within communities.

Fine-Tuning Phase. The teacher network is used to guide the student network in inferring node labels in graph data, as shown in Fig. 5. The teacher obtains embeddings for all nodes through two-layer graph convolution, inputs all node embeddings into the Bayesian neural network to infer labels and confidence scores for each node, and creates pseudo-labels on nodes with high confidence. The labeled node set V_L (including nodes with pseudo-labels and real labels) guides the training of the student network. During fine-tuning, model parameters of the student network are updated based on label information from the labeled node set V_L, which in turn adjusts node embedding information in the teacher network. The teacher network recalculates pseudo-labels and corresponding confidence scores for all nodes based on updated node embeddings, iterating continuously until convergence.

3.3 Synthetic Edge Generation

In the current implementation, generated synthetic nodes are assigned exactly the same neighbor structure as specific nodes v_r, achieved by replicating all adjacency relationships of v_r within their community subgraphs. However, this method may cause generated synthetic nodes to transmit noise to neighbor nodes through v_r, while connections with too many labeled nodes lead to model over-attention to neighbors of v_r, causing overfitting to these neighbor nodes.

Given the above issues, we adopt a more refined approach. Based on node labels derived from self-supervised learning in community subgraphs, connections are established between newly generated synthetic nodes and unlabeled nodes inferred to belong to the same class, thus integrating these synthetic nodes into the graph structure. This method not only avoids noise transmission and overfitting problems but also enhances model robustness and accuracy in handling class-imbalanced graph data.

Based on the gap between currently labeled nodes and the density ratio dr, each synthetically generated node is assigned the same number of connections to establish edges with nodes of the same class in the current community. For example, if dr is set to 3 and the degree of labeled nodes (marked nodes) in the original subgraph is 2, pseudo-labels for nodes in the subgraph are inferred through self-supervised learning to select nodes of the same class as synthetically generated nodes for edge establishment. This ensures that the average degree of labeled nodes in the optimized subgraph (now including nodes with real labels and generated synthetic nodes) increases to 3, meeting hyperparameter requirements, and edge generation stops.

4 Data-Enhanced Mini-Batch Training

SGDA generates synthetic labeled nodes and edges through fitting to ensure topological balance and reduce the loss of label information propagation. However, these synthetic edges exacerbate the phenomenon of redundant node computation during mini-batch training. On the Reddit dataset, when the mini-batch training batch size is 4096, the redundant computation in one iteration increases by nearly 10 times compared to full-batch training. To address this, we include all labeled nodes in each community subgraph as training nodes in the same batch, reducing redundant computation caused by nodes being sampled into subgraphs of multiple batches.

Specifically, when more than 80% of the labeled nodes in the current community belong to the same class, the community is considered to have a dominant community structure for that class. During training, subgraphs are classified according to their dominant categories. For categories with fewer subgraphs, two situations exist: **(1)** nodes of this class belong to the majority class, with consistent and clustered topological positions in the graph; **(2)** nodes of this class belong to the minority class, and due to the small number of nodes, multiple structures cannot be clearly captured in the graph. Regardless of which situation the class nodes belong to, subgraphs of that class are repeatedly sampled as mini-batches to participate in model parameter updates. By ensuring balanced participation of dominant community subgraphs of each class in model training, we implicitly guarantee balanced influence on GNN models under class distribution imbalance. When the average class accuracy on the training set shows no significant improvement after 10 consecutive iterations, the batch construction mode is adjusted, transitioning from the first stage to the second stage. The implementation of this training strategy requires continuous monitoring of model responses and training results to ensure efficiency and accuracy.

5 Experimental Evaluation

Experimental Setup. The experiments are conducted on a server equipped with an Intel Xeon Silver 4110 CPU and an NVIDIA GeForce RTX 2060 Ti GPU, featuring 64GB of RAM and 6GB of GPU memory. The system runs Ubuntu 20.04 with CUDA 11.3. We implement SGDA in Python.

Competitors. We compare SGDA with Reweight [18], GraphSMOTE [20], and ReNode [3]. Reweight is a cost-sensitive approach that assigns higher loss weights to minority samples to mitigate the dominance of samples of the majority class in the loss function. GraphSMOTE samples the minority class by generating new nodes and edges. ReNode reduces the impact of topological imbalance caused by labeled nodes by adjusting the training weights of labeled nodes. We implement these approaches on three commonly used GNN algorithms (GCN [7], GraphSAGE [6], GAT [14]) to compare their node classification performance in semi-supervised training tasks.

Datasets. We evaluate SGDA on five datasets with varying scales, as detailed in Table 1.

Table 1. Dataset Description

Attribute	Cora [10]	Citeseer [5]	PubMed	Reddit	Ogbn-products
# Nodes	2,708	3,327	19,717	232,965	2,449,029
# Edges	10,556	4,732	44,338	114,848,857	126,167,053
Feature Dimension	1,433	3,703	500	602	100
# Classes	7	6	3	41	47
Average Degree	3.89	3.22	4.49	492.99	51.52

5.1 Single Minority Class

We modify the number of labeled nodes in the datasets to ensure there is only one minority class (L0), with the majority classes (L1–L6) having 10 times more labeled nodes than the minority class. The label distribution percentages in the training samples of the Cora dataset are shown in the Cora(S) row of Table 2. We use the Matthews Correlation Coefficient (MCC) as the evaluation metric to analyze the performance of various models under graph imbalance scenarios with a single minority class.

Table 2. Overview of Label Distribution Percentages (%)

Dataset	L0	L1	L2	L3	L4	L5	L6
Cora (S)	1.64%	16.39%	16.39%	16.39%	16.39%	16.39%	16.39%
Cora (M)	29.41%	29.41%	29.41%	2.94%	2.94%	2.94%	2.94%
Citeseer (M)	30.30%	30.30%	30.30%	3.03%	3.03%	3.03%	–
Pumbed (M)	83.33%	8.33%	8.33%	–	–	–	–

As shown in Fig. 6, the graph data augmented by SGDA achieves the best performance across all three GNN algorithms. When L4 and L5 are designated as single minority classes, SGDA shows significant performance improvements. This is because nodes in classes L4 and L5 are located in relatively sparse neighborhood regions with similar neighbor classes, enabling SGDA to generate synthetic node features through neighborhood generation that more closely approximate the actual node features of these classes in the real graph.

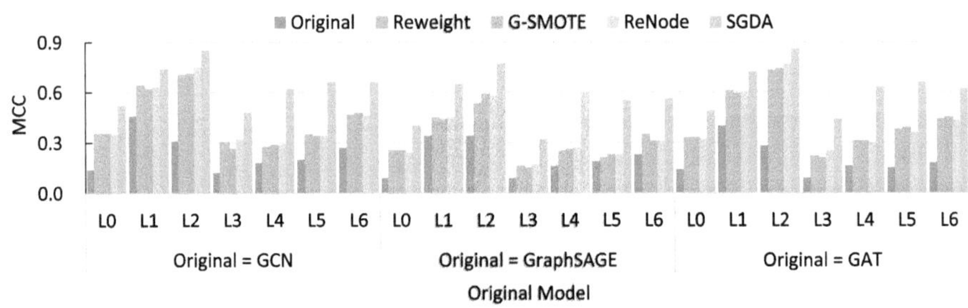

Fig. 6. Comparison for a single minority class on the Cora dataset.

5.2 Multiple Minority Classes

In this experiment, we select GCN as the original model. We designate half of the classes as majority classes and the other half as minority classes, as shown in rows 2–4 of Table 2. The majority classes have 10 times more labeled nodes than the minority classes. We use balanced accuracy (bACC) as the evaluation metric. The bACC computes the average recall across all classes, ensuring equal importance for each class in the overall evaluation.

As shown in Fig. 7a, the data optimized by SGDA achieves significantly better accuracy after training compared to other methods. This is because the SGDA model considers neighborhood information when generating nodes and edges, utilizing graph information more effectively to guide graph data augmentation compared to other approaches. Therefore, even when the number of representative nodes available for expanding training samples is insufficient, the SGDA model can still generate reasonable data based on neighborhood information, thus ensuring the accuracy of model fitting.

5.3 Mini-Batch Training

For the Reddit and Ogbn-products datasets, we construct imbalanced graph data by designating half of the classes as majority classes and the remaining classes as minority classes. We compare the bACC obtained by SGDA mini-batch training with the GCN model, as shown in Fig. 7b. The improvement in balanced accuracy is more pronounced on the Ogbn-products dataset. This is because its topology is relatively sparse, allowing SGDA to achieve better results through data augmentation by fitting nodes and edges.

We also compare the training acceleration performance brought by the deduplication technique, as shown in Fig. 7c. Due to the dense structural characteristics of the Reddit dataset, the acceleration brought by the deduplication technique is relatively significant.

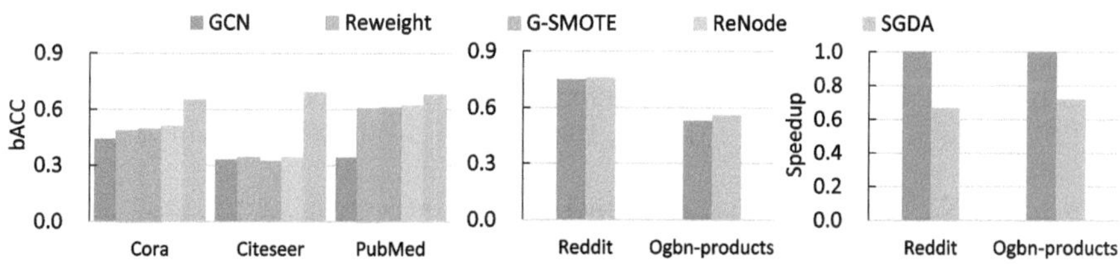

(a) Performance comparison of multiple minority classes **(b) Balanced accuracy in mini-batch training** **(c) Speedup in mini-batch training**

Fig. 7. Performance comparison on multiple minority classes and mini-batch training.

6 Conclusion

This paper presents SGDA, a self-supervised graph data augmentation for addressing class imbalance in large-scale graph learning. By employing graph partitioning to identify nodes requiring augmentation and generating synthetic nodes with topologically-aware edge connections, SGDA ensures numerical and structural balance in imbalanced graph datasets. Additionally, SGDA introduces data-enhanced mini-batch training that maintains class balance while reducing computational overhead. Experimental results show that SGDA achieves up to 0.38 points improvement in balanced accuracy compared to GraphSMOTE and ReNode, while achieving 1.5× speedup in mini-batch training.

Acknowledgments. This work is supported by the National Natural Science Foundation of China (62461146205), the Distinguished Youth Foundation of Liaoning Province (2024021148-JH3/501), and the Fundamental Research Funds for the Central Universities (N25ZLL045).

References

1. Cao, K., Wei, C., Gaidon, A., Arechiga, N., Ma, T.: Learning Imbalanced Datasets with Label-Distribution-Aware Margin Loss. Curran Associates Inc., Red Hook (2019)
2. Chawla, N.V., Bowyer, K.W., Hall, L.O., Kegelmeyer, W.P.: SMOTE: synthetic minority over-sampling technique. J. Artif. Int. Res. **16**(1), 321–357 (2002)
3. Chen, D., et al.: Topology-imbalance learning for semi-supervised node classification. In: Proceedings of the 35th International Conference on Neural Information Processing Systems, NIPS 2021. Curran Associates Inc., Red Hook (2021)
4. Dong, Q., Gong, S., Zhu, X.: Imbalanced deep learning by minority class incremental rectification. IEEE Trans. Pattern Anal. Mach. Intell. **41**, 1367–1381 (2018). https://api.semanticscholar.org/CorpusID:13752387
5. Giles, C.L., Bollacker, K.D., Lawrence, S.: CiteSeer: an automatic citation indexing system. In: Proceedings of the Third ACM Conference on Digital Libraries, DL 1998, pp. 89–98. Association for Computing Machinery, New York (1998). https://doi.org/10.1145/276675.276685
6. Hamilton, W.L., Ying, R., Leskovec, J.: Inductive representation learning on large graphs. In: Proceedings of the 31st International Conference on Neural Information Processing Systems, NIPS 1017, pp. 1025–1035. Curran Associates Inc., Red Hook (2017)
7. Kipf, T.N., Welling, M.: Semi-supervised classification with graph convolutional networks. In: International Conference on Learning Representations (2017). https://openreview.net/forum?id=SJU4ayYgl
8. Liu, Z., Li, Y., Chen, N., Wang, Q., Hooi, B., He, B.: A survey of imbalanced learning on graphs: problems, techniques, and future directions. IEEE Trans. Knowl. Data Eng. **37**(6), 3132–3152 (2025). https://doi.org/10.1109/TKDE.2025.3549299
9. Mao, Z., Ju, W., Qin, Y., Luo, X., Zhang, M.: RAHNet: retrieval augmented hybrid network for long-tailed graph classification. In: Proceedings of the 31st ACM International Conference on Multimedia, MM 2023, pp. 3817–3826. Association for Computing Machinery, New York (2023). https://doi.org/10.1145/3581783.3612360
10. McCallum, A.: Cora (2024). https://doi.org/10.21227/jsg4-wp31
11. Park, J., Song, J., Yang, E.: GraphENS: neighbor-aware ego network synthesis for class-imbalanced node classification. In: International Conference on Learning Representations (2022). https://openreview.net/forum?id=MXEl7i-iru

12. Qu, L., Zhu, H., Zheng, R., Shi, Y., Yin, H.: ImGAGN: imbalanced network embedding via generative adversarial graph networks. In: Proceedings of the 27th ACM SIGKDD Conference on Knowledge Discovery & Data Mining, KDD 2021, pp. 1390–1398. Association for Computing Machinery, New York (2021). https://doi.org/10.1145/3447548.3467334
13. Song, J., Park, J., Yang, E.: TAM: topology-aware margin loss for class-imbalanced node classification. In: Proceedings of the 39th International Conference on Machine Learning. Proceedings of Machine Learning Research, vol. 162, pp. 20369–20383. PMLR (2022)
14. Veličković, P., Cucurull, G., Casanova, A., Romero, A., Liò, P., Bengio, Y.: Graph attention networks. In: International Conference on Learning Representations (2018). https://openreview.net/forum?id=rJXMpikCZ
15. Wang, Y., Zhao, Y., Shah, N., Derr, T.: Imbalanced graph classification via graph-of-graph neural networks. In: Proceedings of the 31st ACM International Conference on Information & Knowledge Management, CIKM 2022, pp. 2067–2076 (2022)
16. Wu, L., Xia, J., Gao, Z., Lin, H., Tan, C., Li, S.Z.: GraphMixup: improving class-imbalanced node classification by reinforcement mixup and self-supervised context prediction. In: Machine Learning and Knowledge Discovery in Databases: European Conference, ECML PKDD 2022, Grenoble, France, 19–23 September 2022, Proceedings, Part IV, pp. 519–535. Springer, Heidelberg (2022). https://doi.org/10.1007/978-3-031-26412-2_32
17. Yi, S.Y., et al.: Towards long-tailed recognition for graph classification via collaborative experts. IEEE Trans. Big Data **9**(6), 1683–1696 (2023). https://doi.org/10.1109/TBDATA.2023.3313029
18. Yuan, B., Ma, X.: Sampling + reweighting: boosting the performance of AdaBoost on imbalanced datasets. In: The 2012 International Joint Conference on Neural Networks (IJCNN), pp. 1–6 (2012https://doi.org/10.1109/IJCNN.2012.6252738
19. Zhang, H., Cisse, M., Dauphin, Y.N., Lopez-Paz, D.: Mixup: beyond empirical risk minimization. In: International Conference on Learning Representations (2018)
20. Zhao, T., Zhang, X., Wang, S.: GraphSMOTE: imbalanced node classification on graphs with graph neural networks. In: Proceedings of the 14th ACM International Conference on Web Search and Data Mining, WSDM 2021, pp. 833–841. Association for Computing Machinery, New York (2021). https://doi.org/10.1145/3437963.3441720

zkANN: Efficient and Verifiable ANN Search with Enhanced Data Privacy

Yang Liu, Linman Li, and Jinchuan Chen(✉)

School of Information, Renmin University of China, Beijing 100872, China
{2019200479,llifem,jcchen}@ruc.edu.cn

Abstract. In recent years, graph-based indexing has emerged as a prominent technique for supporting Approximate Nearest Neighbor Search (ANNS) on massive multimedia datasets. When this data is outsourced to cloud servers, providing verifiable ANNS (vANNS) services presents new challenges. Existing methods typically store intermediate results in Verification Object (VO), enabling users to reproduce the query process and verify the authenticity of the results. However, these intermediate results inadvertently expose excessive data to users, raising privacy concerns. Furthermore, as datasets scale, the VO size grows rapidly, severely impacting efficiency. In this paper, we propose ***zkANN***, a system that combines Merkle trees and zero-knowledge proofs (ZKP) to achieve high-performance query processing while preventing data leakage—a critical capability absent in state-of-the-art (SOTA) methods. Specifically, we employ zk-SNARKs to generate proofs for essential steps in VO construction, eliminating the exposure of raw data. We further propose a monotonicity-aware VO compression technique to enhance efficiency. Experimental results demonstrate that while zkANN incurs additional overhead to ensure data confidentiality, it still outperforms SOTA methods. Without confidentiality constraints, our method achieves 10–100× acceleration.

Keywords: ANNS · Verifiable Query · Zero Knowledge Proof

1 Introduction

Graph data is increasingly essential across various real-world applications, including social networks, recommendation systems, and knowledge graphs. Furthermore, graph-based indexing methods have become popular techniques for supporting ANNS on large-scale multimedia datasets. Well-known examples of such graph-based ANNS indexes include Hierarchical Navigable Small World (HNSW) [4], Hierarchical-Clustering-based Nearest Neighbor Graph (HCNNG) [6], and Locality-Sensitive Hashing-based Approximate Proximity Graphs (LSH-APG) [10], all of which are widely used in multimedia data retrieval. Compared to exact nearest neighbor search (ENNS), ANNS can significantly speed up query processing while maintaining acceptable accuracy [8].

However, as cloud-based graph platforms become more common, verifying the correctness of ANN search results on outsourced data has become increasingly

J. Li et al. (Eds.): APWeb-WAIM 2025, CCIS 2660, pp. 248–260, 2026.
https://doi.org/10.1007/978-981-95-7996-9_22

important. Since users may not fully trust third-party or federated services, ensuring that returned neighbors are correct or sufficiently accurate has emerged as a critical challenge.

Traditional authenticated data structures (ADSs), including Merkle Hash Trees [5], Merkle B-Trees [2], and Merkle R-Trees [9], support verifiable exact queries but cannot handle the inherent uncertainty in ANNS. For example, in ENNS, one can prove that no closer object exists, but this guarantee does not hold in ANNS queries where legal results may exclude some true neighbors.

To address this, recent efforts such as ImageProof [1], ANNProof [3], and GuidedTuple [7] introduce vANNS mechanisms. These systems store graph traversal paths and accessed nodes during query processing in a VO, allowing the user to re-execute the search algorithm and validate results. However, these methods tend to expose significant intermediate data, violating the principle of minimal disclosure and causing large VO size and communication overhead.

Our key insight is that the correctness of ANNS results can be verified without disclosing the entire search path or accessed nodes–by leveraging modern zero-knowledge proof (ZKP) techniques. In this paper, we propose **zkANN**, a privacy-preserving verifiable ANNS framework tailored for graph-derived data. Built atop LSH-APG [10], our system supports efficient query authentication with:

- **Minimal disclosure:** No access to intermediate graph nodes or raw data is required;
- **Compact VOs:** Only the final result and a ZKP are shared;
- **Verifiable correctness:** Clients can independently verify that the result was derived from a specific algorithm on a committed dataset.

2 Problem Formalization and System Overview

Problem Formalization. Given a query object q, an integer $k(\geq 1)$, a dataset $\mathcal{D} = \{o_1, \cdots, o_n\}$ and a set of trusted settings Θ, a verifiable approximate nearest neighbor search (vANNS) returns a result set $\mathbf{R}$ and a verification object ($\boldsymbol{VO}$), such that $\mathbf{R}$ is an approximate kNN set of q and the authenticity of $\mathbf{R}$ can be verified based on $\boldsymbol{VO}$ and Θ.

Safety requirements:

- **Authenticity.** Executing the ANNS algorithm declared by the query service provider on $\mathcal{D}$ guarantees the deterministic generation of result $\mathbf{R}$.
- **Zero Data Leakage.** The query response contains only the result set $\mathbf{R}$ and excludes all other data objects from $\mathcal{D}$.

Other service properties:

- **Efficiency.** Executing a verifiable query incurs not only the cost of query evaluation itself, but also the overhead of constructing and transmitting of VOs, with the latter typically dominating the overall time cost.

- **Error Guarantees.** As an approximate query processing algorithm, the vANNS service provider should give formal error guarantees for its results.

Overview of zkANN. zkANN adopts LSH-APG [10] as its underlying ANNS framework. This selection is motivated by LSH-APG's ability to perform efficient ANN search while offering formal accuracy guarantees.The workflow of zkANN consists of five steps (Fig. 1):

- **1: Data Upload and Root Publication.** The data owner uploads the dataset along with three ADSs: the Merkle LSB-Trees ($MLSBTs$), the MHT over $\mathcal{D}$ ($MHT_{\mathcal{D}}$), and the MHT over the APG (MHT_{APG}). Additionally, the owner publishes: (i) the root digests of all ADSs and (ii) the LSH parameters (e.g., hash functions, z-order function).
- **2: zk-SNARK Parameter Setup.** A trusted third party (TTP, e.g., a designated authority) performs the zk-SNARK circuit setup. The verification key bound to this circuit is published and made publicly available through a secure and authenticated channel to ensure transparency and trust. Note that the involvement of a TTP to generate the circuit constitutes the standard approach in zk-SNARKs [11].
- **3: Query Request.** A user issues a query to the zkANN server. The server initiates the LSH-APG search using the authenticated indexes.
- **4: Result and VO Return.** Upon completing the search, the server returns the result set $\mathbf{R}$ along with a VO.
- **5: Verification.** Client verifies the result's integrity using the root digests and the zk-SNARK parameters through the trusted channel.

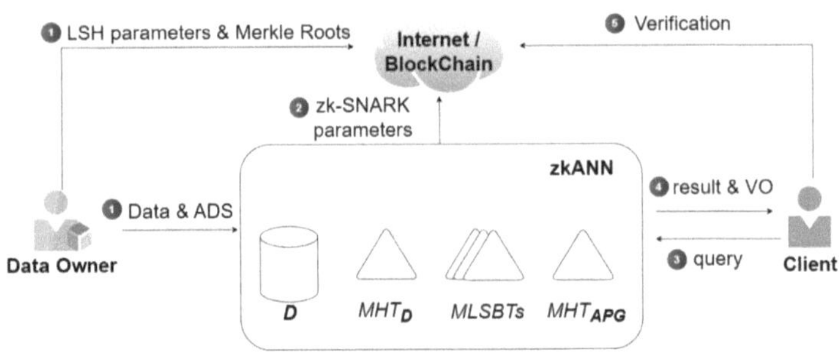

Fig. 1. Workflow of zkANN

3 Verifiable Query Processing

The query processing in zkANN consists of two main steps. First, a set of *entry points* is retrieved via LSH lookup (Sect. 3.1). Subsequently, these entry points are used to iteratively refine the search on an APG, progressively converging toward the query object (Sect. 3.2).

3.1 Generating Entry Points with LSH

The LSH component enables efficient dimensionality reduction and verifiable selection of candidate entry points.

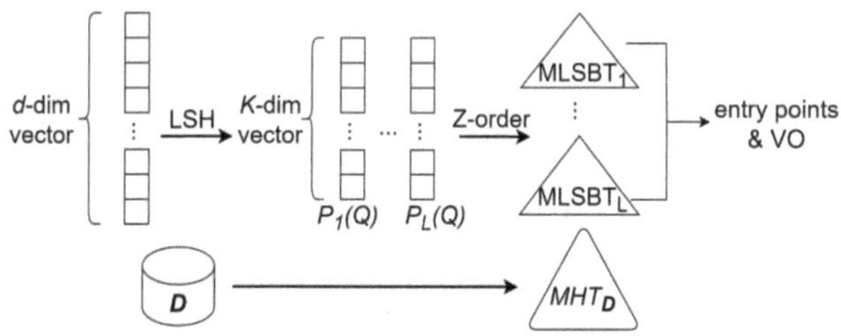

Fig. 2. ADS for LSH

As illustrated in Fig. 2, each data point o is *projected to a hashed value vector* by a sequence of K locality-sensitive hash functions. When evaluating a vANNS, we need first compute the LSH projections of the query object q and conduct nearest neighbor queries over L LSB-trees. The result is a set of pairs $< ID, o >$ which will serve as *entry points* when searching over APG in the next phase. To support integrity verification, we extend the traditional LSB-Tree structure by incorporating MBT-based authentication. Specifically, we transform each LSB tree into an $MLSBT$ by embedding a series of hash values, which are then included in the VO .

During LSB retrieval, when an object o is potentially selected as an entry point, it is necessary to compute the raw distance $||o, q||$. To provide clients with cryptographic proof of this distance, we construct a MHT over $\mathcal{D}$, called $MHT_{\mathcal{D}}$. It serves to verify the integrity of the raw data point o, thereby ensuring the correctness of $||o, q||$. We will revisit this issue and illustrate the whole process of query evaluation and VO generation in Sect. 3.3.

3.2 Searching kNN with APG

The Approximate Proximity Graph (APG) is a graph index structure to support efficient nearest neighbor search. Each node of APG corresponds to a data object o, storing both the raw data and projected data (through LSH functions) of o. During entry point selection, the APG first computes approximate distances using projected data for fast pruning. The algorithm only computes exact distances with raw data for objects that cannot be pruned by this approximation.

The retrieval process in APG starts with a candidate set initialized by LSH-generated entry points, then iteratively: (1) Expand candidates by adding their APG neighbors, (2) Reorder candidates by their distance to q, (3) Prune the set to retain only top-k candidates. This process terminates when no further updates occur.

For cryptographic verification of query result authenticity, we implement a MHT (denoted by MHT_{APG}) built upon the APG. Each leaf node corresponds to an APG node u and stores the hash values of both u and its neighboring nodes' projected data.

During APG retrieval, the VO must contain: (1) All accessed entry points and their neighboring vertices, along with their authenticity proofs derived from MHT_{APG}; (2) The raw data of some data objects which must be pruned by exact distances and their authenticity proofs obtained from $MHT_{\mathcal{D}}$.

3.3 Query Evaluation and Result Verification

Algorithm 1: ANNS Procedure and VO Generation

Input: q, k, $\mathcal{D}$, a set of $MLSBT$s, $MHT_{\mathcal{D}}$, MHT_{APG}
Output: query result and VOs

1 initialize **E**, VO_B, VO_H and VO_G for storing query results and VOs generated with $MLSBTs$, $MHT_{\mathcal{D}}$ and MTH_{APG} respectively;
2 **foreach** $MLSBT$ **do**
3   find the leaf nodes u covering q's search range;
4     $\pi_u \leftarrow calProof(u, MLSBT)$;
5   put $<u,\pi>$ into VO_B;
6   **foreach** $ID \in u$ **do**
7         $o \leftarrow retrieve(ID, \mathcal{D})$;
8         $\pi_o \leftarrow calProof(o, MHT_{\mathcal{D}})$;
9   put $<o, \pi>$ into VO_H;
10   put $<\text{ID}, ||o, q|| >$ into **E**;
11 Prune **E** by preserving only the k-nearest neighbors of q;
12 **while** ***true*** **do**
13     $e \leftarrow pop(\mathbf{E})$;
14     $S \leftarrow$ retrieve e and its neighbors from APG;
15     $\pi_S \leftarrow calProof(S, MHT_{APG})$;
16   put $< S, \pi_S>$ into VO_G;
17   put S into **E**;
18   Prune **E** and break if **E** remains unchanged;
19 **return** $\mathbf{E}, VO_B, VO_H, VO_G$;

Algorithm 1 presents the complete procedure for evaluating a vANNS while constructing VOs. The algorithm begins by identifying candidate points through multiple $MLSBTs$ (Lines 2–6). For each $MLSBT$, it locates the leaf node(s) that cover the query range and calculates corresponding proofs. Then, for each candidate entry point, the algorithm accesses the associated object in the dataset and generates its proof from $MHT_{\mathcal{D}}$ (Line 5).

After obtaining an initial candidate set **E**, the algorithm prunes **E** to retain only the k nearest neighbors (Line 7). It then enters an iterative refinement phase based on the APG (Lines 8–13). In each iteration, the top-ranked element in **E** is expanded by its neighbors in the APG. A proof covering this local neighborhood is generated from MHT_{APG}. The newly revealed neighbors are inserted into **E**, which is subsequently pruned. The process terminates once **E** stabilizes (line 18)–i.e., the current top-k set remains unchanged over consecutive iterations.

Result Verification. The process of verifying ANNS results using VO consists of two steps. First, the verifier recalculates the root digests and compare them with the trusted digests to ensure integrity. Next, it is required to compute the query results via the LSH-APG algorithm and verify them against the server-provided result **R**.

The objective of the first step is to verify the authenticity of intermediate results. These include:

- The leaf nodes containing query objects during LSB retrieval.
- The entry points and their neighbors in each iteration of APG traversal.
- The necessary raw data objects retrieved when searching LSH and APG.

Once integrity is confirmed, the verifier reconstructs the initial candidate set using the data stored in VO_B and VO_H. Subsequently, the verifier attempts to reproduce the multi-iteration k-nearest neighbor search process on APG using intermediate results stored in VO_G. Finally, the verifier compares the locally reconstructed result set with **R** to make the final judgment on the query's correctness.

Discussions

Soundness and Completeness. Our framework ensures **soundness** by verifying that all returned points are part of the raw dataset via VO_H, and by reconstructing the search path using VO_B and VO_G to eliminate incorrect results. **Completeness** is ensured by verifying that the reconstructed results exactly match the expected k points. By reproducing the entire query process, we guarantee both the correctness of results and the integrity of the search procedure.

VO Size Analysis. The VO consists of three components: VO_B, VO_H, and VO_G. VO_B depends on the number of hash tables L and the height of each MLSB-Tree. Since L is typically a small constant and each Merkle path has a height of $O(\log n)$.

VO_H is determined by the number of search iterations m (which is about $k \cdot \log n$) and the data dimensionality d. Hence the space complexity of VO_H is about $O(k \cdot d \cdot \log n)$.

VO_G consists of the projected representations of the m visited anchor points along with their T neighbors in a K-dimensional space, resulting in a total complexity of $O(T \cdot K \cdot \log n)$.

Data Confidentiality. One major limitation of the **Baseline** scheme is its inherent exposure of raw data during verification. The client must access the raw data records in order to recompute distances and confirm result quality. As a consequence, many data points that are not part of the final result set are disclosed.Such privacy risks motivate our adoption of zero-knowledge distance proofs and structure-aware pruning in the enhanced schemes, which effectively reduce unnecessary data disclosure while preserving verifiability.

4 zk-Secured Data Confidentiality Protection and Monotonicity-Aware VO Compression

4.1 Confidentiality Preservation with ZKP

The Baseline approach (Sect. 3) directly expose the raw data in VO_H, which not only increases communication overhead but also compromises user privacy. To address this, we adopt **zk-SNARKs** to validate distance computations without revealing the actual data.

As illustrated in Fig. 3, traditional verification approaches require the server to disclose raw data o_i to the client, who then recomputes the distance $|o_i, q|$, thus risking data leakage. In contrast, **zkANN** allows the server to compute the distance $d_i = |o_i, q|$ and generate a zero-knowledge proof π to attest its correctness, without revealing o_i. This process relies on a zk-SNARK system with a trusted setup phase that generates a proving key (used by the server) and a verification key (shared with clients). To uphold the security guarantees of zkANN, it is crucial that any intermediate artifacts from the setup–commonly referred to as toxic waste–are securely destroyed and never retained after the setup phase.

Efficiency Considerations. Proof generation incurs a complexity of $O(mT)$, where m is the number of search iterations and T is the average node degree. Fully covering the search process with ZKPs is computationally prohibitive for real-time use.

Selective ZK Application. To ensure a balance between privacy and performance, zk-SNARKs are applied only to the distance verification component. Other elements (e.g., Merkle paths, LSH hashes, graph IDs) are non-sensitive and remain outside the circuit. This targeted use ensures confidentiality where most needed without incurring prohibitive overhead.

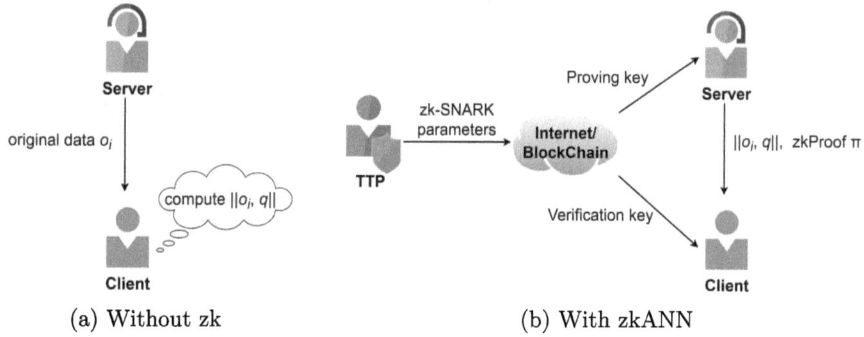

Fig. 3. Comparison of distance verification: (a) Traditional method requiring data disclosure; (b) zkANN approach preserving data confidentiality.

4.2 Monotonic Convergence for VO and ZK Cost Reduction

To further reduce the VO size and the associated computational overhead of zk-SNARKs, we propose a verification strategy based on the *monotonic convergence* property of the search process over the APG. Instead of including *all neighbors and distances* during graph traversal, we demonstrate that verifying *the statistics of an entry point and its neighbors* suffices to guarantee a valid (c, k)-ANN result.

Modifications to ADS. We modify MHT_{APG} to record, for each entry point ep_i:

- w_i: the centroid of the neighbors of ep_i;
- $\bar{r}_i$: the average distance between ep_i and its neighbors.

The hash digest at each leaf node is computed as:

$$H_{\text{Leaf}} = H(P(\text{ep}_i) \mid w_i \mid \bar{r}_i \mid H(N(ep_i))),$$

where $P(\cdot)$ is the projection function, $N(ep_i)$ represents a set containing the IDs of all neighbors that are connected to ep_i.

VO_G Design. Instead of including all neighbors, each VO_G record contains:

$$(P(\text{ep}_i),\ w_i,\ \bar{r}_i,\ H(N(ep_i))),$$

enabling the verifier to compute $\|q, \text{ep}_i\|$ and $\|q, w_i\|$. Only the last VO_G record needs to include $N(ep_i))$ instead of $H(N(ep_i))$ in order to check the convergence condition.

Theorem 1. *Given a query object q, result set $\mathbf{R}$, VO_G, and ZKPs $\{\pi_i\}_{i=0}^m$. The Verifier can be assured that $\mathbf{R}$ satisfies the (c, k)-ANN condition for q, if the provided VO_G sequence satisfies the following conditions:*

1. ***Non-triviality:*** $\forall i \in [0, m], \|q, ep_i\| > \bar{r}_i$;
2. ***Progress:*** $\forall i \in [0, m), \|q, ep_{i+1}\| < \|q, ep_i\|$;

3. ***Sufficiency:*** $\forall i \in [0, m), \|q, ep_{i+1}\| < \|q, w_i\|$;
4. ***Termination:*** $\|q, ep_m\| < (1+\gamma)\bar{r}_m$ *for some constant* $\gamma < 1$, *and every neighbor of* ep_m *lies farther from* q *than all objects in* $\mathbf{R}$.

Proof Sketch. According to LSH-APG [10], assuming the neighborhood is uniformly distributed, given that $\|q, \text{ep}_i\| > \bar{r}_i$ (*Non-triviality*), $\|q, \text{ep}_{i+1}\| < \|q, \text{ep}_i\|$(*Progress*), and a sufficient number of neighbors are explored, the convergence of the query iteration process can be rigorously guaranteed, i.e. there exists a finite number $i > 1$ such that the inequality $\|q, \text{ep}_i\| \leq (1+\gamma)\bar{r}_i$ must hold.

Through the VO_G and $\{\pi_i\}_{i=0}^m$, the verifier (Client) can rigorously confirm that both *Non-triviality* and *Progress* conditions are maintained throughout query execution. By further verifying the *Termination* condition, the verifier obtains cryptographic assurance that: (1) the query has converged within the (c, k)-ANN error bounds, and (2) no further updates to the entry point set are possible. This protocol thus enables verifiable authenticity of the result set $\mathbf{R}$.

The soundness of our optimization also relies on ensuring that a *sufficient number of neighbors* are explored during APG traversal. Under the assumption that neighbors are uniformly distributed around the entry point ep_i, the inequality $||q, \text{ep}_{i+1}|| < ||q, w_i||$ (the *Sufficiency* condition) statistically implies that over half of the neighbors are expected to lie closer to q than ep_i. Thus, by configuring a sufficiently large degree parameter during APG construction, we can guarantee satisfaction of the aforementioned conditions. This ensures that enough neighbors are present such that the traversal can probabilistically identify a closer entry point (thus enabling monotonic progress) and satisfy the convergence guarantee. □

Efficiency Improvement via Monotonic Convergence. By applying the monotonic convergence optimisation, the zk-SNARK proof generation complexity is reduced from $O(mT)$ (as analysed in Sect. 4.1) to $O(m)$, since proofs are required only for m selected distances rather than all mT neighbor comparisons. This yields a T-fold reduction in overhead and substantially improves the practicality of the protocol for real-time deployment.

5 Experimental Study

5.1 Setup

Hardware and Software Environment. All experiments are conducted on a server running Ubuntu 22.04 with an AMD Ryzen CPU (3.8 GHZ) and 128 GB RAM. All codes are implemented with C++ and compiled using g++ version 11.4.0. The zk-SNARK components are constructed using the `libsnark`[1] library. The queries are issued from another PC to evaluate the communication overhead.

Compared Methods. We compare zkANN against two SOTA vANNS systems, i.e. ANNProof [3] and GuidedTuple [7].

[1] https://github.com/scipr-lab/libsnark.

Internal Baselines. To investigate the impact of different verification mechanisms within our framework, we evaluate the following internal variants:

- **Baseline (Sect.** 3**):** The core framework without additional optimization.
- **Baseline + ZK (Sect.** 4.1**):** This version augments the baseline with ZKP.
- **Baseline + Mono (Sect.** 4.2**):** This version augments the baseline by leveraging the monotonic convergence property to reduce VO size.
- **zkANN:** The complete implementation of our framework.

Parameter Settings. The default query setting is to retrieve approximate 10 nearest neighbors, i.e. $k = 10$, $n = 10^5$. For the parameters related to LSH-APG and the compared SOTA methods, we adopt the settings from their papers.

Dataset. We conduct experiments on three widely used real-world datasets for high-dimensional nearest neighbor search: **MNIST**, **SIFT1M**, and **Deep1M** (Table 1).

Table 1. Summary of Datasets Used in Our Experiments

Datasets	Cardinality	Dim.	Datatype	Size (MB)
MNIST	60,000	784	uint8	184
SIFT1M	1,000,000	128	float32	493
Deep1M	1,000,000	256	float32	981

5.2 Evaluation Results

Experiment 1: ADS Construction

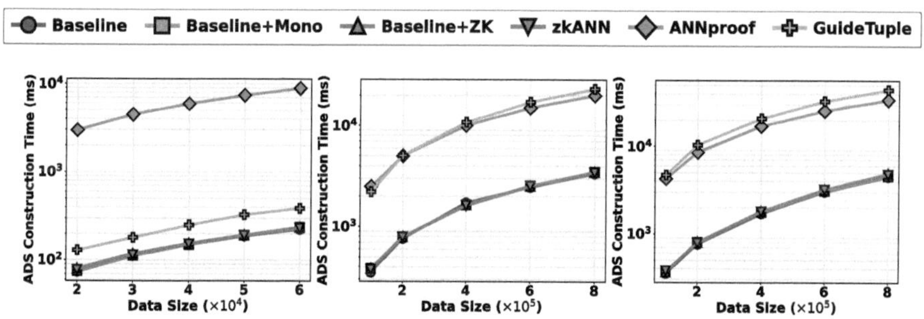

(a) MNIST (b) SIFT1M (c) Deep1M

Fig. 4. ADS construction time

ADS Construction Time. As shown in Fig. 4, our technique outperforms competing methods in ADS construction speed across all datasets. GuidedTuple and ANNProof require 7 to 8 times longer ADS construction time than our approaches, due to their complex clustering and sharding operations.

Experiment 2: Query Efficiency

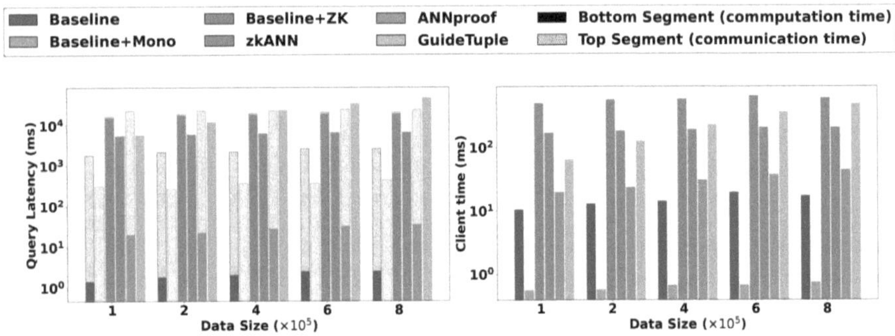

(a) Query Latency (b) Client Time

Fig. 5. Query Efficiency (SIFT1M)

The vANNS query processing pipeline consists of several stages: ANNS query execution, VO generation, VO transmission and result verification. In our evaluation, we report two key performance metrics: **Query Latency**, which includes the time spent on query execution, VO construction, and Communication; and **Client Time**, which refers to the time required by the User to verify the received result.

Query Latency. As shown in Fig. 5(a), zkANN demonstrates the best overall query latency among both SOTA methods. While ANNProof benefits from low VO construction time, its large VO size leads to high transmission overhead. In contrast, GuideTuple produces a compact VO, but incurs significant latency due to expensive VO generation. While zkANN incurs additional overhead to prevent information leakage, it still maintains superior performance compared to the SOTA approaches (which suffer from data exposure risks).

Client Time. As shown in Fig. 5(b), Baseline+Mono maintains the lowest verification time across all datasets, as it avoids distance computation and privacy-preserving validation. Although zkANN introduces ZKP verification, its overhead remains modest, with an overhead of around 100 ms. Our framework ensures privacy-preserving verifiability without compromising too much client-side verification burden.

Experiment 3: Parameter Sensitivity

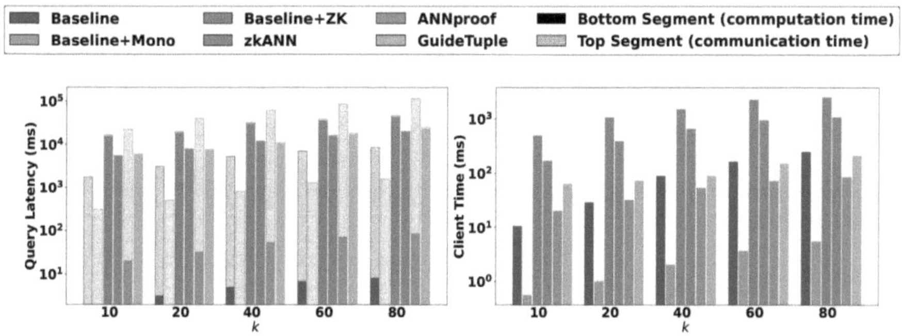

(a) Query Latency (b) Client Time

Fig. 6. Impact of query result size k(SIFT1M)

Impact of k. Figure 6 illustrates the impact of result size k on query latency and Client-side verification time. While all methods incur higher latency as k increases, their growth rates differ. All variants of zkANN demonstrate consistently lower latency than both SOTA methods.

6 Related Works

Graph-Based Approximate Nearest Neighbor Search [8]**.** Graph-based ANNS methods have grown rapidly in recent years, typically relying on structures such as Delaunay Graphs (DG), Relative Neighborhood Graphs (RNG), k-Nearest Neighbor Graphs (KNNG), and Minimum Spanning Trees (MST). DG and RNG provide precise neighborhood guarantees but suffer from exponentially increasing complexity in high-dimensional spaces. In contrast, KNNG and MST offer greater efficiency by simplifying graph construction and storage at a slight accuracy trade-off. Given their superior accuracy in high-precision scenarios, this paper adopts DG and RNG as foundational structures.

Verifiable Nearest Neighbor Search. To ensure result authenticity in outsourced nearest neighbor search, several verifiable frameworks have been developed. In image retrieval, ImageProof [1] combines KD-trees with MHTs and cuckoo filters to balance correctness and VO compactness. Graph-based methods have also been extended. ANNProof [3] authenticates HNSW graphs via MHTs and verifies results by recording search steps and neighbor sets. GuideTuple [7] applies similar principles to HCNNG using prefix trees and MHTs.

These approaches share a common principle: capturing the search path and authenticating index structures is key to verifying ANNS results.

7 Conclusion

This paper introduces zkANN, a high-performance vANNS system designed to minimize data leakage while providing rigorous error bounds for query results. For future work, we will investigate scalability enhancements for efficient handling of multiple concurrent queries. Moreover, we will study trusted collaborative ADS construction between data owners and cloud providers.

Acknowledgments. This work is supported by National Key R&D Program of China (2022YFB2702102).

References

1. Guo, S., Xu, J., Zhang, C., Xu, C., Xiang, T.: ImageProof: enabling authentication for large-scale image retrieval. In: 2019 IEEE ICDE, pp. 1070–1081. IEEE (2019)
2. Li, F., Hadjieleftheriou, M., Kollios, G., Reyzin, L.: Dynamic authenticated index structures for outsourced databases. In: SIGMOD 2006, pp. 121–132. ACM (2006)
3. Lingling, L., Wen, Z., Yuan, Y., He, Q., Chen, J., Liu, Z.: ANNProof: building a verifiable and efficient outsourced approximate nearest neighbor search system on blockchain. Futur. Gener. Comput. Syst. **156**, 206–220 (2024)
4. Malkov, Y.A., Yashunin, D.A.: HNSW. IEEE Trans. Pattern Anal. Mach. Intell. **42**(4), 824–836 (2020)
5. Merkle, R.C.: A digital signature based on a conventional encryption function. In: Conference on the Theory and Application of Cryptographic Techniques, pp. 369–378. Springer (1987)
6. Munoz, J.V., Gonçalves, M.A., Dias, Z., da S Torres, R.: Hierarchical clustering-based graphs for large scale approximate nearest neighbor search. Pattern Recogn. **96**, 106970 (2019)
7. Wang, C., Zhang, J., Liu, X., Wei, K., Feng, B.: Verifiable graph-based approximate nearest neighbor search. In: ADMA, pp. 3–17. Springer (2024)
8. Wang, M., Xu, X., Yue, Q., Wang, Y.: A comprehensive survey and experimental comparison of graph-based approximate nearest neighbor search. arXiv preprint arXiv:2101.12631 (2021)
9. Yang, Y., Papadopoulos, S., Papadias, D., Kollios, G.: Authenticated indexing for outsourced spatial databases. VLDB J. **18**(3), 631–648 (2009)
10. Zhao, X., Tian, Y., Huang, K., Zheng, B., Zhou, X.: Towards efficient index construction and approximate nearest neighbor search in high-dimensional spaces. PVLDB **16**(8), 1979–1991 (2023)
11. Zhou, L., Diro, A., Saini, A., Kaisar, S., Hiep, P.C.: Leveraging zero knowledge proofs for blockchain-based identity sharing: a survey of advancements, challenges and opportunities. J. Inf. Secur. Appl. **80**, 103678 (2024)

Text Classification Method Based on Hypergraph Convolutional Network and Attention Mechanism

Yifeng Xu[1,2], Xiaojie Chu[1], Xiaoli Lin[1,2], Quanxin Li[1], Ying Song[1], Shuang Li[1], and Jun Pang[1,2](✉)

[1] School of Computer Science and Technology, Wuhan University of Science and Technology, Wuhan, China
{xu1feng,chuxiaojie,linxiaoli,202113704098,song_ying,lishuang, pangjun}@wust.edu.cn

[2] Hubei Province Key Laboratory of Intelligent Information Processing and Real-Time Industrial System, Wuhan, China

Abstract. Existing text classification methods based on graphs and hypergraphs have issues such as incomplete capture of text feature information, insufficient expression of multiple relationships, and inadequate attention to important information. Therefore, we propose a text classification model based on hypergraph convolutional networks and attention mechanisms, i.e., HCWA. Firstly, we construct three types of textual hyperedges for the text, including sequential hyperedges, thematic hyperedges, and syntactic hyperedges, to form multi-layered textual expressions and enrich the captured textual feature information. Secondly, we process the data using a hypergraph convolutional network to transmit and aggregate different relational information between nodes, enhancing the understanding of textual information. Then, attention is extracted from the nodes to improve the representation weight of important features, obtaining more critical features within the hypergraph. Finally, multi-channel pooling can capture overall features while preserving important features, reducing the loss of textual information. Extensive experimental results on real datasets verify that HCWA can effectively improve the performance of text classification.

Keywords: knowledge hypergraph · text classification · hypergraph convolutional network · attention mechanism

1 Introduction

Text classification refers to the task of automatically assigning one or multiple category labels to input text using computational algorithms. As an essential task in the field of natural language processing, it has extensive and important applications in sentiment analysis [1], news categorization [2], and question-answering systems [3].

Existing text classification methods are mainly divided into two categories based on the different text feature extraction methods: traditional machine learning-based text classification methods and deep learning-based text classification methods. The first category

J. Li et al. (Eds.): APWeb-WAIM 2025, CCIS 2660, pp. 261–271, 2026.
https://doi.org/10.1007/978-981-95-7996-9_23

uses machine learning techniques such as BOW (bag of words) [4] and TF-IDF (term frequency-inverse document frequency) [5] to represent text as feature vectors, which are then input into classifiers such as Naïve Bayes [6], support vector machine (SVM) [7], and random forests [8] for classification prediction. Due to the typically lengthy feature engineering required by these methods, the design and analysis of text features are difficult to scale to new tasks, resulting in subpar performance in classification.

The second category of methods uses deep embedding models to represent text as feature vectors, replacing manually designed features. This category can be further divided into two types based on the neural network models used: methods based on convolutional neural networks (CNNs) [9], and methods based on graph neural networks (GNNs) [10]. The former extracts local features from text through local perception mechanisms, significantly improving the performance of text mining. Although these models can well capture semantic and syntactic associations from continuous word sequences, they struggle to capture non-continuous and long-range text information. The latter uses graph neural networks to obtain non-continuous word relationships in documents. However, most graph neural network methods focus on pairwise interactions between words. In natural language, word relationships are not always second order. They are often third-order, fourth-order, or higher-order. The learning of binary relationship text representations limits the richness and relevance of text information representation. Ding et al. [12] then transformed text into hypergraphs to deepen understanding and learning of text information. However, their method is insufficient in mining semantic relationships and considering contextual relationships. Li et al. [13] proposed a method that fuses BERT [14] with hypergraph convolutional networks based on attention mechanisms, effectively modeling context semantics and learning deep embedding representations, thereby improving text representation learning performance. However, there are the following limitations to this type of method: (1) The text hyperedge structure is relatively simple, making it difficult to build multi-level text representations. (2) The attention mechanism's ability to focus on key information is insufficient, affecting further improvement in model performance.

Therefore, we propose a new text classification model, called HCWA. Firstly, by constructing three types of text hyperedges, including sequence hyperedges, topic hyperedges, and syntactic hyperedges, a multi-layer text representation is formed, enriching the captured text feature information. Secondly, a hypergraph convolutional network is used to process, propagate, and aggregate different association information between nodes, enhancing the understanding of text information. Then, attention is extracted from nodes to increase the representation weight of important features and obtain more critical features in the hypergraph. Finally, multi-channel pooling can capture overall feature information while retaining important features. The main contributions of this paper are summarized as follows:

1. We propose a text classification model based on hypergraph convolutional networks and attention mechanisms. Three types of text hyperedges is constructed to combine the multi-dimensional structural information of the text for enhancing the richness and hierarchy of text representations.

2. We integrate hypergraph convolutional networks with attention mechanisms to effectively capture complex high-order relationships between nodes. By further optimizing the feature extraction process through multi-channel pooling strategies, we significantly enhance the model's ability to capture key information and classification performance.
3. Extensive experimental results on real datasets verify the effectiveness of the model proposed in this paper.

2 Related Work

In this section, we introduce text classification models, including text classification methods based on traditional machine learning and text classification methods based on deep learning.

2.1 Based on Traditional Machine Learning Methods

In text representation learning, common models include BOW [4], TF-IDF [5], and Word2vec [15]. For example, Word2vec generates word vector representations using local contextual information, primarily consisting of two components: CBOW and Skip-gram. CBOW predicts the middle word based on the surrounding words in the context, while Skip-gram predicts its neighboring words by given words. After obtaining the text representation, the extracted features are input into the classifier for classification, with representative methods including Naive Bayes, k-Nearest Neighbors (KNN) [16], and Support Vector Machines (SVM) [7]. Machine learning models learn from predefined features and can achieve better classification results than deep learning methods when the amount of data is small and computational complexity is constrained. However, feature engineering is complex and time-consuming, typically requiring a significant amount of time and professional knowledge and experience to complete. Therefore, traditional models have more advantages in fields with smaller sample sizes.

2.2 Based on Deep Learning Methods

These methods can learn more potential effective feature representations from training data, thereby achieving better classification performance than traditional methods. Among the research methods relevant to this work are those based on convolutional neural networks, pre-trained models, and graph neural networks.

Convolutional neural network-based methods extract local features from text through local perception mechanisms, offering advantages such as strong parallel computing capabilities and high training efficiency. For example, CNNs [9] can effectively extract local features from text through local perception mechanisms and shared weight structures. A typical approach involves concatenating word vectors in a sentence into a matrix, inputting it into convolutional and pooling layers, and ultimately obtaining a text representation vector. However, CNNs primarily focus on local context and struggle to capture global semantic correlations.

Graph neural networks can model complex structural information in text and have made significant progress in text classification in recent years. TextGCN [11] is one of the representative models in this category. This method constructs a heterogeneous graph containing word and document nodes and uses graph convolution to learn node representations based on word co-occurrence frequency and TF-IDF weighted edges, achieving text classification. TextGCN can effectively capture global corpus relationships, but its fixed and non-updatable edge weights limit representation flexibility, and it also suffers from high computational overhead on large-scale datasets. The IBHC [13] model combines pre-trained model BERT with hypergraph neural networks and introduces an attention mechanism to better capture the deep semantic representation of text context. However, IBHC still faces challenges in constructing hierarchical text structures, and its attention mechanism's ability to focus on key information still has room for optimization. Zheng [17] et al. established text structure hypergraphs and semantic-enhanced hypergraphs, leveraging hypergraph neural networks to propagate and aggregate their features, and fused the dual graph representations through an attention mechanism to enhance the discriminative power of text representations. However, when facing multilingual scenarios or long texts, the scalability and generalization ability of the hypergraph construction strategy in this model still have further room for optimization.

Emerging pre-trained language model methods such as CINO-Large [18] and prompt learning [19] demonstrate strong language understanding and transfer capabilities in low-resource language text classification. CINO-Large, combining a denoising loss function and graph attention mechanisms, proposes a text classification model suitable for Tibetan long texts, significantly enhancing the model's ability to model long-range dependencies and semantic global relationships. Chen [19] et al. propose a model that integrates prompt learning with adaptive loss weighting strategies, reconstructing the classification task as a semantic relevance recognition task, guiding the training model to utilize few-shot data for cross-lingual text semantic alignment and knowledge transfer, particularly suitable for few-shot classification tasks. Although it possesses strong generalization capabilities at the semantic expression level, there is still room for further enhancement in deep modeling of structured information.

3 The Proposed Model

In this section, we will provide a comprehensive description of HCWA. As shown in Fig. 1, the HCWA architecture consists of three basic components: hypergraph convolutional layers, attention extraction layers, and output layers. The HCWA model first constructs three types of hyperedges based on the input text. The sequence hyperedge helps capture dependencies between words in a sentence, the topic hyperedge provides semantic information between words, and the syntactic hyperedge provides information about the text structure, including the grammatical structure of sentences and long-distance dependencies. After processing by the hypergraph convolutional layer, it can more comprehensively consider the relationships between multiple nodes, aggregate hyperedge features and node information, improve the modeling ability of the entire hypergraph, and enhance node feature representation. The attention extraction layer further refines the model's node representation, assigning different weights to different nodes to retain more useful information and obtain a model with higher generalization ability. Finally, through multi-channel pooling in the output layer, it focuses on information at different scales and uses the final text representation for label prediction. The following section will provide a detailed description of each component of the model.

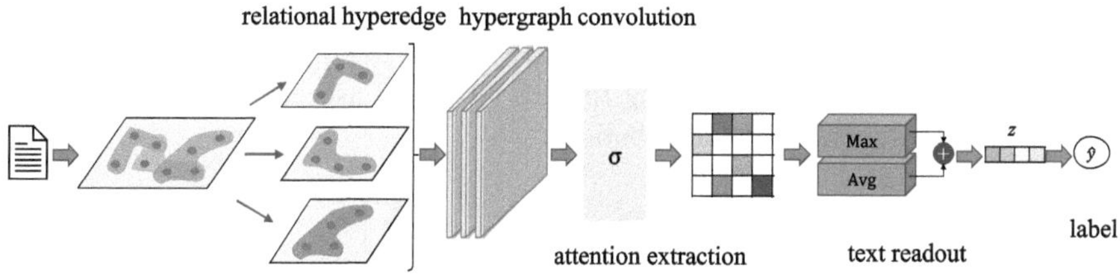

Fig. 1. Overall framework of HCWA.

3.1 Hypergraph Convolutional Layer

To obtain expressive text representations in the hypergraph, we process the hyperedges using hypergraph convolution, utilizing one-hot vectors as the initialization attributes for nodes, as shown in Eq. 1.

$$X^{(l)} = HyperCov(E^{(l-1)}, X^{(l-1)}) \tag{1}$$

where $X^{(l)}$ is the node feature representation of the l-th layer, $HyperCov()$ is the hypergraph convolution operation, and E is the hyperedge representation. Through the "node-hyperedge-node" information propagation process, the node features are aggregated.

3.2 Attention Extraction Layer

After obtaining the node feature representation, the attention method was used to extract the key nodes with salient features. Firstly, the features output by the hypergraph convolution are processed nonlinearly to highlight the expressive ability of the features and deepen the depth of expression, as shown in Eq. 2. Then the attention weight score is calculated by Eq. 3, MLP_2 helps to introduce the contextual features into the hidden layer, adjust the attention to different features, and avoid the complete loss of features, where the σ is the activation function, and finally the output features are obtained by Eq. 4.

$$h' = tanh(MLP_1(X)) \tag{2}$$

$$score = \sigma(MLP_2(X)) \tag{3}$$

$$h = h' \odot score \tag{4}$$

where MLP is a multi-layer perceptron, $\odot$ is the Hadamard product, h is the node feature after the attention layer processing.

3.3 Readout Layer

By using max pooling to make key features contribute more, enhance sensitivity to important information, while avoiding the loss of low significance features to obtain global information, and ensuring each feature plays its own role in text classification, max pooling and average pooling are combined as shown in Eq. 5. Max pooling focuses on significance, while average pooling focuses on global aspects. Their combination better adapts to different feature distributions.

$$z = Maxpool(h_1, ..., h_v) + \frac{1}{|\mathcal{V}|}\sum_{v \in \mathcal{V}} h_v \tag{5}$$

where h_v is the feature representation of node v, and z is the document representation of the output. Finally, *softmax*() is used to classify the labels, with cross-entropy as the loss function, as shown in Eq. 6 and Eq. 7.

$$\hat{y} = softmax(Wz + b) \tag{6}$$

$$\mathcal{L} = -\sum_i y_i \log(\hat{y}_i) \tag{7}$$

where W and b are the weights and biases, respectively, and y_i is the true label.

4 Experiments

This section verifies the text classification capability of the proposed model from multiple perspectives on real datasets.

4.1 Experiment Settings

Datasets. The datasets used include R8, R52, Ohsumed, MR, 20NG, and SST, and the statistical information of the datasets is shown in Table 1.

Table 1 shows the detailed information of these datasets, where #Doc, #Word, and #Class represent the number of texts, unique words, and categories, respectively, and #train and #test represent the number of texts in the training set and test set, respectively. "Ave Len" indicates the average length of the texts.

Experiment Setup. The experiments were conducted on a CentOS8 operating system with a 2.60 GHz CPU and a GeForce RTX 3070 GPU. Development and debugging were performed on the PyCharm platform, and the HCWA model was implemented using the PyTorch framework. For all datasets, 90% of the data in the training set is randomly selected for model training, while the remaining 10% is used as the validation set. Accuracy is used as the performance evaluation metric for the model. Adam is used for optimizing model parameters, with the optimizer's initial learning rate set to 0.001. Dropout is used during training to prevent overfitting, with the Dropout rate set to 0.5 and the L2 regularization rate set to 0.001.

Table 1. Datasets statistics

Datasets	R8	R52	Ohsumed	MR	20NG	SST
# Doc	7674	9100	7400	10662	18846	11855
# Train	5485	6532	3357	7108	11314	9465
# Test	2189	2568	4043	3554	7532	2210
# Word	7688	8892	14157	18764	42757	19524
# Class	8	52	23	2	20	5
Ave Len	65.72	69.82	135.82	20.39	221.26	20.17

Baseline Model and Evaluation Indicators. This section introduces baseline models, which mainly include 4 categories for verifying model performance.

(1) Word embedding-based models: Use pre-trained embedding models to obtain text representations, then classify the text, including fastText [20] and LEAM [21].
(2) Sequence-based deep learning models: Capture text contextual features from local continuous text sequences for classification, including CNN [9] and Bi-LSTM [22].
(3) Graph neural network-based models: Model text as a graph structure, utilize GNN to learn node representations and capture text features, including Graph-CNN [23], the inductive and deductive versions of TextGCN [11], and Text-level GNN [24].
(4) Models based on hypergraphs: Introduce hypergraphs to allow more complex and multi-layered representation relationships between nodes and hyperedges, including HyperGAT [12], HyperGAT(no sem), IBHC, where HyperGAT(no sem) is HyperGAT without introducing semantic features.

The performance evaluation metrics for text classification algorithms include Accuracy, macro-F1, and micro-F1. All three have a range of [0,1], with higher values indicating better model performance.

4.2 Experimental Results and Analysis

This section compares the performance of the model we proposed with other baseline models on six datasets and analyzes the experimental results. The classification accuracy results are shown in Table 2.

From the results, it can be seen that our proposed model HCWA achieved the highest improvements of 3.3% and 7.3% compared to word embedding class models on the shorter text length datasets MR and SST. On the R8 and R52 datasets, the HCWA model achieved the highest improvements of 4.4% and 10.1% compared to sequence-based deep learning models. On the longer text length datasets Ohsumed and 20NG, our model HCWA improved by 3.0%–14.7% and 1.3%–6.3% compared to GNN class models, respectively. Additionally, on both these long text datasets, GNN class and hypergraph class models outperformed sequence-based deep learning models, demonstrating the significant advantages of graph and hypergraph-based methods in capturing features of long text content, interacting contextual information, and handling more complex global relationships.

Table 2. Classification accuracy results on different datasets

Model	R8	R52	Ohsumed	MR	20NG	SST
fastText	0.9613	0.9281	0.5770	0.7514	0.7938	0.3608
LEAM	0.9331	0.9184	0.5858	0.7695	0.8191	0.4293
CNN	0.9402	0.8537	0.4387	0.7498	0.7693	0.4230
Bi-LSTM	0.9638	0.9054	0.4927	0.7767	0.7318	0.4263
Graph-CNN	0.9699	0.9275	0.6386	0.7722	0.8142	0.3523
TextGCN(transductive)	0.9707	0.9356	0.6836	0.7674	0.8643	0.4065
TextGCN(inductive)	0.9578	0.8820	0.5770	0.7480	0.8331	0.3905
Text-level GNN	0.9783	0.9462	0.6940	0.7596	0.8416	0.4302
HyperGAT	0.9797	0.9498	0.6990	0.7832	0.8662	0.4196
HyperGAT(no sem)	0.9714	0.9415	0.6848	0.7811	0.8602	0.3985
IBHC	0.9812	0.9510	0.7027	**0.7846**	0.8676	**0.4408**
HCWA	**0.9841**	**0.9549**	**0.7235**	0.7843	**0.8778**	0.4339

In hypergraph-based methods, HyperGAT (no sem) without semantic features performs worse across all six datasets compared to the HyperGAT method, while the model proposed in this paper, HCWA, outperforms both types of hypergraph models. Compared to the IBHC model, on long-text datasets Ohsumed and 20NG, the model proposed in this paper achieves significant improvements. On short-text datasets MR and SST, the IBHC model introduces the pre-trained language model BERT to enhance contextual semantic modeling, which has a natural advantage in feature extraction that relies on local semantics for short texts. In contrast, HCWA emphasizes the construction of multi-type hyperedges and high-order modeling of structural information, demonstrating stronger representation capabilities for complex structures and long-range dependencies, thus performing more outstandingly in long-text tasks and generally in short-text tasks. However, overall, the HCWA model still shows superior text classification performance across multiple datasets. The HCWA model introduces sequential hyperedges, thematic hyperedges, and syntactic hyperedges to more comprehensively describe the relationships between nodes, enabling the model to learn richer hypergraph structural information. Different hyperedges capture distinct textual features, and through hypergraph convolutional information propagation and processing, the model deepens its understanding of complex textual relationships. The attention extraction layer helps the model improve its ability to capture and utilize important information, allowing it to more effectively focus on expressive features and enhance the model's classification performance.

4.3 Ablation Experiment

To further verify the effectiveness of the proposed model, we explored the impact of different types of hyperedges on the model's classification performance by conducting classification experiments on datasets R52, Ohsumed, SST, and 20NG. The experimental results are shown in Table 3, where "w/o sequential" refers to the removal of sequential hyperedges, "w/o topic" refers to the removal of topic hyperedges, and "w/o syntax" refers to the removal of syntactic hyperedges.

Table 3. Results of ablation experiments

Model	R52	Ohsumed	20NG	SST
w/o sequential	0.9236	0.6339	0.7959	0.3965
w/o topic	0.9447	0.6855	0.8583	0.4107
w/o syntax	0.9463	0.7021	0.8651	0.4163
HCWA	**0.9549**	**0.7235**	**0.8778**	**0.4339**

The experimental results in Table 3 can verify that by constructing different hyperedges and integrating multiple relationships, text expression can be enriched, and the ability to understand different contexts can be improved. Sequence hyperedges have a greater impact on text classification, but the three relationships can complement each other. Sequence relationships involve the order of words, topic hyperedges focus on the association of word meanings, and syntactic dependency relations describe the structure between words. By utilizing information from different levels, the meaning of the text can be better understood, the adaptability to complex texts can be improved, and it plays an important role in enhancing the performance of text classification.

4.4 Parametric Analysis

To investigate the impact of sliding windows of different sizes on the accuracy of text classification, experiments were conducted on the R52 and 20NG datasets, with the results shown in Fig. 2.

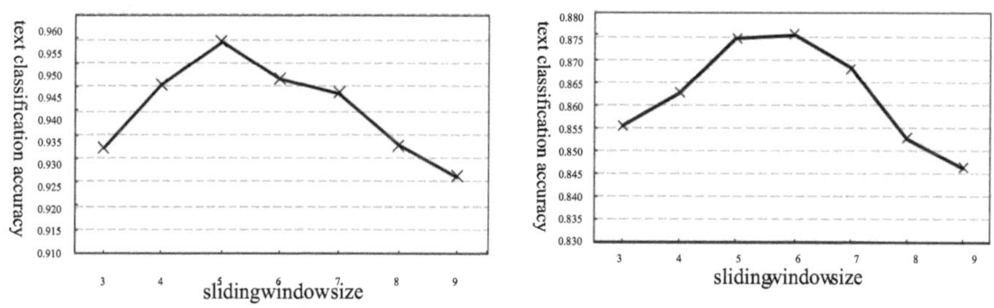

(a) R52 (b) 20NG

Fig. 2. The effect of sliding window size on classification performance

Figure 2, shows that an appropriate sliding window size can effectively improve classification accuracy. When the sliding window is too small, it causes data sparsity, leading to insufficient information utilization. While a too large sliding window increases irrelevant textual information, making the nodes contained in the super edge redundant, reducing information expression, and simultaneously increasing memory consumption.

Although our model has achieved good results in multiple text classification tasks, it still has limitations. The current hyperedge construction method is mainly based on fixed grammatical, semantic, and sequential relationships, and have not yet captured semantic associations in finer detail, which may lead to insufficient expression of complex linguistic structures.

5 Conclusion

In this paper, we propose a text classification model based on hypergraph convolutional networks and attention mechanisms, called HCWA. The model constructs hyperedges of three types: sequences, topics, and syntax, to achieve multi-level text structure modeling. It combines hypergraph convolution with attention mechanisms to enhance node representations and further improves global feature extraction through multi-channel pooling. Extensive experimental results on real datasets validate that the proposed method can effectively enhance the performance of text classification tasks. In the future, we plan to further explore finer-grained semantic relationships to extend the hyperedge construction method and enhance the ability to express complex language structures.

Acknowledgments. This work is supported by the National Natural Science Foundation of China (No. 62372342), the General Program of National Natural Science Foundation of Hubei Province (No. 2024AFB865).

References

1. Tao, L., Hua, G., Li, B.: Aspect-level sentiment analysis based on location-enhanced word embeddings and GRU-CNN model. Comput. Eng. Appl. 212–218 (2024)
2. Hu, J., Lv, X., You, X., et al: Fund news classification based on clustering annotation and multi granularity feature fusion. Small Micro Comput. Syst. 257–264 (2024)
3. Li, Q., Peng, H., Li, J., et al.: A survey on text classification: from traditional to deep learning. ACM Trans. Intell. Syst. Technol. (TIST) **13**(2), 1–41 (2022)
4. Zhang, Y., Jin, R., Zhou, Z.: Understanding bag-of-words model: a statistical framework. Int. J. Mach. Learn. Cybern. 43–52 (2010)
5. Jain, S., Jain, S., Vasal, S.: An effective TF-IDF model to improve the text classification performance. In: 2024 IEEE 13th International Conference on Communication Systems and Network Technologies (CSNT), pp. 1–4. IEEE (2024)
6. Ruan, S., Chen, B., Song K, et al.: Weighted naïve Bayes text classification algorithm based on improved distance correlation coefficient. Neural Comput. Appl. 1–10 (2022)
7. Palanivinayagam, A., El-Bayeh, C., Damaševičius, R.: Twenty years of machine-learning-based text classification: a systematic review. Algorithms **16**(5), 236 (2023)
8. Chen, H., Wu, L., Chen, J., et al.: A comparative study of automated legal text classification using random forests and deep learning. Inf. Process. Manage. **59**(2), 102798 (2022)

9. Kim, Y.: Convolutional neural networks for sentence classification. In: Proceedings of the 2014 Conference on Empirical Methods in Natural Language Processing (EMNLP). Doha, pp. 1746–1751. Association for Computational Linguistics (2014)
10. Chen, Y.: Convolutional neural network for sentence classification. University of Waterloo (2015)
11. Yao, L., Mao, C., Luo, Y.: Graph convolutional networks for text classification. In: Proceedings of the AAAI Conference on Artificial Intelligence, vol. **33, no.** 01, pp. 7370–7377 (2019)
12. Ding, K., Wang, J., Li, J., et al.: Be more with less: Hypergraph attention networks for inductive text classification (2020)
13. Li, Q., Pang, J., Zhu, F.: Text classification method based on integration of bert and hypergraph convolutional network. J. Comput. Eng. Appl. **59**(17) (2023)
14. Devlin, J.: Bert: pre-training of deep bidirectional transformers for language understanding (2018)
15. Church, K.: Word2Vec. Nat. Lang. Eng. **23**(1), 155–162 (2017)
16. Abu Alfeilat, H.A., et al.: Effects of distance measure choice on k-nearest neighbor classifier performance: a review. Big Data **7**(4), 221–248 (2019)
17. Zheng, C., Li, P.: Text classification based on feature fusion of dual hypergraph neural networks. Comput. Eng. **51**(06), 127–135 (2025)
18. Jing, R., Wan, F., Huang, R.: Tibetan long text classification by fusing denoising fine-tuning and graph attention mechanism. Comput. Eng. Sci. **47**(06), 1133–1140 (2025)
19. Chen, L., Ma, L., Zhang, Y., et al.: Industrial text classification for Chinese and vietnamese based on prompt learning and adaptive loss weighting. Comput. Sci. 1–16 (2025)
20. Joulin, A., Grave, E., Bojanowski, P., et al.: Bag of tricks for efficient text classification, pp. 427–431 (2017)
21. Wang, G., Li, C., Wang, W., et al.: Joint embedding of words and labels for text classification (2018)
22. Jiang, W., Jin, Z.: Integrating bidirectional LSTM with inception for text classification. In: 2017 4th IAPR Asian Conference on Pattern Recognition (ACPR) (2017)
23. Defferrard, M., Bresson, X., Vandergheynst, P.: Convolutional neural networks on graphs with fast localized spectral filtering. In: Advances in Neural Information Processing Systems, pp. 3844–3852 (2016)
24. Ma, D., Li, S., Zhang, X., et al.: Text level graph neural network for text classification. In: Proceedings of the 2019 Conference on Empirical Methods in Natural Language Processing and the 9th International Joint Conference on Natural Language Processing (EMNLP-IJCNLP). Hong Kong, pp. 3442–3448. Association for Computational Linguistics (2019)

Enhancing Traffic Sign Detection with EEM-YOLO for Complex Scenarios

Jian Wang[1,2(✉)], Xiulan Hao[1,2], and Qiang Gao[1,2]

[1] School of Information Engineering, Huzhou Normal University, Zhejiang 313000, China
{hxl2221_cn,02794}@zjhu.edu.cn

[2] Zhejiang-French Digital Monitoring Lab for Aquatic Resources and Environment, Huzhou Normal University, Zhejiang 313000, China
2023388324@stu.zjhu.edu.cn

Abstract. Traffic sign recognition (TSR) is a crucial application of computer vision, using image recognition to identify traffic signs in road scenes. Accurate detection is vital for autonomous driving, especially in complex scenarios with occlusion or environmental interference. Traditional models often suffer from low accuracy, missed detections, and false positives. To address these issues, we propose EEM-YOLO (Efficient-former Enhanced Multi-Detection YOLO). It uses the EfficientFormerV2 backbone, which efficiently extracts spatial features and reduces redundant computations. In the Neck module, eSE attention is integrated with MBConv to enhance feature fusion, and a new C2f_EC design improves regression prediction accuracy. Additionally, the Wisev3-SIoU loss function replaces CIoU to address missing/false detections and gradient issues. Extensive experiments on TT100K and CCTSDB2021 datasets show that EEM-YOLO achieves mAP of 83.1% and 85.9%, respectively, outperforming benchmark models by 9.4% and 7.8%. This demonstrates its superior detection performance and feature extraction capability, making it highly competitive among existing models.

Keywords: Computer vision · Object detection · Traffic signs detection · Feature fusion · YOLOv8

1 Introduction

TSR is crucial for autonomous driving and intelligent driving assistance systems. However, TSR faces challenges like small targets, lighting variations, obstructions, and background interference. This leads to difficulties in balancing real-time performance and accuracy, posing safety risks. Thus, developing algorithms that ensure both accuracy and real-time performance is urgently needed.

Traffic sign detection methods are divided into traditional approaches and deep learning-based techniques. Traditional methods use manual feature extraction based on color, texture, and shape, with candidate regions generated via thresholding or sliding windows. They rely on rule-based validation, template

J. Li et al. (Eds.): APWeb-WAIM 2025, CCIS 2660, pp. 272–284, 2026.
https://doi.org/10.1007/978-981-95-7996-9_24

matching, or edge detection for classification. However, these methods have limitations in feature design, adaptability to lighting and viewpoint changes, and robustness against occlusions, resulting in poor real-time performance

Deep learning, especially CNNs, has transformed image feature extraction, improving traffic sign detection accuracy and efficiency. Deep learning-based detection methods are categorized into two-stage and one-stage approaches.

Representative two-stage methods like Faster R-CNN [13] first generate region proposals before performing convolutional classification, achieving high accuracy at the cost of computational speed. One-stage methods, such as YOLO [19] and SSD [3], simultaneously locate and classify objects, and position coordinates and classification probability of the object are directly regressed. They go after detection speed other than accuracy. Given the strict real-time requirements in traffic sign detection, YOLO and SSD frameworks have caught substantial research attention.

While current traffic sign recognition mainly targets simple scenarios, challenges like low accuracy and false detections in complex scenes remain. To tackle these, we propose EEM-YOLO—an improved YOLOv8-based method for detecting small and medium-sized traffic signs. Our key contributions include:

1) The EfficientFormerV2 network is adopted as the feature extraction framework to reduce redundant computations and enhance detection speed.
2) A novel C2f_EC module is proposed by integrating the eSE with MBConv, effectively reducing model computational complexity while improving accuracy.
3) A Wise-v3-SIoU loss function is designed to replace conventional CIoU, addressing the issues of introduced by low-quality anchor boxes.
4) Extensive experiments are done on TT100K and CCTSDB2021, demonstrate the effectiveness of EEM-YOLO through evaluations.

2 Related Works

Deep learning-based detection methods have become the leading approach in traffic sign recognition, overcoming limitations of traditional algorithms that rely on shape and color features. Traditional methods struggle with accuracy under distorted images, small signs, or complex lighting, posing safety risks. In contrast, deep learning offers robust feature extraction and recognition capabilities. We categorize and analyze the latest advancements in traffic sign recognition and object detection technologies from various technical perspectives.

2.1 Scenario-Specific Improvements

The main challenges in traffic sign detection arise from environmental complexity and the need for real-time accuracy. Autonomous driving systems require robust traffic sign recognition that can adapt to diverse operational scenarios, including small targets, varying lighting conditions, adverse weather (e.g., fog), and low-light nighttime environments. Many studies have developed specialized solutions to address these environmental challenges.

Saxena et al. [15] proposed a multi-stage framework integrating nighttime image enhancement, mosaic-style data augmentation, and dedicated filter banks for disentangled feature learning. The architecture uses Generalized Intersection over Union (GIoU) for anchor clustering and enhances PANet convolutional layers with extra feature scales to improve small traffic sign detection.

To address detection under complex weather conditions, Qu et al. [12] incorporated the Coordinate Attention (CA) mechanism into the backbone network. The CA module uses positional encoding to help the network capture extended contextual features, improving object localization accuracy. For small traffic signs, they augmented the architecture with specialized small-object detection layers, utilizing high-resolution feature maps to preserve fine-grained details and reduce missed detections.

2.2 Model-Specific Improvements: YOLO Models and Beyond

In recent years, YOLO models have dominated traffic detection due to their real-time performance.

YOLO-Based Approaches. YOLOv8, released in 2023, is a target detection model in the YOLO series. It has four main components: Input, Backbone, Neck (Feature Fusion Layer), and Head (Output Layer). For input processing, YOLOv8 uses Mosaic data augmentation and other techniques.

The backbone extracts features, the fusion layer integrates them, and the detection head makes predictions. YOLOv8 uses a C2f module, replacing YOLOv5's C3 module and inspired by YOLOv7's ELAN design, to improve feature fusion. The SPPF (Spatial Pyramid Pooling Fast) module captures local and global feature information.

YOLOv8's Neck combines FPN [9] and PAN [10] for cross-level multi-scale perception and comprehensive feature fusion, enhancing network expression and detection accuracy.

Zhou et al. [23] developed a traffic sign detection algorithm using the Selective Kernel (SK) attention mechanism, Explicit Visual Center (EVC), and YOLOv5 architecture. It fuses small-scale feature maps with those from other scales, with SK balancing cross-scale features and EVC capturing local features within layers.

Zhao et al. [22] proposed YOLOv7-TS, which uses sub-pixel convolution and feature fusion. It enhances channel upsampling and reduces information loss with a feature mapping extraction module.

Alternative Approaches. While YOLO-series models dominate traffic sign recognition research, alternative neural architectures and computer vision techniques continue to show promising potential.

Zhang et al. [20] developed RTS R-CNN, an enhanced instance segmentation network based on Mask R-CNN, to improve small target detection accuracy.

Gang et al. [4] proposed a lightweight SSD-based algorithm for complex road scenarios. They replaced the VGG16 backbone with MobileNetV3_large to reduce parameters and accelerate detection.

3 EEM-YOLO Model

YOLOv8's detection head uses a decoupled structure for separate classification and localization, improving precision in complex scenarios. It employs an anchor-free strategy for better alignment with ground truth boxes and a task-aligned assigner for dynamic task allocation, enhancing generalization and performance.

3.1 EEM-YOLO Network

As the application scope expands, the requirements for model size and inference accuracy have become increasingly demanding. To further optimize YOLOv8's traffic sign detection performance in complex environments, EEM-YOLO is proposed. The framework is illustrated in Fig. 1.

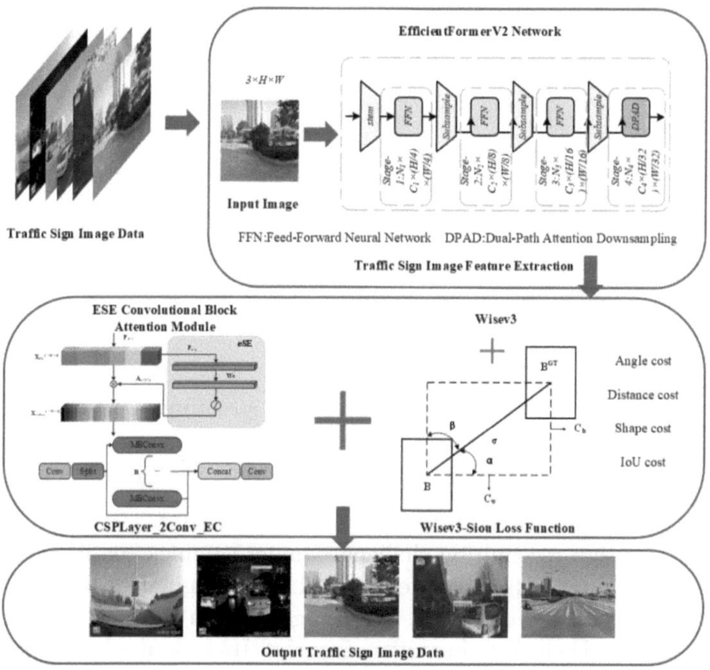

Fig. 1. EEM-YOLO Framework.

EfficientFormerV2 [8] is introduced as the backbone network for feature extraction. Through efficient token mixing, optimized feed-forward networks, and a dual-path attention mechanism for downsampling, our approach achieves more effective spatial feature extraction while reducing redundant computations and improving detection efficiency.

For the neck section of the network, the eSE [7] attention mechanism is integrated with EfficientNet's MBConv [14] to facilitate the full integration of shallow and deep features. A novel C2f_EC module is designed to reduce the model's computational complexity while enhancing the accuracy of regression predictions.

To address the challenges of missed and false detections in complex scenarios, a new Wisev3-SIoU loss function is designed to replace the original CIoU loss function. The overall architecture of our EEM-YOLO network is illustrated in Fig. 2.

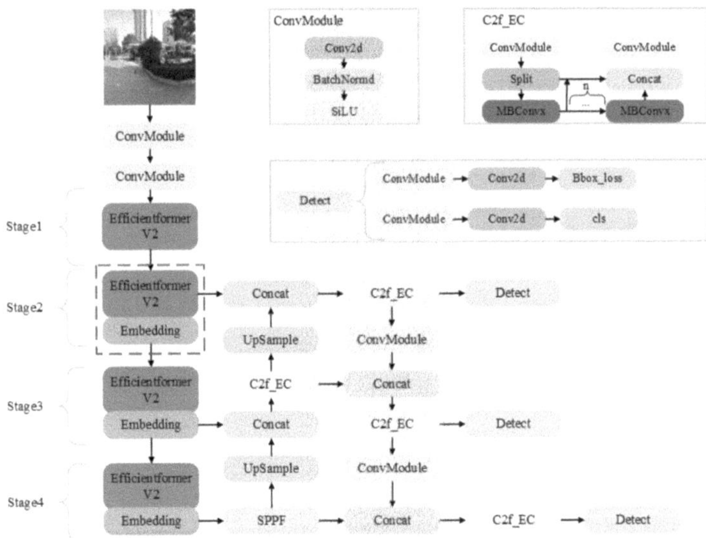

Fig. 2. EEM-YOLO network structure.

3.2 EfficientFormerv2

To tackle traffic sign recognition challenges under complex conditions, we focus on selecting an improved backbone network. The original YOLOv8 backbone excels in general object detection but struggles with illumination variations, occlusions, and scale changes in traffic scenarios. Its lack of a background interference suppression mechanism results in suboptimal feature extraction.

EfficientFormerV2 is used as the backbone network to replace the original backbone of YOLOv8. Figure 3 shows its network architecture. Following the design philosophy of EfficientFormer, this architecture introduces a novel supernet design that focuses on low latency and high parameter efficiency, aiming to improve the overall performance of the Transformer based models [18].

EfficientFormerV2 employs four hierarchical stages to capture multi-scale features at input resolutions of 1/4, 1/8, 1/16, and 1/32. The first two stages utilize uniform feed-forward networks (FFN) to capture high-resolution local information, while the latter two stages combine local FFN with global multi-head self-attention (MHSA) blocks.

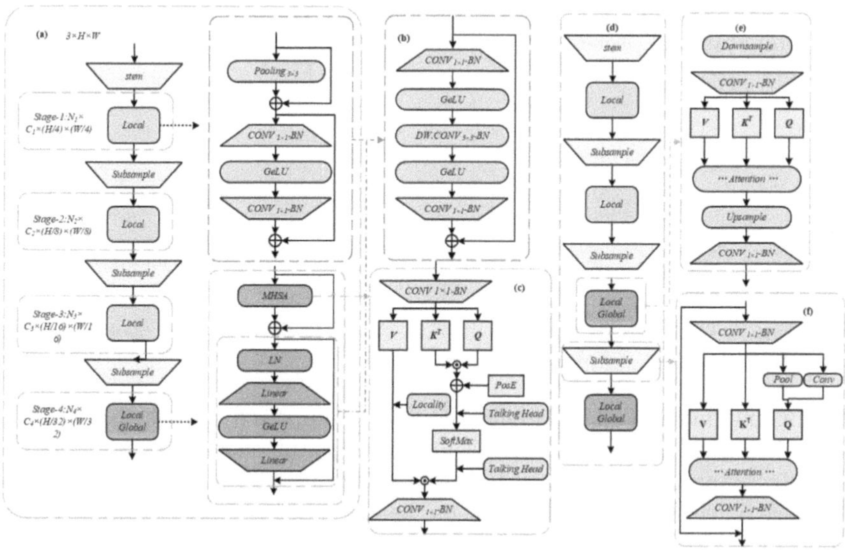

Fig. 3. EfficientFormerV2 network structure.

By replacing traditional average pooling layers in local token mixers with depthwise separable convolution layers within the FFN, EfficientFormerV2 enhances model performance without increasing latency while providing superior feature extraction and scene adaptation capabilities.

Additionally, EfficientFormerV2 introduces a fine-grained joint search strategy and defines a new evaluation metric, MES (Multi-dimensional Efficiency Score), which enables flexible configuration of each module. These improvements significantly enhance real-time performance and accuracy in complex scenarios, ensuring reliable and rapid traffic sign recognition under various challenging environmental conditions.

3.3 C2f_EC

In the design of the Neck structure, the C2f module employed in YOLOv8 integrates concepts from both C3 and ELAN architectures. It incorporates a supplementary branch structure to enhance auxiliary flow during gradient backpropagation.

C2f_EC, as shown in Fig. 4, is an improved version of C2f, which is replaced the traditional standard residual block (Bottleneck) with MBConvx module. It fuses eSE attention mechanism with MBConv (Mobile Inverted Bottleneck Convolution), where eSE is used to replace the original SE attention mechanism [6]. eSE can focus on important feature channels by decreasing parameters and optimize computational process, thus, improving the model's feature representation capability. Architecture of MBConvx is shown in Fig. 5.

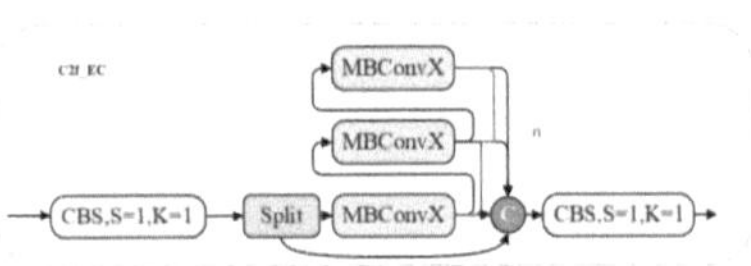

Fig. 4. C2f_EC structure.

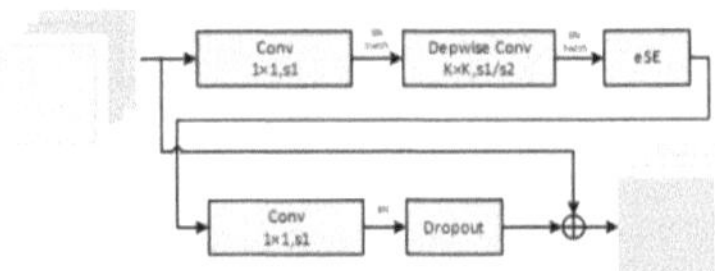

Fig. 5. MBConvx structure.

MBConv, introduced in MobileNetV2, is often paired with the SE attention mechanism to enhance feature sensitivity by adjusting channel weights. SE uses two fully connected layers to reduce dimensionality and computation, but this can cause information loss. The eSE mechanism addresses this by using a single layer to preserve channel information. Additionally, residual connections in the OSA module maintain gradient flow, improving training stability and network robustness

3.4 Wisev3-SIoU Loss Function

In the YOLOv8 framework, the regression loss is jointly constituted by DFL (Distribution Focal Loss) and CIoU (Complete Intersection over Union) loss functions. The CIoU formulation is expressed as follows:

$$\alpha = \frac{v}{(1 - IoU) + v} \tag{1}$$

$$v = \frac{4}{\pi^2}\left(\arctan\frac{w_g}{h_g} - \arctan\frac{w}{h}\right)^2 \tag{2}$$

$$CIoULoss = 1 - IoU + \frac{d^2}{c^2} + \alpha v \tag{3}$$

Here, IoU is the Intersection over Union loss, which measures the degree of overlap between the predicted bounding box and the ground truth box; d is the Euclidean distance between the centers of the predicted box and the ground truth box; c is the diagonal length of the smallest enclosing rectangle that bounds both the predicted box and the ground truth box; v is a similarity measure that evaluates the consistency of the aspect ratio between the predicted box and the ground truth box; w, h, w_g, h_g are the width and height of the predicted box and the ground truth box, respectively. The parameter α is a factor that adjusts the influence of v.

CIoU comprehensively considers multiple factors between the ground truth box and the predicted box, including overlap area, center point distance, and aspect ratio. Figure 6 visually illustrates some key elements of CIoU.

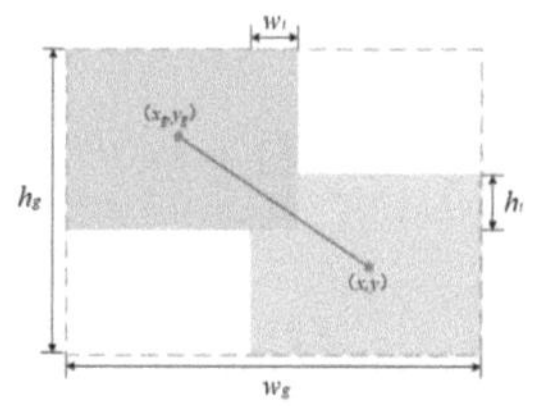

Fig. 6. Visualization of CIoU elements in ground truth and predicted boxes.

In complex scenarios, traffic signs often suffer from degraded image quality due to lighting, occlusion, and background clutter, causing large aspect ratio variations. This leads geometric metrics like CIoU to over-penalize low-quality samples, reducing generalization. CIoU also struggles with modeling widthheight relationships, especially when aspect ratios are similar, making the penalty ineffective. Its gradient formulation prevents simultaneous optimization of width and height, hindering accuracy and convergence.

To overcome CIoU's limitations, we propose the Wisev3-SIoU loss, combining WIoU's dynamic non-monotonic focusing [17] and SIoU's vector angle [5]. This hybrid reduces unnecessary penalties when boxes are well-aligned, improving generalization and mitigating training interference. Designed for bounding box regression, Wisev3 builds on IoU with a refined two-stage attention mechanism, as follows.

$$\mathcal{L}_{WIoUv1} = \mathcal{R}_{WIoU}\mathcal{L}_{IoU} \tag{4}$$

$$\mathcal{R}_{WIoU} = \exp\left(\frac{(x - x_g)^2 + (y - y_g)^2}{(w_g^2 + h_g^2)^*}\right) \tag{5}$$

Wisev1 reduces the R_{WIoU} of high-quality anchor boxes by lowering their sensitivity to center distance when they are well-aligned with targets. To avoid gradient issues affecting convergence, W_g and H_g are detached from the computational graph (indicated by $*$).

Building on this, Wisev3 introduces a dynamic non-monotonic focusing mechanism by replacing IoU with an "outlier score" β to evaluate anchor quality. It employs an optimized gradient gain allocation strategy, adaptively balancing gradient pressure—reducing it for high-quality anchors and mitigating harmful gradients from low-quality data. The Wisev3 loss function is defined as:

$$\beta = \frac{\mathcal{L}_{IoU}^{*}}{\mathcal{L}_{IoU}} \in [0, +\infty) \tag{6}$$

$$\mathcal{L}_{WIoUv3} = r\mathcal{L}_{WIoUv1}, r = \frac{\beta}{\delta\alpha^{\beta-\delta}} \tag{7}$$

SIoU loss function considers the vector angle between the ground truth box and the predicted box, a penalty metric is redefined. It is composed of four components: angle cost, distance cost, shape cost and IoU cost.

$$L_{box} = 1 - IoU + \frac{\Delta + \Omega}{2}, \Delta = \sum_{t=x,y} \left(1 - e^{-\gamma\rho_t}\right), \Omega = \sum_{t=w,h} \left(1 - e^{-\omega t}\right)^{\theta} \tag{8}$$

$$\Lambda = 1 - 2 \times \sin^2\left(\arcsin\left(\frac{c_h}{\sigma}\right) - \frac{\pi}{4}\right) \tag{9}$$

$$\rho_x = \left(\frac{b_{g_{c_x}} - b_{c_x}}{c_w}\right)^2, \rho_y = \left(\frac{b_{g_{c_y}} - b_{c_y}}{c_h}\right)^2, \gamma = 2 - \Lambda \tag{10}$$

$$\omega_w = \frac{|w - w_g|}{\max(w, w_g)}, \omega_h = \frac{|h - h_g|}{\max(h, h_g)} \tag{11}$$

where L_{box} is SIoU loss, Δ is distance cost, Ω is shape cost, Λ is angle cost, and γ is the Sensitivity to distance.

The final Wisev3-SIoU loss function is defined as:

$$\mathcal{L}_{Wisev3SIoU} = rL_{box} \tag{12}$$

The loss combines Wisev3's weighting with SIoU's geometry to suppress low-quality gradients, prioritize average-quality anchors, and incorporate box directionality, improving accuracy and convergence.

4 Experimental Results and Analysis

The experiments are primarily designed to address the following Research Questions (RQs):

RQ1: Are all modules in EEM-YOLO necessary, and do they contribute to improving detection performance?

RQ2: Does EEM-YOLO exhibit performance improvements compared to the current state-of-the-art models?

4.1 Experiments Settings

The experimental software is: operating system Linux, Python 3.8.16, PyTorch 2.3.0, CUDA 12.1; the hardware is NVIDIA GeForce RTX 4090.

The experimental parameter settings are: the optimizer uses Stochastic Gradient Descent (SGD), initial learning rate 0.01, weight decay 5e-4, momentum coefficient 0.937, NMS threshold is set to 0.7, batch size 16, training for 400 epochs, and the default value of workers is 8.

4.2 Datasets

Experiments were conducted on two Chinese traffic sign datasets: TT100K and CCTSDB2021. CCTSDB2021 has 17,856 images split into 16,356 for training and 1,500 for testing, covering indicative, warning, and prohibitory signs under diverse weather conditions. TT100K contains 10,000 high-resolution images from multiple cities, with 45 subcategories having over 100 samples each, split into 6,793 training, 1,949 validation, and 996 testing images.

4.3 Evaluation Metrics

The evaluation metrics used include Precision, Recall, mean Average Precision (mAP50, mAP50:95), F1 Score, and Parameter count. The specific calculation methods for Precision, Recall, and mean Average Precision are as follows:

$$Precision = \frac{TP}{TP + FP}, Recall = \frac{TP}{TP + FN} \tag{13}$$

$$mAP = \frac{1}{N}\sum_{i=1}^{N} AP_i, \qquad AP_i = \int_0^1 (P_i \cdot R_i) \tag{14}$$

Among these, TP is True Positive samples, FP is False Positive samples, and FN is False Negative samples. N is the total number of categories in the dataset, while i is the index of a specific category.

$$F_1 = \frac{2PR}{P + R} \tag{15}$$

4.4 Comparative Experiments with SOAT Models

To verify EEM-YOLO's advancement (answering RQ2), we conducted comparative experiments with some baseline models. The experimental results are shown in Table 1 and Table 2, where results in bold are best and results in italic are second best.

The results on two datasets show EEM-YOLO is more competitive. Due to the constraints of paper length, further interpretation is omitted.

Table 1. Comparative experiments on CCTSDB2021

Model	P/%	R/%	Params/M	GFLOPs	mAP50
FasterR-CNN	–	–	41.6	195.8	0.566
YOLOv5s	87.8	71.4	9.1	16.1	0.798
YOLOv8n	86.9	70.4	3.01	8.1	0.781
YOLOv10n	89.6	70.1	**2.3**	*6.5*	0.81
YOLOv11n	88.2	71.7	*2.59*	**6.3**	0.808
RT-DERT-1	–	–	32	103.6	0.821
Zhao [22]	*92.6*	78.3	34.4	–	0.86
Luo [11]	89.4	71.9	5.2	6.9	0.829
Du [2]	91.8	*78.6*	8.8	65.6	**0.865**
Shi [16]	**93.8**	76.8	6.1	–	0.843
Ours	89.3	**79.2**	5.3	11.3	*0.859*

Table 2. Comparative experiments on TT100K

Model	P/%	R/%	Params/M	GFLOPs	mAP50
FasterR-CNN	–	–	41.6	195.8	0.566
YOLOv5s	87.8	71.4	9.1	16.1	0.798
YOLOv8n	86.9	70.4	3.01	8.1	0.781
YOLOv10n	**89.6**	70.1	**2.3**	*6.5*	*0.81*
YOLOv11n	*88.2*	71.7	*2.59*	**6.3**	0.808
RT-DERT-1	–	–	32	103.6	0.796
Zhang [21]	81.9	**77.2**	–	–	0.82
Chen [1]	–	–	4.51	6.9	0.751
–					
–					
Ours	84.8	*74.1*	5.3	11.3	**0.831**

4.5 Ablation Experiments

To verify how well the proposed model addresses RQ1, we did ablation experiments. We include/exclude EfficientFormerv2, C2f_EC, and Wisev3-SIoU respectively, to demonstrate the effectiveness and necessity of these methods. The results are shown in Table 3 and Table 4.

Table 3. Ablation Experiments on CCTSDB2021

Efficientv2	C2f_EC	Wisev3-SIoU	Precision	Recall	Params/M	GFLOPs	mAP50	mAP50:95	APs	APm	F1
			0.869	0.704	3.01	8.1	0.781	0.493	0.442	0.596	0.778
✓			0.905	0.774	5.11	11.7	0.848	0.562	0.496	0.630	0.834
✓	✓		0.898	0.779	5.30	11.3	0.851	0.563	0.501	0.617	0.834
✓	✓	✓	0.893	0.792	5.30	11.3	0.859	0.566	0.520	0.630	0.839

Table 4. Ablation Experiments on TT100K

Efficientv2	C2f_EC	Wisev3-SIoU	Precision	Recall	Params/M	GFLOPs	mAP50	mAP50:95	APs	APm	F1
			0.724	0.676	3.01	8.1	0.737	0.553	0.314	0.680	0.699
✓			0.839	0.644	5.11	11.7	0.789	0.602	0.385	0.773	0.728
✓	✓		0.805	0.743	5.30	11.3	0.817	0.612	0.411	0.731	0.772
✓	✓	✓	0.848	0.741	5.30	11.3	0.831	0.621	0.411	0.785	0.791

In Ablation experiment 1, after adding EfficientFormerv2, the performance has been improved. On CCTSDB2021, mAP50 is improved from 78.1% to 84.8%; while mAP50:95 is improved from 49.3% to 56.2%. On TTK100, mAP50 is improved from 73.1% to 81.7%; while mAP50:95 is improved from 55.3% to 61.2%. These results are achieved at the expense of an extra 2.18M parameters.

In Ablation experiment 2, after replacing C2f with C2f_EC, the computational cost is reduced. The results show that this improvement has improved all indicators of the model except precision, and even recall has increased by 9.9% on TT100K, while the floating-point calculation has decreased.

Ablation experiment 3 introduces Wisev3-SIoU. On CCTSDB2021, mAP50 is improved from 85.1% to 85.9%; while mAP50:95 is improved from 56.3% to 56.6%. On TTK100, mAP50 is improved from 81.7% to 83.1%; while mAP50:95 is improved from 61.2% to 62.1%.

Due to space constraints, certain experimental results, such as backbone network comparisons and detection result visualizations, are omitted.

5 Conclusion

We propose the EEM-YOLO model to tackle issues like background noise and poor anchor boxes in traffic sign detection. It significantly reduces false/missed detection rates, improves accuracy, and shrinks model size. Experiments on two datasets validate its potential for edge devices. However, it currently performs best in normal weather, with extreme conditions like heavy snow/rain needing further investigation. Future work will focus on enhancing robustness and generalization in complex environments.

Acknowledgments. This work is supported by Natural Science Foundation of Huzhou (Grant Nos. 2022YZ20).

References

1. Chen, B., Fan, X.: MSGC-YOLO: an improved lightweight traffic sign detection model under snow conditions. Mathematics **12**(10), 1539 (2024)
2. Du, S., et al.: TSD-YOLO: small traffic sign detection based on improved YOLO v8. IET Image Proc. **18**(11), 2884–2898 (2024)
3. Fu, C.Y., Liu, W., Ranga, A., Tyagi, A., Berg, A.C.: DSSD: deconvolutional single shot detector. arXiv preprint arXiv:1701.06659 (2017)
4. Gang, Z., Yunming, W., Chaoliang, P.: Traffic sign detection algorithm based on lightweight SSD. Exp. Technol. Manage. **41**(1) (2024)
5. Gevorgyan, Z.: Siou loss: more powerful learning for bounding box regression. arXiv preprint arXiv:2205.12740 (2022)
6. Hu, J., Shen, L., Sun, G.: Squeeze-and-excitation networks. In: Proceedings of the IEEE Conference on Computer Vision and Pattern Recognition, pp. 7132–7141 (2018)
7. Lee, Y., Park, J.: Centermask: real-time anchor-free instance segmentation. In: Proceedings of the IEEE/CVF Conference on Computer Vision and Pattern Recognition, pp. 13906–13915 (2020)
8. Li, Y., et al.: Rethinking vision transformers for mobilenet size and speed. In: Proceedings of the IEEE/CVF International Conference on Computer Vision, pp. 16889–16900 (2023)

9. Lin, T.Y., Dollár, P., Girshick, R., He, K., Hariharan, B., Belongie, S.: Feature pyramid networks for object detection. In: Proceedings of the IEEE Conference on Computer Vision and Pattern Recognition, pp. 2117–2125 (2017)
10. Liu, S., Qi, L., Qin, H., Shi, J., Jia, J.: Path aggregation network for instance segmentation. In: Proceedings of the IEEE Conference on Computer Vision and Pattern Recognition, pp. 8759–8768 (2018)
11. Luo, Y., Ci, Y., Jiang, S., Wei, X.: A novel lightweight real-time traffic sign detection method based on an embedded device and YOLOv8. J. Real-Time Image Proc. **21**(2), 24 (2024)
12. Qu, S., Yang, X., Zhou, H., Xie, Y.: Improved YOLOv5-based for small traffic sign detection under complex weather. Sci. Rep. **13**(1), 16219 (2023)
13. Ren, S., He, K., Girshick, R., Sun, J.: Faster R-CNN: towards real-time object detection with region proposal networks. In: Advances in Neural Information Processing Systems, vol. 28 (2015)
14. Sandler, M., Howard, A., Zhu, M., Zhmoginov, A., Chen, L.C.: Mobilenetv2: inverted residuals and linear bottlenecks. In: Proceedings of the IEEE Conference on Computer Vision and Pattern Recognition, pp. 4510–4520 (2018)
15. Saxena, S., Dey, S., Shah, M., Gupta, S.: Traffic sign detection in unconstrained environment using improved YOLOv4. Expert Syst. Appl. **238**, 121836 (2024)
16. Shi, Y., Li, X., Chen, M.: SC-YOLO: a object detection model for small traffic signs. IEEE Access **11**, 11500–11510 (2023)
17. Tong, Z., Chen, Y., Xu, Z., Yu, R.: Wise-IoU: bounding box regression loss with dynamic focusing mechanism. arXiv preprint arXiv:2301.10051 (2023)
18. Vaswani, A., et al.: Attention is all you need. In: Advances in Neural Information Processing Systems, vol. 30 (2017)
19. Wang, C.Y., Bochkovskiy, A., Liao, H.Y.M.: YOLOv7: trainable bag-of-freebies sets new state-of-the-art for real-time object detectors. In: Proceedings of the IEEE/CVF Conference on Computer Vision and Pattern Recognition, pp. 7464–7475 (2023)
20. Zhang, G., Peng, Y., Wang, H.: Road traffic sign detection method based on RTS R-CNN instance segmentation network. Sensors **23**(14), 6543 (2023)
21. Zhang, R., Zheng, K., Shi, P., Mei, Y., Li, H., Qiu, T.: Traffic sign detection based on the improved YOLOv5. Appl. Sci. **13**(17), 9748 (2023)
22. Zhao, S., Yuan, Y., Wu, X., Wang, Y., Zhang, F.: YOLOv7-TS: a traffic sign detection model based on sub-pixel convolution and feature fusion. Sensors **24**(3), 989 (2024)
23. Zhou, F., Zu, H., Li, Y., Song, Y., Liao, J., Zheng, C.: Traffic-sign-detection algorithm based on SK-EVC-YOLO. Mathematics **11**(18), 3873 (2023)

Knowledge Hypergraph Link Prediction Based on Attention and Relational Contrast Learning

Ying Song[1,2], Zhifen Ma[1], Xiaoli Lin[1,2], Xiaojie Chu[1], Yifeng Xu[1], Shuang Li[1], and Jun Pang[1,2](✉)

[1] School of Computer Science and Technology, Wuhan University of Science and Technology, Wuhan, China
{song_ying,linxiaoli,chuxiaojie,xu1feng,lishaung,pangjun}@wust.edu.cn
[2] Hubei Province Key Laboratory of Intelligent Information Processing and Real-time Industrial System, Wuhan, China

Abstract. Knowledge Hypergraph (KHG) is a specialized form of knowledge graph that uses a hypergraph structure to represent n-ary relationships. Link prediction in KHGs aims to infer missing entities and relations from existing ones, and is an essential task for many knowledge-driven applications. However, current neural approaches often struggle to model interfact dependencies and maintain semantic consistency across complex relational patterns. To address these challenges, we propose a hypergraph link prediction method based on attention and convolutional neural network, called HPAC. Firstly, our refined attention module assigns distinct weights to nodes and hyperedges, enabling effective aggregation of local and global contexts to capture intricate relational dependencies. Secondly, inspired by contrastive learning, we propose a context-level contrastive model to capture the latent semantics among related entities and refine their embeddings. By constructing multiple contextual views through cross-perspective contrastive strategies, HPAC enriches node and edge representations with diverse semantic cues, thereby boosting its discriminative power. Finally, we apply regularization techniques to mitigate neuron overreliance, enhancing prediction accuracy and generalization. Extensive experimental results on real-world datasets demonstrate that HPAC consistently outperforms established baselines in link-prediction tasks.

Keywords: knowledge hypergraph · link prediction · attention mechanism · contrast learning

1 Introduction

The knowledge graph represents real-world entities and their relationships in the form of a graph [1]. However, traditional triple-based knowledge graphs struggle to represent n-ary relations. For instance, in the Freebase dataset, over 61% of the relations are n-ary [2]. Knowledge hypergraph represents facts through n-ary relations (relations involving three or more entities) [3]. A typical example

J. Li et al. (Eds.): APWeb-WAIM 2025, CCIS 2660, pp. 285–296, 2026.
https://doi.org/10.1007/978-981-95-7996-9_25

of knowledge hypergraph is shown as Fig. 1 [4]. Entities and n-ary relationships correspond to nodes and hyperedges in the hypergraph, respectively.

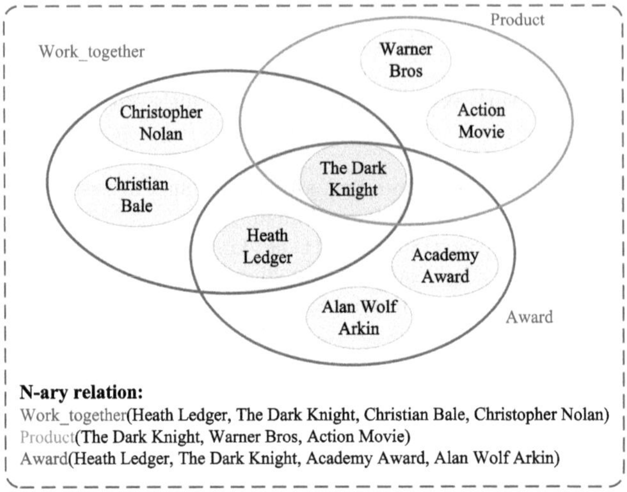

Fig. 1. Example of knowledge hypergraph.

Because of incomplete knowledge acquisition and other reasons, the knowledge hypergraphs constructed in real life are usually incomplete. Therefore, how to complete the knowledge hypergraph has become one of the hot issues in the research field of knowledge graph. As an effective means of knowledge hypergraph completion, knowledge hypergraph link prediction aims to predict missing entities or relationships based on the existing structure, and has extensive research value and application prospects, including recommender systems [5], social network analysis [6], and so on.

It is worth noting that knowledge hypergraphs are different from hyper-relational knowledge graphs. Hyper-relational knowledge graphs augment binary triples (h, r, t) with auxiliary attribute-value pairs. Knowledge hypergraphs model relational facts as n-ary tuples of the form $r(e_1, e_2, ..., e_n)$, without the need for triple decomposition or auxiliary qualifiers, e.g., HJE [7], and HyCubE [8]. This work focuses solely on knowledge hypergraphs for link prediction.

Existing knowledge hypergraph link prediction methods fall into three categories: (1) Translation-based methods model interactions through node translations but have limited expressiveness. (2) Tensor decomposition methods reveal low-dimensional patterns in high-dimensional data but lack deep semantic understanding. (3) Neural network methods aggregate entity interactions to learn richer representations. For example, the current HPGC model [4] with better performance combines Graph Attention Networks (GAT) with convolutional layers to capture complex associations. It still struggles to capture relational dependencies and maintain semantic consistency.

To address the above challenges, this paper proposes HPAC, a knowledge hypergraph link prediction model that combines an optimized attention mechanism with relational contrast learning. The primary contributions are summarized as follows: (1) A link prediction method HPAC based on attention mechanism and relation comparison learning is proposed to solve the problem of knowledge hypergraph link prediction. The optimized attention mechanism is used to extract the key features of nodes and relations, and assign weights to different nodes and relations, so as to effectively fuse local and global information, capture the complex dependencies between entities and relations, and enhance the modeling capability. (2) A context-level contrast model is proposed to capture the potential semantic features of entities. Multiple context patterns are also constructed by using the cross-view contrastive learning strategy to enhance the representation of nodes and edges. Further introduces a regularization mechanism to improve the model's link pre-diction performance and generalization capability. (3) Extensive experimental results on the real-word data sets demonstrate the effectiveness and advantages of our HPAC.

2 Related Work

This section focuses on the research works related to this paper, including translation-based modeling, tensor decomposition-based modeling, and neural network-based modeling.

2.1 Translation-Based Modeling

Translation-based methods for link prediction in knowledge hypergraphs extend binary to multivariate relationships by modeling translations between entities, thereby enhancing prediction accuracy. m-TransH [9] extends TransH [10] by projecting tuple entities onto a relation-specific hyperplane and computing confidence via a weighted sum of projections. RAE [11] further extends m-TransH by adding the correlation of the values to the loss function. However, due to the weak expressiveness of the translation model, better prediction cannot be achieved.

2.2 Methods Based on Tensor Decomposition

Tensor decomposition-based methods can discover potential low-dimensional structures in high-dimensional data [21], which enables faster and more accurate knowledge graph querying and reasoning. GETD+ [12] is an extension of GETD [13], which combines the Tucker decomposition with a fully connected tensor decomposition network (FCTN) to establish multilinear operations and connections between any two factor tensors obtained from the tensor decomposition. HypE [2] reduces parameter count by leveraging hyperedge-level parameter sharing. However, all such methods have shallow model structures, which may limit the extraction of deep or implicit knowledge.

2.3 Neural Network Based Methods

Neural network-based approaches can model complex nonlinear interactions between entities and effectively extract representative features from knowledge hypergraphs. tNaLP+ [14] is an extension of NaLP [15], further introducing the type embedding to enhance the semantic representation and improving the negative sampling method to enhance the training effect and model performance. HINGE [16] is a hyper-relational knowledge graph embedding model that uses CNNs to learn relational feature representations and uses a fully connected layer to score facts. NeuInfer [17] represents multivariate relational facts as main triples and auxiliary attribute pairs and realizes effective reasoning about multivariate relational facts through fully connected networks. HyperMLN [18] is a multivariate hybrid interpretable prediction framework based on first-order logic. The model enhances logic rule representation by introducing domain knowledge to increase its weight and role in reasoning. RD-MPNN [19] incorporates the positional information of entities within hyper relations and performs representation learning by dynamically updating hyper relations using entity information. The HPGC model [4] integrates the strengths of graph neural networks and convolutional neural networks to capture intricate associations among entities. In addition, there have been efforts focusing on temporal knowledge hypergraphs [20] and distributed processing of hypergraphs [11]. It is worth noting that all these studies differ from text-based approaches. Nevertheless, existing models still exhibit limitations in capturing deep semantic dependencies, which constrains their overall performance.

3 The HPAC Method

In this section, the proposed HPAC method is presented. We first introduce the overall framework of the HPAC model, and then detail its specific components, including attention-based feature fusion, comparative optimization learning, and global enhancement contrast. The proposed HPAC is built upon an attention mechanism and relational contrastive learning. The overall framework of HPAC is shown in Fig. 2.

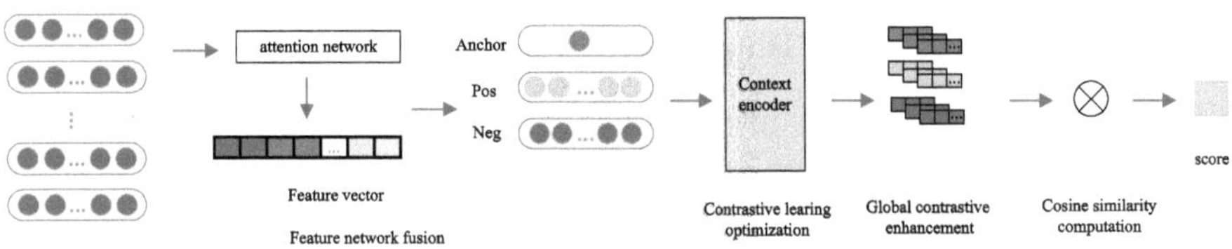

Fig. 2. A figure caption is always placed below the illustration. Short captions are centered, while long ones are justified. The macro button chooses the correct format automatically.

3.1 Attention Network Fusion Module

Inspired by Graph Attention Network (GAT) [22], we introduce an optimized attention mechanism for multi-level feature extraction of nodes and relations in the knowledge hypergraphs. Constructing the initialized features of nodes based on the dependent correlations between sequence nodes, a single-layer feed-forward neural network is used to compute the attention coefficients, and the graph structure information is injected into the attention computation process. The calculation process is as follows:

$$\mathrm{b_{ijk}}{=}W[\overrightarrow{h_i}||\overrightarrow{h_j}||\mathrm{r_k}] \tag{1}$$

where W is the weight matrix, $W \in \mathbb{R}^{d' \times d}$. $\overrightarrow{h_i} \in \mathbb{R}^d$ denotes the feature vector of node i, and d denotes the dimensionality of each node. r_k is the relation feature vector.

Then, the LeakyReLU activation function is applied to nonlinearly transform the computed attention weights e_{ijk}. The calculation process is as follows:

$$\mathrm{e_{ijk}} = LeakyReLU(b_{ijk}) \tag{2}$$

To measure the importance of different neighboring nodes within the same hyper-edge to the central node, e_{ijk} is normalized using the softmax function. The calculation process is as follows:

$$a_{ij} = soft\max_{N_i}(e_{ijk}) = \frac{\exp(\overrightarrow{a}(W\overrightarrow{h_i}, W\overrightarrow{h_j}, r_k))}{\sum_{k \in N_i} \exp(\overrightarrow{a}(W\overrightarrow{h_i}, W\overrightarrow{h_j}, r_k))} \tag{3}$$

For node i, due to its small initial feature dimension, a shared weight matrix is first used for feature enhancement. Then, the features of each neighboring node are spliced with the features of node i and mapped to a real number (i.e., it is the correlation coefficient between the neighboring nodes and node i), which is then normalized to the coefficient of interest using the softmax function. The calculation process is as follows:

$$a_{ij} = \frac{\exp(Leaky\mathrm{ReL}U(\overrightarrow{a}^T[W\overrightarrow{h_i}||W\overrightarrow{h_j}||r_k]))}{\sum_{k \in N_i} \exp(Leaky\mathrm{ReL}U(\overrightarrow{a}^T[W\overrightarrow{h_i}||W\overrightarrow{h_j}||r_k]))} \tag{4}$$

Finally, using the attention coefficients as weights, the neighborhood features are weighted and operated, and the new feature vector of the node h_i is obtained by the nonlinear activation function σ. The calculation process is as follows:

$$h_i = \sigma(\sum_{j \in N_i} a_{ij} W \overrightarrow{h_j}) \tag{5}$$

where $||$ denotes the connection operation, $\overrightarrow{a}^T$ the transpose for $\overrightarrow{a}$, and denotes the learnable parameter. and $\overrightarrow{h_i}$ and $\overrightarrow{h_j}$ are the feature vectors of the node and its neighbor nodes h_j. a_{ij} is the attention coefficient of node h_j for h_i. W is the weight matrix.

Furthermore, to enhance the robustness of the link prediction results and ensure that the embedding reliably represents node i, we employ a multihead attention mechanism by independently applying Equation (5) K times. The resulting node embeddings are then concatenated and passed through a multi-head attention layer to derive new features for node h_i, yielding the final output embedding. The calculation process is as follows:

$$\overrightarrow{h_i'} = \|_{k=1}^{K} \sigma\left(\sum_{j \in N_i} a_{ij}^k W^k \overrightarrow{h_j}\right) \tag{6}$$

where a_{ij}^k is the normalized attention coefficient of the k-th attention mechanism a^k. W^k is the weight matrix of the corresponding input linear transformation. σ is the nonlinear activation function.

3.2 Contrastive Learning Optimization Module

The context-level contrastive module is designed to capture the latent semantic information and interrelations among contextual entities, thereby enhancing the semantic expressiveness of embedding representations. By reinforcing the positive and negative sample selection mechanism in contrastive learning, the model's robustness and generalization ability are further improved. Given an anchor entity s, the positive samples of anchor entities are defined as follows.

$$P_S = \{u | (s, r, u) \in T_c(s), u \notin s\} \tag{7}$$

where $T_c(s)$ is the set of middle triples $g_c(s)$.

To match two anchor entities of the same type, and from the generated context subgraphs $g_c(u)$ and $g_c(v)$ can be projected to the same context pattern, the $g_c(s)$ inner neighbor entities are defined as the negative samples between the two. Given a batch of anchor entities ε_B, the negative samples of entities are defined as follows:

$$N_s^{cur} = \{P_i | i \in \varepsilon_B \backslash \{s\}, S_c(i) = S_c(s), t_i = t_s\} \tag{8}$$

This paper uses a dynamic sequence storage mechanism for storing entity embeddings from previous batches. By storing embeddings from previous batches, the model can select negative samples from historical batches, not just limited to the current batch. Specifically, the pre-batch negative samples of entity s are defined as follows:

$$N_s^{pre} = \{P_j | j \in \varepsilon_B^{-1} \cup \varepsilon_B^{-2} ... \cup \varepsilon_B^{-n}, S_c(j) = S_c(s), t_j = t_s\} \tag{9}$$

where ε_B^{-n} denotes the entity of the nth pre-batch. To ensure that the embedding of the prior batch is consistent with the negative samples in the individual, the total pattern negative samples of anchor entities in the context level are defined as follows:

$$N_s^{ia} = N_s^{cur} \cup N_s^{pre} \tag{10}$$

Finally, for the context embedding of the entity, the contrast estimation is performed using the information loss function. The calculation process is as follows:

$$L_s^c = -\log \frac{\sum_{t \in P_s} \exp(\emptyset(c_s, c_t)/\tau)}{\sum_{k \in \{P_s \cup N_s^{ia}\}} \exp(\emptyset(c_s, c_k)/\tau)} \tag{11}$$

where τ is a hyperparameter that controls the sensitivity of the scoring function. Different from previous contrast methods, this paper considers multiple positive samples when calculating the contrast loss.

3.3 Global Enhanced Contrast Module

This section constructs multiple context modes, compares nodes and edges from different perspectives. Suppose u and v are two anchor entities corresponding to two different contextual models. The model can learn a more distinguishable embedding representation by comparing the relationship between anchor entities and their contexts. The calculation process is as follows:

$$N_s^{ie} = \{\varepsilon_c^i | i \in \varepsilon_B \backslash \{s\}, S_c(i) \neq S_c(s)\} \tag{12}$$

A multilayer perceptron (MLP) is used to align embeddings and compute the loss. Firstly, the contextual embedding representation of an entity s is generated based on its neighbors, and contrastive learning is applied to enhance local representation. Then, a global embedding mapping is performed. The specific calculation formulas are as follows:

$$h_s^p = W^2\sigma(W^1 h_s + b^1) + b^2 \tag{13}$$

$$c_s^p = W^2\sigma(W^1 c_s + b^1) + b^2 \tag{14}$$

where σ is the ELU activation function, which plays the role of nonlinear transformation in the mapping process.

After obtaining the contextual and global embeddings, the model applies an entity-level, context-structure cross-temporal contrastive learning strategy by measuring the similarity of embedding pairs. The calculation is as follows:

$$L_s^g = -\log \frac{\sum_{t \in P_s} e^{(\emptyset(c_s^p, c_t^p)/\tau)}}{\sum_{k \in P_s} e^{(\emptyset(c_s^p, c_k^p)/\tau)} + \sum_{k \in N_s^{ie}} e^{(\emptyset(c_s^p, h_k^p)/\tau)}} \tag{15}$$

where $\emptyset$ is the cosine similarity score, a common measure of similarity between two vectors.

3.4 Training Process

The overall workflow of HPAC is summarized in Algorithm 1. Initially, all entity and relation embedding vectors are randomly initialized (Line 1). The iterative procedure defined in Lines 2–12 then executes repeatedly until the specified number of training epochs is reached. Finally, as detailed in Lines 5–11, HPAC employs a cross-entropy loss to compute the training objective and updates the embedding parameters via backpropagation.

Algorithm 1. HPAC model

Inputs: training T, loss function L, number of training rounds n_{epoch}, embedding dimension d, batch size $batch$.
Output: embedded representation of entities and relations.

1: Initialize entities and relations;
2: **for** $t = 1, \ldots, n_{epoch}$ **do**
3:     Divide T into batches of size n_{epoch}, denoted as T_{batch};
4:     $L \leftarrow 0$;
5:     **for** each batch T_{batch} of t **do**
6:         $T^{+}_{batch} \leftarrow$ The set of positive samples generated;
7:         $T^{*} \leftarrow T^{+} \cup T$;
8:         $Score \leftarrow$ Calculate the final score;
9:         $L \leftarrow$ Calculate the loss;
10:         Backpropagation updates the embedding vectors of entities and relations, updating all parameters of the model;
11:     **end for**
12: **end for**

4 Performance Evaluation

In this section, we compare the link prediction ability of the HPAC algorithm proposed in this paper with that of the baseline models on two commonly used real-world datasets. To further assess the impact of the model's components and hyperparameters, we conduct ablation studies and parameter sensitivity analyses.

4.1 Experimental Setting and Baseline Models

In this paper, we use two publicly available and commonly used knowledge hypergraph link prediction datasets for our experiments: the JF17K [9], and the FB-AUTO [2]. Table 1 shows statistics.

The baseline models used in this paper include the translation-based m-TransH [9] and RAE [11] methods and the neural network-based HINGE [16], NeuInfer [17], HypE [2], tNaLP+ [14], HyperMLN [18], RD-MPNN [19], and HPGC [4] methods. The operating system used in the experimental environment is Ubuntu, the CPU is Intel Core i5-11400F, and the GPU is NVIDIA GeForce

Table 1. Datasets statistics

Datasets	$\|E\|$	$\|R\|$	#train	#valid	#test
JF17K	29177	327	61104	-	24195
FB-AUTO	3388	8	6778	2255	2180

RTX 3090. The HPAC model is implemented using the PyTorch framework. For all datasets, the embedding dimension size is set to 128 and the learning rate is set to 0.01; the batch size is set to 128; and the epoch value is set to 500. The training process uses Dropout to prevent over-fitting, where the Dropout ratio is set to 0.2 and the L2 regularization ratio is set to 0.001. In this paper, we use the evaluation criteria commonly used for link prediction in knowledge hypergraphs, mean reverse rank (MRR) [2] and Hit@n [2]. The larger the value of MRR and Hit@n, the better the link prediction performance. In this paper, we experimentally take MRR, Hit@1, Hit@3, and Hit@10 as performance index criteria to evaluate the effectiveness of the model proposed in the text.

4.2 Comparison of Link Prediction Performance

This section compares the MRR, Hit@1, Hit@3, and Hit@10 of the HPAC and baseline models proposed in this paper on two datasets, respectively. The experimental results are shown in Table 2, where the bolded data are the optimal results for each evaluation metric. The experimental results show that HPAC outperforms other baseline models. This is because traditional models like m-TransH and RAE maintain translation invariance but fail with higher-order relations. NeuInfer and HINGE lose structural and semantic details, while HypE emphasizes local structure, ignoring global semantics. tNaLP+ uses key-value pairs for

Table 2. Comparison of link prediction performance of HPAC and baseline methods

Model	JF17K				FB-AUTO			
	MRR	Hit@1	Hit@3	Hit@10	MRR	Hit@1	Hit@3	Hit@10
m-TransE	0.458	0.383	0.490	0.596	0.645	0.549	0.710	0.809
RAE	0.315	0.215	0.433	0.651	0.703	0.612	0.764	0.712
HINGE	0.449	0.424	0.490	0.618	0.679	0.630	0.710	0.765
NeuInfer	0.415	0.436	0.553	0.675	0.730	0.621	0.755	0.805
HypE	0.491	0.404	0.535	0.655	0.734	0.661	0.799	0.839
tNaLP	0.449	0.370	0.484	0.598	0.729	0.645	0.748	0.826
HyperMLN	0.556	0.472	0.581	0.617	0.765	0.645	0.801	0.837
RD-MPNN	0.512	0.445	0.573	0.680	0.795	0.670	0.818	0.879
HPGC	0.603	0.512	0.601	0.682	0.801	0.672	0.821	0.885
HPAC	**0.611**	**0.523**	**0.616**	**0.693**	**0.811**	**0.684**	**0.828**	**0.887**

multivariate relations, causing information loss. HyperMLN focuses on symbolic reasoning but underutilizes neighbor features. RD-MPNN captures structural dynamics but lacks semantic depth, resulting in inferior performance.

4.3 Ablation Experiments

To further validate the effectiveness of the proposed model, we perform ablation studies on three components: attention-based feature fusion (w/o AT), contrastive learning (w/o CL), and global enhancement modules (w/o GL). Table 3 shows that HPAC can effectively represent the implicit dependencies and semantic consistency between different relational facts, thereby enhancing model discrimination. By integrating attention mechanisms and relational contrastive learning, HAPC better models complex node-relation dependencies and improves semantic understanding of hypergraphs.

Table 3. Comparison of link prediction performance of HPAC and ablated variants

Model	JF17K				FB-AUTO			
	MRR	Hit@1	Hit@3	Hit@10	MRR	Hit@1	Hit@3	Hit@10
w/o AT	0.480	0.452	0.522	0.553	0.614	0.533	0.702	0.743
w/o CL	0.491	0.464	0.542	0.571	0.637	0.558	0.741	0.761
w/o GL	0.552	0.501	0.552	0.631	0.759	0.645	0.781	0.836
HPAC	**0.611**	**0.523**	**0.616**	**0.693**	**0.811**	**0.684**	**0.828**	**0.887**

4.4 Parameter Analysis

In order to investigate the effect of embedding dimensions on the model, experiments using different embedding dimensions are conducted on the JF17K and FB-AUTO datasets, with results shown in Fig. 3. Performance improves as embedding dimension increases but plateaus after a certain point.

The experimental results in Fig. 3 shows that performance improves with increasing embedding dimension but eventually stabilizes. When the dimension is too small, the model lacks sufficient capacity to capture complex structural patterns. However, beyond a certain threshold, the benefit of additional dimensions diminishes, as essential feature information has already been encoded. Moreover, excessively high dimensions may lead to overfitting, where the model captures noise or irrelevant patterns, reducing its generalization to unseen data.

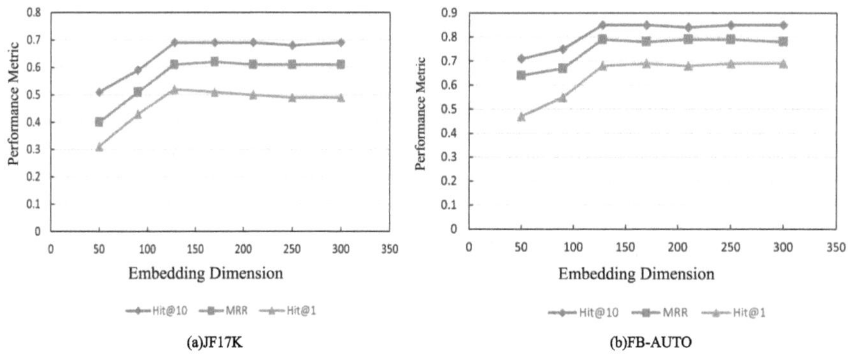

Fig. 3. Effect of embedding dimensions on model performance.

5 Conclusion

In this paper, a knowledge hypergraph link prediction method called HPAC is proposed based on attention mechanism and relational contrast learning. HPAC employs an optimized attention mechanism to extract key features of nodes and relations, assign importance weights, and integrate both local and global information to capture complex dependencies. Building on contrastive learning principles, we design a context-hierarchical contrast model to capture latent semantic information among contextual entities. By constructing multiple contextual patterns using cross-view contrastive strategies, HPAC enhances node and edge representations, and captures diverse semantic features. Extensive experimental results on real-world datasets demonstrate the superior performance of the proposed HPAC model.

Acknowledgements. This work is supported by the National Natural Science Foundation of China (No. 62372342), the General Program of National Natural Science Foundation of Hubei Province (No. 2024AFB865).

References

1. Du, X., Liu, M., Shen, L.: A review of knowledge graph representation learning methods for link prediction. J. Softw. **35**(1), 87–117 (2024)
2. Fatemi, B., Taslakian, P., Vázquez, D., et al.: Knowledge hypergraphs: prediction beyond binary relations. In: Proceedings of the Twenty-Ninth International Joint Conference on Artificial Intelligence, pp. 2191–2197 (2020)
3. Pang, J., Xu, H., Qin, H.C., et al.: Link prediction in knowledge hypergraphs via joint attention and convolutional networks. J. Front. Comput. Sci. Technol. **17**(11), 2734–2742 (2023)
4. Pang, J., Ma, Z., Lin, X., et al.: Link prediction in knowledge hypergraphs combining GAT and convolutional neural networks. Comput. Eng. Appl. **61**(9), 194–201 (2025)

5. Burke, R.: Knowledge-based recommender systems. Encyclopedia Libr. Inf. Syst. **69**(Supplement 32), 175–186 (2000)
6. Antonio, P., Gianvito, P., Domenico, R., et al.: SAIRUS: spatially-aware identification of risky users in social networks. Inf. Fusion **92**, 435–449 (2023)
7. Li, Z., Wang, C., Wang, X., Chen, Z., Li, J.: HJE: joint convolutional representation learning for knowledge hypergraph completion. IEEE Trans. Knowl. Data Eng. **36**(8), 3879–3892 (2024)
8. Li, Z., Wang, X., Zhao, J., Guo, W., Li, J.: HyCubE: efficient knowledge hypergraph 3D circular convolutional embedding. IEEE Trans. Knowl. Data Eng. **37**(4), 1902–1914 (2025)
9. Wen, J., Li, J., Mao, Y.: On the representation and embedding of knowledge bases beyond binary relations. CoRR abs/1604.08642 (2016)
10. Wang, Z., Zhang, J., Feng, J., et al.: Knowledge graph embedding by translating on hyperplanes. In: Proceedings of the AAAI Conference on Artificial Intelligence, vol. 28, no. 1 (2014)
11. Zhang, R., Li, J., Mei, J., et al.: Scalable instance reconstruction in knowledge bases via relatedness affiliated embedding. In: Proceedings of the 2018 World Wide Web Conference, pp. 1185–1194 (2018)
12. Pang, J., Qin, H.C., Liu, Y., et al.: Two birds with one stone: a link prediction model for knowledge hypergraph based on fully-connected tensor decomposition. In: International Conference on Advanced Data Mining and Applications, pp. 78–90. Springer, Cham (2023)
13. Liu, Y., Yao, Q., Li, Y.: Generalizing tensor decomposition for N-ary relational knowledge bases. In: Proceedings of the Web Conference 2020, pp. 1104–1114 (2020)
14. Guan, S., Jin, X., Guo, J., et al.: Link prediction on N-ary relational data based on relatedness evaluation. IEEE Trans. Knowl. Data Eng. **35**(1), 1 (2021)
15. Guan, S., Jin, X., Wang, Y., et al.: Link prediction on N-ary relational data. In: The World Wide Web Conference, pp. 583–593 (2019)
16. Rosso, P., Yang, D., Cudré-Mauroux, P.: Beyond triplets: hyper-relational knowledge graph embedding for link prediction. In: Proceedings of The Web Conference 2020, pp. 1885–1896 (2020)
17. Guan, S., Jin, X., Guo, J., et al.: Neuinfer: knowledge inference on N-ary facts. In: Proceedings of the 58th Annual Meeting of the Association for Computational Linguistics, pp. 6141–6151 (2020)
18. Chen, Z., Wang, X., Wang, C.: Explainable link prediction in knowledge hypergraphs. In: Proceedings of the 31st ACM International Conference on Information & Knowledge Management, pp. 262–271 (2022)
19. Zhou, X., Hui, B., Zeira, I., et al.: Dynamic relation learning for link prediction in knowledge hypergraphs. Appl. Intell. **53**(22), 26580–26591 (2023)
20. Wang, G., Zeng, Y., Li, R.H., et al.: Temporal graph cube. IEEE Trans. Knowl. Data Eng. **35**(12), 13015–13030 (2023)
21. Zhang, W.H., Xu, Z.S., Liu, N., et al.: Overview of knowledge graph completion methods. Comput. Eng. Appl. 1–16 (2024)
22. Velickovic, P., Cucurull, G., Casanova, A., et al.: Graph attention networks. arXiv preprint arXiv:1710.10903 (2017)

GDMMA Workshop

Multi-hop Reasoning Method Oriented Question Answering in Time-Constrained Temporal Knowledge Graph

Yingping Su(✉) and Dechang Pi

College of Computer Science and Technology, Nanjing University of Aeronautics and Astronautics, Nanjing, China
{sypcs16,pinuaa}@nuaa.edu.cn

Abstract. Knowledge Graph Question Answering (KGQA) has gained significant attention in recent years, but many existing methods struggle to effectively reason over temporal constraints. This limitation reduces accuracy in temporal contexts and creates challenges in maintaining consistency and detecting conflicts within temporal knowledge graphs. To address this, a temporal constraint-based multi-hop reasoning approach is proposed, consisting of four modules: temporal information parsing, fact retrieval, constraint extraction and subgraph temporal reasoning. By incorporating temporal constraints, the model identifies implicit time-related information, thereby improving answer accuracy. Unlike conventional methods that focus on entities, this approach applies the attention mechanism at the relationship level to enhance reasoning performance. Experiments conducted on two public datasets, CronQuestions and ComplexCronQuestions, demonstrate the method's superiority over state-of-the-art models. On CronQuestions, the model improves Hits@1 by 4.8% for complex questions and 3.9% for entity answers. On ComplexCronQuestions, it achieves Hits@1 gains of 10.4%, 6.8%, and 2.0% for temporal precedence, extremum, and conjunction questions, respectively. Additionally, a 7.9% improvement in Hits@10 is observed for temporal precedence questions, with an overall Hits@1 increase of 8.2%, highlighting its effectiveness in temporal KGQA tasks.

Keywords: Question Answering · Attention Mechanism · Temporal Knowledge Graph

1 Introduction

With the rapid growth of information technology, there is increasing demand for efficient and accurate retrieval from large-scale data [1]. Question answering (QA) systems have emerged as a key solution in artificial intelligence and natural language processing, aiming to understand user intent and return precise, natural language answers. Compared with traditional keyword-based retrieval, QA systems improve efficiency by reducing irrelevant information and user effort. Advances in deep learning and large-scale language models have further

J. Li et al. (Eds.): APWeb-WAIM 2025, CCIS 2660, pp. 299–311, 2026.
https://doi.org/10.1007/978-981-95-7996-9_26

broadened QA applications across domains such as customer service, education, healthcare, and law [2].

Among QA methods, knowledge graph-based QA (KGQA) has gained prominence due to its structured representation of entities and relations. Knowledge graphs (KGs) enable semantic reasoning beyond surface-level text and support fact-based queries using large datasets. While KGQA performs well on simple "single-hop" questions, it struggles with complex multi-hop reasoning, which requires traversing multiple relations [3].

Embedding-based methods support multi-hop reasoning by learning dense representations of entities and relations. However, they often ignore the temporal nature of real-world facts. Traditional static KGs miss temporal constraints, limiting their ability to handle time-sensitive queries. Temporal Knowledge Graphs (TKGs) solve this by attaching timestamps to facts, enabling reasoning over dynamic events and causal relationships.

Temporal KGQA (TKGQA) systems aim to answer questions requiring semantic and temporal reasoning. However, current models face key challenges such as insufficient use of temporal relational semantics, poor handling of temporal logic, and reliance on static embeddings that ignore evolving contexts. To address these challenges, a novel temporal knowledge graph question answering model is proposed, integrating temporal logic constraints with relation-aware reasoning. Our contributions are as follows:

(1) A novel model is proposed to address temporal knowledge graph-based question answering. This model improves the capability to identify the temporal information embedded in the question.
(2) Temporal constraints are introduced to ensure temporal consistency and detect conflicts in knowledge graphs, improving answer accuracy.
(3) The model applies attention to relationships using a transformer framework, capturing long-range dependencies and distinguishing relationships, thereby improving reasoning accuracy.

The remainder of this paper is organized as follows. Section 2 reviews related work. Section 3 introduces our proposed model. Section 4 describes the experimental setup and results. Finally, Sect. 5 concludes the paper and outlines future work.

2 Related Work

Temporal knowledge graph question answering (TKGQA) poses unique challenges. Building on this context, our model incorporates specialized modules that improve temporal understanding and constraint extraction in QA tasks.

EaE [4] combines entity embeddings with Transformer representations to improve entity awareness but lacks temporal modeling. Our model adds explicit temporal parsing and constraint handling for more accurate answers.

TempoQR [5] uses temporal vectors and TComplEx embeddings to enhance temporal awareness but doesn't support dynamic constraints. Our model constructs temporal and logical constraints during inference, offering better flexibility.

CTRN [6] applies attention and temporal GCNs to extract temporal features but overlooks relational structures. Our model captures both temporal and relational dependencies using Transformer-based relation encoding.

TSQA [7] leverage LSTMs, GNNs, and contrastive learning for entity and timestamp modeling but are limited by static architectures. Our modular design supports joint reasoning across entities, relations, and time.

SubGTR [8] transforms implicit constraints into explicit ones using rule matching but depends on pre-trained embeddings and fixed scoring. Our model builds constraints dynamically, enhancing interpretability and adaptability.

TMA [9] generates temporal features via a multi-directional module with pre-trained embeddings. Our model further incorporates dynamic temporal parsing and constraint extraction to improve accuracy and generalization.

GATQR [10] uses GATs and a temporal knowledge base to detect temporal cues, aided by RoBERTa-based relation representations, but lacks constraint modeling. Our model fills this gap by constructing constraints dynamically during inference.

Prog-TQA [11] combines temporal operators and self-boosting with draft generation to improve temporal understanding. Our model complements this with fine-grained parsing and constraint modules for better time-sensitive reasoning.

RTA [12] applies reinforcement learning and path matching to capture evolving relations but omits temporal constraints. Our model explicitly incorporates them, enabling unified reasoning over time, entities, and relations.

3 The Proposed Model

3.1 Overview of the Model

The overall architecture of our model is shown in Fig. 1. The proposed model consists of four modules: a) Temporal Information Parsing Module, b) Fact Retrieval Module, c) Constraint Extraction Module, and d) Temporal Reasoning Module. First, the implicit temporal information in the question is explicitly rewritten by leveraging background knowledge from the Temporal Knowledge Graph (TKG) to simplify the problem. Then, embeddings are employed to retrieve candidate entities from the TKG and compute initial scores. Subsequently, frequent candidate constraints are mined from the knowledge graph, and the confidence scores of these constraints are calculated. These constraints are refined using dynamically adjusted strategies, resulting in an accurate set of temporal constraints. Finally, the answer entity is obtained by constructing a subgraph corresponding to the question, and temporal constraints are applied for pruning and logical reasoning.

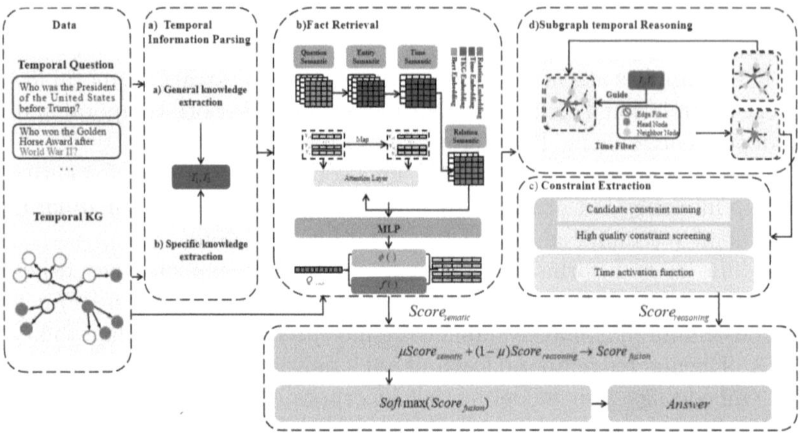

Fig. 1. The Architecture of the Model Proposed.

3.2 Temporal Information Parsing Module

This module includes two parts: basic knowledge extraction, which retrieves common event-based knowledge, and specific knowledge extraction, which focuses on temporal details from domain knowledge bases.

Basic Knowledge Extraction. Many questions use well-known events as temporal constraints, and temporal graphs convert them into explicit data. Queries in the TKG retrieve the event's start time T_1 and end time T_2, replacing the event text with these time expressions [13].

Specific Knowledge Extraction. Implicit time information can also appear in complex questions, such as "Who was the President of the United States before Trump?" Simply querying "Trump" lacks the necessary time context. Key information like "President of the United States" and "Trump" must be combined to retrieve the relevant temporal details.

For a given question involving multiple entities e_1 ande_2, we first perform a joint query on the knowledge graph to extract the relevant tuples G_r corresponding to the pair(e_1, e_2) from the TKG, the calculation process is shown in Eq. (1).

$$\mathrm{Gr} = \{(e_h, r, e_t, t_1, t_2) \mid e_h \in [e_1, e_2], e_t \in [e_1, e_2]\} \tag{1}$$

Then we merge and sort all tuples $t1$ and $t2$ in G_r based on their values, with the smallest value marked as start time $t1$ and the largest value marked as end time $t2$. Finally, regular expressions are used to replace the entity text q_{entity} in the natural language question q_{text} with the corresponding time expressions T_1 and T_2.

3.3 Fact Retrieval Module

In the fact retrieval module of the temporal knowledge graph question answering model, the natural language question is first transformed into a vector representation that encapsulates the question semantics, temporal knowledge graph semantics, and time information. This process combines pre-trained language models with temporal knowledge graph embeddings. Subsequently, the TKG scoring function and the time weighting function are applied to evaluate candidate entities and calculate their corresponding semantic scores.

3.3.1 Question Embedding

A pre-trained language model is used to extract the semantic information of the natural language question. Specifically, the question is transformed into a semantic matrix using the pre-trained RoBERTa [14] model.

Semantic Information of Entities. To capture entity knowledge in the question, entity and timestamp tokens Q_R are replaced with their pre-trained TKG embeddings, as shown in Eq. (2).

$$q_{Ei} = \begin{cases} W_E e_\xi & \text{if token is linked to an entity } \xi, \\ W_E t_\tau & \text{if token is linked to a timestamp } \tau, \\ q_R & \text{otherwise.} \end{cases} \tag{2}$$

where W_E is a projection matrix, e_ξ and t_τ are pre-trained entity embeddings and time embeddings, respectively.

Semantic Information of Time. The temporal information parsing module explicitly transforms time information in the question. To avoid increasing model parameters, this paper uses a split strategy that divides the timestamp into three parts: year, month, and day. Then, the t_y, t_m, and t_d are combined using vector concatenation to form a timestamp vector, as shown in Eq. (3).

$$T_t = T_y \oplus T_m \oplus T_d \tag{3}$$

where t_y, t_m, and $t_d \in R^k$denote vectors of years, months, and days, respectively.

In this paper, the construction of the mapping space W involves calculating the transformation matrix based on v_s and T_t^T. This process is detailed in Eq. (4).

$$W = v_s T_t^T \tag{4}$$

where v_s denotes the vectors for the entity s, T_t^T denotes the vectors for the timestamp t, and T denotes the transpose.

To enhance the perception of temporal information, temporal semantic embeddings extracted from the knowledge graph are integrated into the question representation, as shown in Eq. (5).

$$q_{T_i} = \begin{cases} q_{E_i} & \text{if token } i \in \text{entity}, \\ q_{E_i} + T_1 + T_2 & \text{if token } i \in \text{entity}. \end{cases} \tag{5}$$

where $Q_T = [q_{T_1}, q_{T_2}, ..., q_{T_N}]$.

Semantic Information of Relation. A relation-focused attention layer is introduced, using Scaled Dot-Product and Multi-head Attention to compute attention among relations. The query v'_r represents relation r, with keys and values from K_{A_e} and V_{A_e}, based on sequence A_e. The output is v''_r, as shown in Eq. (6).

$$v''_r = \text{Attn}(v'_r, K_{A_e}, V_{A_c}) \tag{6}$$

where Attn () represents the procedure for computing attention.

Following the "Scaled Dot-Product Attention" approach, we compute the attention value for each $r_i \in A_e$. The attention value represents the importance of each relationship r_i to the source relationship r, denoted as α_i. The relative attention value λ_i is calculated by the Softmax function over all attention values, as described in Eq. (7).

$$\lambda_i = \text{Softmax}(\alpha_i) = \frac{\exp(\alpha_i)}{\sum_{r_k \in \mathcal{A}_e} \exp(\alpha_k)} \tag{7}$$

where α_i denotes the attention value and λ_i denotes the relative attention value.

To capture various relational information, a multi-head attention mechanism [15] is introduced, using H parallel attention heads to compute embedding vectors, each focusing on different aspects of the input. The outputs are concatenated to integrate the information, as shown in Eq. (8).

$$v''_r = \prod_{h=1}^{H} \sum_{r_i \in \mathcal{A}_e} \lambda_i^k v'^k_{r_i} \tag{8}$$

where $\prod$ denotes tandem operation.

Semantic Fusion. The information fusion layer integrates context, entity, temporal, and relational information into a unified query representation q using a Transformer encoder with two layers. A relation-aware attention mechanism adjusts attention weights, and the final token embedding matrix Q_E combines all information.

3.3.2 Semantic Scoring Function

The semantic score of the entity $\xi \in E$ as an answer is given by Eq. (9),

$$score_{entity} = \max\left(\phi(e_s, P_E q, e_\xi, t_\tau), \phi(e_o, P_E q, e_\xi, t_\tau)\right) \tag{9}$$

where P_E is a matrix that can be learned specific to the prediction of entities.

In addition, the correctness of timestamp t is determined based on a specific scoring criterion, which is defined and computed in detail in Eq. (10).

$$score_{time} = \max\left(\phi(e_s, P_T q, e_o, t_\tau), f_{sentity}(P_T q, t_\tau)\right) \tag{10}$$

where s and o are the entities with attributes in the query and $score_{time}$ represents the score of the timestamp candidate answer.

Finally, the semantic scores of the candidate answers, including entity and timestamp candidates, are computed as detailed in Eq. (11).

$$score_{semantic} = score_{entity} \oplus score_{time} \tag{11}$$

where $\oplus$ denotes vector concatenation.

3.4 Constraint Extraction Module

The constraint extraction module detects frequent constraint patterns in the knowledge graph by analyzing entities and relations. It involves two steps: mining and evaluation. Starting from a structural pattern (SP), it generates graph patterns (GP) by matching entities and filling attributes using local data, then adds temporal predicates to form temporal constraints. Patterns with high support and confidence are kept; others are discarded.

The quality of a temporal constraint tc is evaluated on $IG(tc)$, the set of facts matching its graph pattern. We use entity-level confidence, defined in Eq. (12), to measure this quality.

$$\text{confidence} = \frac{\#\text{entities}_\text{p}\text{os}}{\#\text{entities}_\text{p}\text{os} + \#\text{entities}_\text{n}\text{eg}} \tag{12}$$

where $entities_{pos}$ and $entities_{neg}$ are subsets of subject entities in $IG(tc)$.

The logical value is computed based on the time interval $T = (t.s, t.e)$, where $t.s$ and $t.e$ represent the start and end times of the event.

When the confidence of a certain constraint c, does not meet the strict threshold θ_{c_1} but exceeds the relaxed threshold θ_{c_2}, it attempts to limit the scope to a specific combination of entity categories [16]. Ultimately, refined constraints that satisfy the support $s \geq \theta_{freq}$ and confidence $c \geq \theta_{c_1}$ will be added to the final result set.

3.5 Subgraph Temporal Reasoning Module

The subgraph temporal reasoning module constructs a local temporal graph of relevant entities and time nodes for each question. It filters out irrelevant elements using temporal constraints, then scores the rest with a temporal activation function to enhance reasoning accuracy.

3.5.1 Local Graph Extraction

To construct the local graph, a relevant subgraph is extracted from the temporal knowledge graph based on the question. Let $G = (V, R)$ be the knowledge graph, with v_{center} as the central entity. The local graph includes the L-hop neighborhood $N_L(v_{center})$, containing all nodes and edges reachable within L steps [17].

All relevant facts in the TKG are queried, and the associated entities and temporal information are extracted to construct the temporal local graph. Let $N_L(v_{center})$ denote the set of L-hop neighborhood nodes of v_{center} in the knowledge graph, and $R_L(v_{center})$ represent the corresponding edges. The calculation process is detailed in Eq. (13).

$$G_L(v_{\text{center}}) = \{(e, r) \mid e \in N_L(v_{\text{center}}), r \in R_L(v_{\text{center}})\} \tag{13}$$

where v_{center} is the central node, and $G_L(v_{center})$ denotes the corresponding L-hop local graph of v_{center}.

Paths P are extracted from neighboring nodes to the central node to capture temporal relationships and semantic associations. Each path P consists of a sequence of nodes and edges, as defined in Eq. (14). Each edge e_k in the path is checked against temporal constraints. Edges that fail are pruned, paths are updated, and isolated nodes are removed.

$$P = \{v_0, e_1, v_1, e_2, v_2, ..., e_k, v_k\} \tag{14}$$

where v_0 is the central node, v_k is its neighboring node, and e_k is the edge.

3.5.2 Answer Scoring

In temporal constraint evaluation, specific scoring mechanisms measure the relevance of entities to the given time range Time for different question types. Below are the scoring methods for each type:

1) Temporal Conjunction Type. For temporal conjunction questions, entities with time between $Time_1$ and $Time_2$ receive a positive score; others receive 0, as defined in Eq. (15). The C is a constant.

$$Score_1(e_i) = C \cdot \mathbf{1}_{\{Time_1 < \text{time}_i < Time_2\}} \tag{15}$$

2) Temporal Extremity Type. For "First" or "Last" questions, scores are assigned based on temporal precedence: earlier times for "First" and later times for "Last" receive higher scores, as shown in Eq. (16).

$$Score_2(e_i) = -a_i \cdot \mathbf{1}_{\{\text{first}\}} + a_i \cdot \mathbf{1}_{\{\text{last}\}} \tag{16}$$

3) Temporal Precedence Type. For "Before" or "After" questions, entities closer in time to T_1 or T_2 receive higher scores, as detailed in Eq. (17).

$$Score_3(e_i) = (T_1 - t_i) \cdot \mathbf{1}_{\{\text{before}\}} + (t_i - T_2) \cdot \mathbf{1}_{\{\text{after}\}} \tag{17}$$

4) Temporal Matching Type. For "Equal" questions, entities whose time exactly matches the specified Time receive a positive score, while others receive 0, as shown in Eq. (18).

$$Score_4(e_i) = M \cdot \mathbf{1}_{\{t_i=\text{Time}\}} \tag{18}$$

4 Experiments

4.1 Experimental Setup

In this section, the datasets and evaluation metrics are first introduced. Then, baseline methods for comparison are presented, followed by a description of the implementation details.

Datasets and Evaluation Metrics. This study uses two datasets: CronQuestions and ComplexCronQuestions [18]. CronQuestions contains 125k entities and 328k facts, covering various temporal queries but includes irrelevant pseudo-temporal questions. ComplexCronQuestions refines CronQuestions by removing irrelevant questions, focusing on those requiring temporal reasoning with at least five relevant facts. Evaluation for each test event (s, r, o, t) is based on Hits@1 and Hits@10 [19]. Hits@i reflects the proportion of correct predictions in the top-i, with higher values signifying better accuracy, as shown in Eq. (19).

$$\text{Hit@10} = \frac{\sum_{(h, r_{\text{tem}}, t) \in \text{test}} val\left(\text{rank}(h, r_{\text{tem}}, t) \leq 10\right)}{|\text{test}|} \tag{19}$$

Baselines and Implementation Details. The proposed model is compared with recent TKGQA methods across four categories: knowledge graph embedding-based models (EaE [4], T53B [20]), temporal embedding-based models (CronKGQA [18], TSQA [7]), and specialized models for the TimeQuestions dataset (TempoQR [5], CTRN [21], TMA [9], GATQR [10], Prog-TQA [11] and RTA [12]). The model is implemented in PyTorch, with TComplEx for temporal KG embeddings and a pre-trained RoBERTa for semantic search; both components remain fixed during training.

4.2 Overall Results in Cronquestions

Table 1 presents a comparison of our method with several baseline models from 2020 to 2025 on the CronQuestions dataset, focusing on Hits@1 and Hits@10 as key evaluation metrics. The best scores are marked in bold, while the second-best scores are underlined. The proposed model achieves a Hits@1 score of 0.967, making it the best-performing model. It outperforms the second-best model,

Table 1. Overall Results.

Model	Hits@1					Hits@10				
	Overall	Question Type		Answer Type		Overall	Question Type		Answer Type	
		Complex	Simple	Entity	Time		Complex	Simple	Entity	Time
EaE	0.288	0.257	0.329	0.318	0.231	0.678	0.623	0.725	0.850	0.341
T53B	0.252	0.240	0.251	0.283	0.180	–	–	–	–	–
CronKGQA	0.647	0.392	0.987	0.699	0.549	0.884	0.802	0.990	0.898	0.857
TSQA	0.831	0.713	0.987	0.829	0.836	0.980	0.968	0.997	0.981	0.978
TempoQR	0.918	0.864	0.990	0.926	0.903	0.978	0.967	0.993	0.980	0.974
CTRN	0.920	0.869	0.900	0.921	0.917	0.980	0.970	0.993	0.982	0.976
TMA	0.784	0.632	0.987	0.792	0.743	0.943	0.904	0.995	0.947	0.936
GATQR	0.848	0.738	0.988	0.937	0.681	0.957	0.927	0.997	0.986	0.940
Prog-TQA	0.937	0.898	0.985	0.914	0.982	0.973	0.960	0.993	0.968	0.982
RTA	0.930	0.885	0.987	0.876	**0.988**	0.989	0.973	**0.999**	0.989	0.981
Ours	**0.967**	**0.936**	**0.994**	**0.971**	0.968	**0.990**	**0.987**	0.990	**0.993**	**0.991**

RTA (2025), which scores 0.930, showing an improvement of 4.0%. In Hits@10, the proposed model scores 0.990, slightly surpassing RTA's 0.989.

For complex questions, ours model scores 0.936 in Hits@1, which is a 5.8% improvement over RTA's 0.885. In Hits@10, it scores 0.987, outperforming RTA's 0.973 by 1.4%. These results highlight the model's strong ability in multi-faceted temporal reasoning tasks.

For simple questions, ours model scores 0.994 in Hits@1, ahead of RTA's 0.987, showcasing excellent precision in simpler queries. In Hits@10, it scores 0.990, slightly trailing RTA's 0.999 but still maintaining high performance. In entity-type answer tasks, ours model achieves 0.971 in Hits@1, a 10.8% improvement over RTA's 0.876. In Hits@10, it scores 0.993, outperforming RTA's 0.989. For time-type answers, the model reaches 0.968 in Hits@1, surpassing RTA's 0.988, although in Hits@10, it scores 0.991, slightly behind RTA's 0.999. Nonetheless, the model shows strong temporal reasoning capabilities.

Table 2. Results on the ComplexCronQuestions Dataset

Model	Hits@1				Hits@10			
	Overall	BA	FL	TJ	Overall	BA	FL	TJ
CronKGQA	0.266	0.252	0.235	0.458	0.767	0.645	0.777	0.851
EntityQR	0.425	0.473	0.365	0.733	0.859	0.855	0.845	0.951
TempoQR	0.792	0.683	0.789	0.937	0.959	0.923	0.960	0.990
Prog-TQA	0.766	0.618	0.693	0.954	0.826	0.893	0.911	0.994
Ours	0.857	0.754	0.843	0.973	0.970	0.995	0.963	0.953
Improvement (%)	**8.2%**	**10.4%**	**6.8%**	**2.0%**	**1.2%**	**7.9%**	**0.4%**	**-0.6%**

Table 2 shows that the proposed model outperforms the RTA(2025) baseline on the ComplexCronQuestions dataset. It achieves 4.0% higher overall Hits@1, with notable improvements on complex questions, increasing Hits@1 by 5.8%, and on entity-type queries, with a 10.8% gain. These results highlight the model's strength in handling complex and entity-focused temporal reasoning tasks.

4.3 Ablation Study

To assess the contribution of each module within the proposed model, an ablation study was conducted on the ComplexCronQuestions dataset. The key components were individually removed: the Temporal Information Parsing Module ("Time"), the Constraint Extraction Module ("CE"), the Fact Retrieval Module ("Relation") and time weighting function("f_{sensity}"). Table 3 shows that removing the Temporal Parsing module causes the largest drop, with Time_Join Hits@1 decreasing by 7.8%, highlighting its importance. Further ablation of the time weighting function f_{sensity} leads to a 4.0% decrease in Time_Join Hits@1, indicating its independent contribution to modeling temporal relevance. Removing the Constraint Extraction module also significantly lowers performance, with First/Last Hits@1 dropping by 9.5%, while excluding the Fact Retrieval module results in smaller but consistent declines, such as a 5.4% drop in Before/After Hits@1.

Table 3. Ablation Results on the ComplexCronQuestions Dataset.

Model	Hits@1			Hits@10		
	Before/After	First/Last	Time_Join	Before/After	First/Last	Time_Join
Ours(w/o Time)	0.665	0.723	0.897	0.916	0.895	0.918
Ours(w/o CE)	0.684	0.748	0.922	0.944	0.902	0.935
Ours(w/o Relation)	0.713	0.793	0.943	0.973	0.942	0.941
Ours(w/o f_{sensity})	0.707	0.781	0.933	0.965	0.938	0.939
Ours	**0.754**	**0.843**	**0.973**	**0.995**	**0.963**	**0.953**

5 Conclusions and the Future Work

This paper introduces a time-aware model for answering complex questions in temporal knowledge graphs. Experiments show the model outperforms existing methods, effectively handling time and constraints to provide accurate answers. However, limitations include the ComplexCronQuestions dataset's limited relationship types, which affects diversity and generalization, and challenges in handling large-scale knowledge graphs, impacting reasoning speed and accuracy. Future research will focus on improving scalability with more efficient temporal functions and exploring diverse datasets to enhance generalization and real-world performance.

Funding Information. This work was supported by the National Science and Technology Major Project under Grant 2021ZD0113103.

References

1. Chen, X., Jia, S., Xiang, Y.: A review: knowledge reasoning over knowledge graph. Expert Syst. Appl. **141**, 112948 (2020)
2. Qamar, F., Latif, S., Shah, A.: Techniques, datasets, evaluation metrics and future directions of a question answering system. Knowl. Inf. Syst. **66**(4), 2235–2268 (2024)
3. Bai, L., Zhang, H., An, X., et al.: Few-shot multi-hop reasoning via reinforcement learning and path search strategy over temporal knowledge graphs. Inf. Process. Manage. **62**(3), 104001 (2025)
4. Thibault, F., Livio, B.S., Nicholas, F., Eunsol, C., Tom, K., et al.: Entities as experts: sparse memory access with entity supervision. computing research repository, pp. 4937–4951 (2020)
5. Mavromatis, C., Subramanyam, P.L., Ioannidis, V.N., et al.: Tempoqr: temporal question reasoning over knowledge graphs. In: Proceedings of the AAAI Conference on Artificial Intelligence, vol. 36, no. 5, pp. 5825–5833 (2022). https://doi.org/10.1609/aaai.v36i5.20526
6. Jiao, S., Zhu, Z., Wu, W., et al.: An improving reasoning network for complex question answering over temporal knowledge graphs. Appl. Intell. **53**(7), 8195–8208 (2023)
7. Chao, S., Guangtao, W., Peng, Q., Jing, H., et al.: Improving time sensitivity for question answering over temporal knowledge graphs. In: Proceedings of the 60th Annual Meeting of the Association for Computational Linguistics, (Volume 1 Long Papers), pp. 8017–8026 (2022)
8. Chen, Z., Zhao, X., Liao, J., et al.: Temporal knowledge graph question answering via subgraph reasoning. Knowl.-Based Syst. **251**, 109134 (2022)
9. Liu, Y., Liang, D., Fang, F., et al.: Time-aware multiway adaptive fusion network for temporal knowledge graph question answering. In: ICASSP 2023 – IEEE International Conference on Acoustics, Speech and Signal Processing, pp. 1–5. IEEE, Rhodes Island (2023). https://doi.org/10.1109/ICASSP49357.2023.10095395
10. Jiang, W., Guo, Y., Fu, J.: Temporal knowledge graph question answering models enhanced with GAT. In: BigData 2023 – IEEE International Conference on Big Data, pp. 1162–1167. IEEE, Sorrento (2023). https://doi.org/10.1109/BigData59044.2023.10386891
11. Zhuo, C., et al.: Self-improvement programming for temporal knowledge graph question answering. arXiv **2024**, 14579–14594 (2024)
12. Zhu, Y., Ma, T., Sun, S., et al.: RTA: a reinforcement learning-based temporal knowledge graph question answering model. Neurocomputing **617**, 128994 (2025)
13. Yifu, G., et al.: Two-stage generative question answering on temporal knowledge graph using large language models. *arXiv* (Cornell University), abs/2402.16568, 6719–6734 (2024)
14. Liu, Y., Ott, M., Goyal, N., et al.: RoBERTa: a robustly optimized BERT pretraining approach. *CoRR*, abs/1907.11692 (2019)

15. Dan, B., Dane, S., Pietro, C., Nils, Y.H., et al.: Relational Graph Attention Networks. *CoRR*, abs/1904.05811 (2019)
16. Peng, W., Xin, X., Wang, X., Zhang, N., et al.: Reasoning through memorization: nearest neighbor knowledge graph embeddings. In: Liu, L., et al. (eds.) NLPCC 2023, LNCS, vol. 14323, pp. 111–122. Springer, Cham (2023)
17. Bai, L., Chen, M., Xiao, Q.: Multi-hop temporal knowledge graph reasoning with multi-agent reinforcement learning. Appl. Soft Comput. **160**, 111727 (2024)
18. Saxena, A., Chakrabarti, S., Talukdar, P.: Question answering over temporal knowledge graphs. *Computer Science*, abs/2106.01515 (2021)
19. Zhu, L., Chai, D., Bai, L.: CRmod: context-aware rule-guided reasoning over temporal knowledge graph. Inf. Sci. **664**, 120343 (2024)
20. Raffel, C., Shazeer, N., Roberts, A., et al.: Exploring the limits of transfer learning with a unified text-to-text transformer. J. Mach. Learn. Res. **21**(140), 1–67 (2020)
21. Dai, R., Das, S., Bremond, F.: CTRN: class-temporal relational network for action detection. In: British Machine Vision Conference (BMVC), pp. 224. BMVA Press, Online (2021)

M4-Greedy: Visualization-Oriented Representative Selection of Time Series Data

Duosi Jin[1], Shuai Xu[1(✉)], Minhan Ju[2], Ziqian Zhang[3], and Feng Lin[3]

[1] Nanjing University of Aeronautics and Astronautics, Nanjing 211106, China
{jinduosi,xushuai7}@nuaa.edu.cn
[2] The University of Sydney, Sydney, NSW 2006, Australia
miju0633@uni.sydney.edu.au
[3] State Grid Electric Power Research Institute, Nanjing 211106, China
{zhangziqian,linfeng}@sgepri.sgcc.com.cn

Abstract. To explore time series data patterns, understand data and enable communication, visualization is crucial in various fields. The visual analysis of large-scale time series faces two fundamental challenges: (i) individual series often contain excessive data points that lead to visual clutter when rendered, and (ii) simultaneous display of multiple series results in overlapping patterns that obscure key trends. To address these issues, we propose selecting representative time series that can effectively capture overall trends while maintaining visual clarity. This work investigates the representative selection problem with dual objectives: maximizing both diversity among selected series and their coverage of the complete dataset. We formally prove this bi-objective optimization problem to be NP-hard. Our solution combines a preprocessing technique with a similarity-based greedy algorithm, achieving a 10× speedup with minimal memory overhead across four real-world datasets. Experimental results demonstrate that the proposed method maintains strong performance in both diversity and coverage metrics while significantly improving computational efficiency.

Keywords: Time Series · Reduction · Representative Selection · Visualization

1 Introduction

Modern applications generate massive multiple time series (MTS) data in finance, healthcare, and environmental monitoring. Visual analytics plays a crucial role in identifying patterns, trends, and anomalies within this data. Tools like Tableau, PowerBI and Python are widely used for time series visualization, they face significant challenges in effectively handling large-scale data. A time series with 10^5 data points exceeds the 10^3 to 10^4 pixel capacity of standard displays, causing detail loss and cognitive overload [1,10,11]. Simultaneously

J. Li et al. (Eds.): APWeb-WAIM 2025, CCIS 2660, pp. 312–323, 2026.
https://doi.org/10.1007/978-981-95-7996-9_27

visualizing 100+ correlated series creates clutter, obscuring patterns [2]. Computational costs are also prohibitive. For instance, rendering a single time series with 10^5 data points using Python takes approximately 16 s, and visualizing 100 series fails in 5 h. Despite reducing the data scale to 10^4 points per series, visualization takes over 30 s and the display remains overwhelming. Consequently, selectively displaying a small subset of representative time series is a clearer and more practical approach for effective analysis. Figure 1 illustrates the selection of 3 representative time series from dataset.

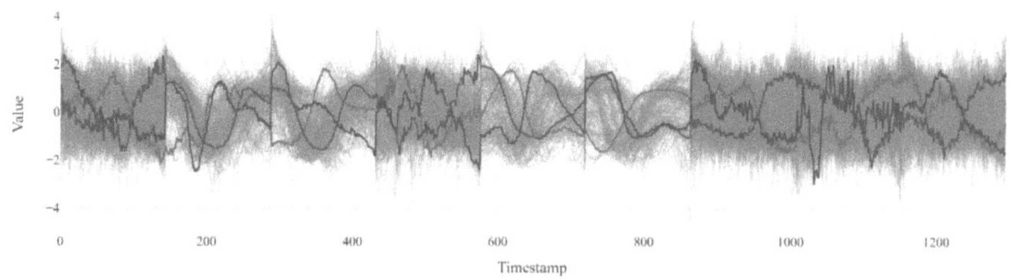

Fig. 1. The grey lines: the original time series data; the red, blue, purple lines: the selected representatives of the time series data. (Color figure online)

Many existing time series clustering and representative selection methods use machine learning or deep learning techniques to extract temporal features before selecting cluster centers as representatives [6], but often overlook ensuring sufficient diversity among the selected representatives by maintaining low inter-series similarity. Unsupervised learning-based approaches typically incur high training costs [4] and rely heavily on large volumes of high-quality data. Current research tackles temporal compression and representative selection independently, each with notable shortcomings. In temporal compression, fixed window aggregation methods such as M4 representation [10] can preserve local extremes and global trends. However, their potential for enabling shape-preserving representative selection in large-scale time series visualization remains underexplored. The approach in reference [3] for representative selection focuses on diversity and coverage, but it is mainly for spatial rather than time series data.

These approaches lack a unified framework for optimizing temporal compression and cross - series representation selection under visual fidelity constraints. To bridge this gap, we propose a formulation of the representative selection problem with two key objectives: *(i) diversity preservation and (ii) visual representativeness.* Specifically, we define these objectives via two quantifiable metrics - diversity and coverage - with the goal of simultaneously maximizing both properties. This Bi-Objective Representative Selection Problem (BORSP) for time series data is proven to be NP-hard. To address this challenge, we develop a unified single-objective optimization framework solved via a greedy algorithm, coupled with M4 preprocessing for large-scale data. The M4 preprocessing preserves essential data features while significantly accelerating similarity computations.

In this paper, we make the following contributions:

1. We study BORSP and prove that its two independent subproblems are NP-hard. Since BORSP is a linear combination of these subproblems, it is also NP-hard.
2. We develop a unified single-objective optimization framework that can be solved via a greedy algorithm, and accelerate the selection speed by preprocessing the data with M4.
3. We conduct experiments on four real-world datasets, achieving a 10× speedup with minimal memory overhead, and its adjustable parameters better meet visualization diversity and coverage requirements.

The rest of the paper is organized as follows. Section 2 defines the problem and proves it is NP-hard. Section 3 presents the greedy algorithm, along with preprocessing and precomputation techniques for acceleration. Section 4 reports and analyzes experimental results. Section 5 reviews related work. Section 6 concludes the paper and outlines future work.

2 Preliminaries

Definition 1 (Multiple Time Series). *A multiple time series dataset $\mathcal{D}$ is a collection of $|\mathcal{D}|$ individual time series, where each time series is represented as a sequence of L points, formally expressed as:*

$$\mathcal{D} = \{X_1, X_2, \ldots, X_{|\mathcal{D}|}\}, \tag{1}$$

where each time series X_i ($i \in \{1, 2, \ldots, |\mathcal{D}|\}$) is defined as:

$$X_i = \{x_{i,1}, x_{i,2}, \ldots, x_{i,L}\}, \tag{2}$$

and $x_{i,j}$ denotes the j-th point in the i-th time series. Here, $|\mathcal{D}|$ represents the total number of time series in the dataset, and L is the length of each time series.

Definition 2 (Representative Subset Diversity). *The diversity of a representative subset $\mathcal{D}_r$ is defined as the minimum inter-representative distance:*

$$\mathcal{F}_{div}(\mathcal{D}_r) = \min_{\substack{X_i, X_j \in \mathcal{D}_r \\ X_i \neq X_j}} \mathrm{DTW}(X_i, X_j) \tag{3}$$

where $\mathrm{DTW}(X_i, X_j)$ denotes the Dynamic Time Warping distance [5] between representatives X_i and X_j.

Lemma 1. *Maximizing $\mathcal{F}_{div}(\mathcal{D}_r)$ is NP-hard.*

Proof. We prove that maximizing $\mathcal{F}_{\text{div}}(\mathcal{D}_r)$ is NP-hard via a reduction from the Max-Min Facility Dispersion (MMFD) problem, which is known to be NP-hard [8]. Given an MMFD instance with node set $V = \{v_1, \ldots, v_n\}$, distance metric $w : V \times V \to \mathbb{R}^+$, and integer p $(2 \leq p \leq n)$, we construct an equivalent instance of our problem by corresponding each node $v_i \in V$ to a time series $X_i \in \mathcal{D}$, and setting $\text{DTW}(X_i, X_j) = w(v_i, v_j)$. Let the target subset size be $k = p$. It follows that a subset $P \subseteq V$ achieves dispersion δ if and only if the corresponding subset $\mathcal{D}_r \subseteq \mathcal{D}$ satisfies $\mathcal{F}_{\text{div}}(\mathcal{D}_r) = \delta$. The reduction is polynomial, hence the problem of maximizing $\mathcal{F}_{\text{div}}$ is NP-hard. □

Definition 3 (Representative Subset Coverage). *The diversity of a representative subset $\mathcal{D}_r$ is defined as the average distance from original data to their nearest representatives:*

$$\mathcal{F}_{cov}(\mathcal{D}_r) = \frac{1}{|\mathcal{D}|} \sum_{X \in \mathcal{D}} \min_{X' \in \mathcal{D}_r} \text{DTW}(X, X') \tag{4}$$

Lemma 2. *Minimizing $\mathcal{F}_{cov}(\mathcal{D}_r)$ is NP-hard.*

Proof. We prove the NP-hardness of coverage minimization via a reduction from the k-Medoids Problem [9], which is known to be NP-hard. Given a set of points $P = \{a_1, \ldots, a_n\}$ and a distance metric $\delta(a_i, a_j)$, the goal in k-Medoids is to select k points $M \subseteq P$ minimizing $\sum_{a_i \in P} \min_{m \in M} \delta(a_i, m)$. We construct an equivalent instance of the coverage minimization problem by mapping each point $a_i \in P$ to a time series $X_{a_i} \in \mathcal{D}$, and setting $\text{DTW}(X_{a_i}, X_{a_j}) = \delta(a_i, a_j)$. A subset $\mathcal{D}_r \subseteq \mathcal{D}$ with $|\mathcal{D}_r| = k$ achieves minimal coverage if and only if the corresponding $M \subseteq P$ achieves minimal assignment cost. Since this reduction is polynomial, coverage minimization is NP-hard. □

Definition 4 (Bi-Objective Representative Selection Problem). *Given a multiple time series dataset $\mathcal{D}$, we aim to select a representative subset $\mathcal{D}_r \subseteq \mathcal{D}$ of size $|\mathcal{D}_r| = k$ that simultaneously maximizes the weighted combination:*

$$\mathcal{F}(\mathcal{D}_r) = \alpha \cdot \mathcal{F}_{div}(\mathcal{D}_r) - \beta \cdot \mathcal{F}_{cov}(\mathcal{D}_r),$$

where $\alpha, \beta \geq 0$ are trade-off coefficients balancing the importance of diversity ($\mathcal{F}_{div}$) and coverage ($\mathcal{F}_{cov}$).

Theorem 1. *The bi-objective representative selection problem is NP-hard.*

Proof. We prove the NP-hardness of the bi-objective representative selection problem by observing that it generalizes both the diversity maximization problem and the coverage minimization problem, which are each proven to be NP-hard in Lemmas 1 and 2, respectively. Specifically, any instance of the diversity maximization problem can be embedded into the bi-objective framework by assigning full weight to the diversity objective, effectively reducing the bi-objective problem to a known NP-hard problem. Similarly, setting full preference on coverage minimization recovers the k-Medoids problem, another NP-hard task.

Therefore, since the bi-objective problem subsumes two distinct NP-hard problems as special cases, it is itself at least as hard as either one. Hence, the bi-objective representative selection problem is NP-hard. □

3 The M4-Greedy Algorithm

To address BORSP, we propose **M4-Greedy**. As outlined in Fig. 2, our approach has three steps. In Step 1, $\mathcal{D}$ is reduced to a feature-preserving subset $\mathcal{D}'$ to facilitate distance computations. In Step 3, an approximate solution $\mathcal{D}'_r$ is selected using a greedy algorithm that optimizes for both diversity and coverage. Step 2 involves computing the pairwise DTW distance matrix M and the average distance vector $\bar{\boldsymbol{d}}$ for all $X_i, X_j \in \mathcal{D}'$. Key notations are in Table 1.

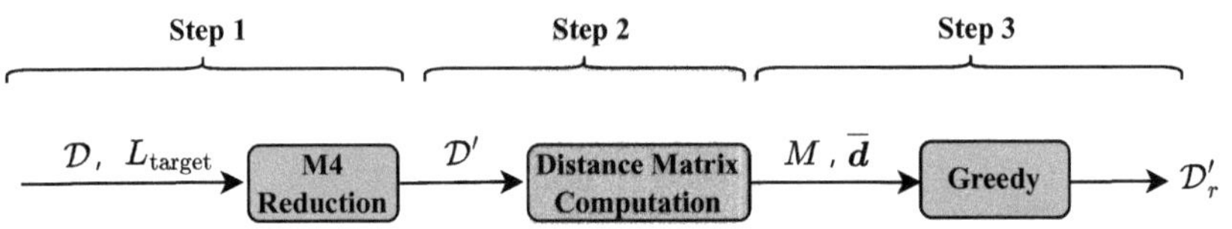

Fig. 2. M4-Greedy workflow.

Table 1. Notation Summary.

Notation	Description
D,D'	multiple time series
D_r	representatives subset
X	a time series
α	diversity-coverage trade-off parameter
DTW($\cdot$)	dynamic time warping distance

3.1 M4 Preprocessing

Algorithm 1 compresses time series by preserving critical points in sliding windows. For each window, as depicted in Fig. 3, it extracts four features in every time span: first and last timestamps, min and max values. The number of time spans L_{target}, corresponding to the number of pixel columns in a line chart. The method guarantees the output length of $\leqslant 4L_{\text{target}}$ for an input of length L.

Complexity Analysis. The time complexity of Algorithm 1 is $O(|\mathcal{D}|(L + L_{\text{target}} \log L_{\text{target}}))$. For $L_{\text{target}} \ll L$, it reduces to $O(|\mathcal{D}|L)$.

Algorithm 1 M4 Reduction

Input: $\mathcal{D}$: original dataset, $\mathcal{D} = \{X_1, ..., X_{|\mathcal{D}|}\}$, $X_i \in \mathbb{R}^L$;
L_{target}: the number of time spans in $\mathbb{N}^+$;
Output: $\mathcal{D}'$:reduced dataset, $\mathcal{D}' = \{X'_1, ..., X'_{|\mathcal{D}'|}\}$, $X'_i \in \mathbb{R}^{L'}$, $L' \leq 4L_{\text{target}}$

for each $X = \{x_1, ..., x_L\} \in \mathcal{D}$ **do**
  $\mathcal{P} \leftarrow \emptyset$ ▷ Initialize feature set
  $s \leftarrow \max\left(1, \left\lfloor \frac{L}{L_{\text{target}}} \right\rfloor\right)$
  for $\tau = 0$ **to** $L-1$ **step** s **do**
    $\tau_{\text{end}} \leftarrow \min(\tau + s - 1, L - 1)$
    if $\tau \leq \tau_{\text{end}}$ **then**
      $\mathcal{P} \leftarrow \mathcal{P} \cup \{(\tau, x_\tau)\}$ ▷ Window start
      $\mathcal{P} \leftarrow \mathcal{P} \cup \{(\tau_{\text{end}}, x_{\tau_{\text{end}}})\}$ ▷ Window end
      $\mathcal{P} \leftarrow \mathcal{P} \cup \{(\hat{t}, \max_{t \in [\tau, \tau_{\text{end}}]} x_t)\}$ ▷ Local maximum
      $\mathcal{P} \leftarrow \mathcal{P} \cup \{(\check{t}, \min_{t \in [\tau, \tau_{\text{end}}]} x_t)\}$
  Sort $\mathcal{P} = \{(t_1, v_1), ..., (t_{|\mathcal{P}|}, v_{|\mathcal{P}|})\}$ with $t_1 \leq \cdots \leq t_{|\mathcal{P}|}$
  $X' \leftarrow (v_1, ..., v_{|\mathcal{P}|})$
  $\mathcal{D}' \leftarrow \mathcal{D}' \cup \{X'\}$
return $\mathcal{D}'$

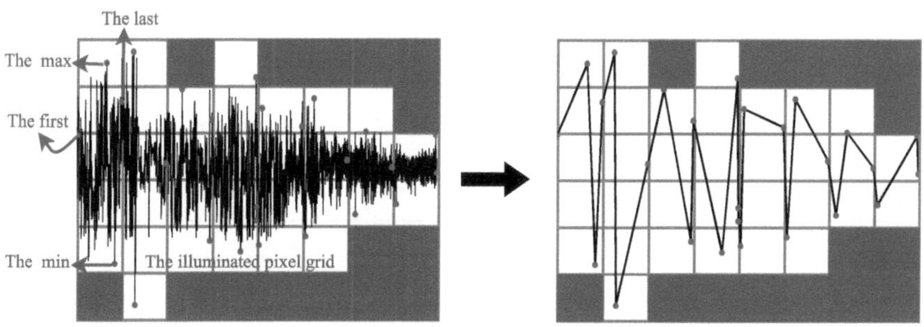

(a) 107904 timestamps in 8 pixel columns **(b) after M4 aggregation**

Fig. 3. After M4 aggregation, the data points for transmission and rendering are limited to four times the horizontal pixel columns of the target device.

3.2 Diversity-Coverage Greedy

In Algorithm 2, the greedy selection criterion then optimizes a dual objective. It combines representative distinctiveness and dataset coverage through a composite score:

$$c^{(t)} = \underset{c \in R}{\arg\max} \left[\alpha \cdot \underbrace{\log \left(\min_{j \in \mathcal{I}_r^{(t-1)}} M(c, j) \right)}_{\text{Log-Diversity}} - (1 - \alpha) \cdot \underbrace{\log \left(\frac{1}{|\mathcal{D}'|} \sum_{i=1}^{|\mathcal{D}'|} \Delta_i(c) \right)}_{\text{Log-Coverage}} \right] \quad (5)$$

The logarithmic transformation balances scale disparities, ensuring balanced contributions to the objective function. Here, $\mathcal{I}_r^{(t-1)}$ denotes the set of indices of

Algorithm 2 Greedy Representative Selection

Input: $\mathcal{D}'$: reduced dataset;
$M \in \mathbb{R}^{|\mathcal{D}'| \times |\mathcal{D}'|}$: pairwise DTW distance matrix;
$\overline{\boldsymbol{d}} \in \mathbb{R}^{|\mathcal{D}'|}$: average distance vector;
k: target size;
α: trade-off parameter in [0,1].
Output: $\mathcal{D}'_r$: Representative subset
1: $r_1 \leftarrow \arg \min_{1 \leq i \leq |\mathcal{D}'|} \overline{d}_i$ ▷ Initial representative index
2: $\mathcal{I}_r \leftarrow \{r_1\},\ R \leftarrow \{1, \ldots, |\mathcal{D}'|\} \setminus \{r_1\}$ ▷ Index sets
3: $\boldsymbol{d}_{\text{min}} \leftarrow \boldsymbol{m}_{r_1}$ ▷ Initialize with $\boldsymbol{m}_{r_1} := [M(i, r_1)]_{i=1}^{|\mathcal{D}'|}$
4: **for** $t = 2$ **to** k **do**
5: Initialize $s^*,\ c^*$ ▷ Reset scores for current iteration
6: **for each** $c \in R$ **do**
7: $S_{\text{div}} \leftarrow \alpha \cdot \log \left(\min_{j \in \mathcal{I}_r} M(c, j) \right)$ ▷ Diversity term
8: Update $\Delta_i(c)$ via Equation 6 ▷ Coverage improvement
9: $S_{\text{cov}} \leftarrow -(1-\alpha) \cdot \log \left(\frac{1}{|\mathcal{D}'|} \sum_{i=1}^{|\mathcal{D}'|} \Delta_i(c) \right)$
10: $S_{\text{com}} \leftarrow S_{\text{div}} + S_{\text{cov}}$
11: **if** $S_{\text{com}} > s^*$ **then**
12: $s^* \leftarrow S_{\text{com}},\ c^* \leftarrow c$
13: $\mathcal{I}_r \leftarrow \mathcal{I}_r \cup \{c^*\}$
14: $R \leftarrow R \setminus \{c^*\}$
15: $\boldsymbol{d}_{\text{min}} \leftarrow \min(\boldsymbol{d}_{\text{min}},\ \boldsymbol{m}_{c^*})$ ▷ $\boldsymbol{m}_{c^*} := [M(i, c^*)]_{i=1}^{|\mathcal{D}'|}$
16: $\mathcal{D}'_r \leftarrow \{X_j \in \mathcal{D}' \mid j \in \mathcal{I}_r\}$
17: **return** $\mathcal{D}'_r$

previously selected representatives, and the updated minimum distance metric $\Delta_i(c)$ is defined as:

$$\Delta_i(c) = \min \left(d_{\text{min}}^{(t-1)}(i),\ M(i, c) \right) \tag{6}$$

Here, i indexes a time series in $\mathcal{D}'$, $d_{\text{min}}^{(t-1)}(i)$ is its minimum distance to selected representatives, and $M(i, c)$ is the pairwise DTW distance to candidate c.

The historical minimum distance vector $\boldsymbol{d}_{\text{min}}$ evolves iteratively through element-wise minimization:

$$\boldsymbol{d}_{\text{min}}^{(t)} = \min \left(\boldsymbol{d}_{\text{min}}^{(t-1)},\ \boldsymbol{m}_{c^{(t)}} \right) \tag{7}$$

where superscripts $^{(t)}$ and $^{(t-1)}$ mark variables at iteration t and $t-1$ and $\boldsymbol{m}_{c^{(t)}} = [M(1, c^{(t)}), \ldots, M(|\mathcal{D}'|, c^{(t)})]^\top$ denotes the $c^{(t)}$-th column vector of matrix M, and the min operator acts element-wise:

$$\min(\boldsymbol{u}, \boldsymbol{v}) \triangleq [\min(u_1, v_1),\ \ldots,\ \min(u_n, v_n)]^\top \tag{8}$$

At Line 5 of Algorithm 2, s^* is the composite score at iteration t, and c^* is the representative index at iteration t. The algorithm iteratively selects the

Algorithm 3 Distance Matrix Computation

Input: $\mathcal{D}'$: Reduced dataset.
Output: $\mathbf{M}$: Distance matrix in $\mathbb{R}^{|\mathcal{D}'|\times|\mathcal{D}'|}$;
$\overline{\mathbf{d}}$: Average distance vector in $\mathbb{R}^{|\mathcal{D}'|}$

Initialize $\mathbf{M}$ ▷ Zero matrix
for $i = 1$ **to** $|\mathcal{D}'| - 1$ **do** ▷ Upper triangular computation
  for $j = i + 1$ **to** $|\mathcal{D}'|$ **do**
    $\mathbf{M}(i, j) \leftarrow \mathrm{DTW}(X_i, X_j)$ ▷ $X_i, X_j \in \mathcal{D}'$
    $\mathbf{M}(j, i) \leftarrow \mathbf{M}(i, j)$ ▷ Enforce symmetry
for $k = 1$ **to** $|\mathcal{D}'|$ **do**
  $\overline{d}_k \leftarrow \frac{1}{|\mathcal{D}'|-1} \sum_{\substack{m=1 \\ m\neq k}}^{|\mathcal{D}'|} \mathbf{M}(k, m)$
return $\mathbf{M}$, $\overline{\mathbf{d}}$

next representative $c^{(t)}$ that maximizes the composite score $s^{(t)}$. The algorithm terminates when the number of selected representatives reaches k.

The behavior of the method is governed by the trade-off parameter $\alpha \in [0, 1]$. When α is close to 1, the algorithm prioritizes diversity by maximizing the log-scaled minimum distance to previously selected representatives $\mathcal{I}_r^{(t-1)}$. When α is close to 0, it emphasizes coverage by minimizing the log-scaled average coverage gap.

Complexity Analysis. The algorithm exhibits $O(k|\mathcal{D}'|^2)$ time complexity, driven by k iterative selections where each evaluates $O(|\mathcal{D}'|)$ candidates with $O(|\mathcal{D}'|)$-scale coverage score computations per candidate; space complexity is $O(|\mathcal{D}'|^2)$ due to storage of the pairwise distance matrix $\mathbf{M} \in \mathbb{R}^{|\mathcal{D}'|\times|\mathcal{D}'|}$.

3.3 Precompute Matrix

Algorithm 3 computes pairwise DTW similarities to form a symmetric distance matrix and calculate the average. It employs nested loops for DTW computations and exploits matrix symmetry to optimize storage.

Complexity Analysis. The time complexity of Algorithm 3 is $O(|\mathcal{D}'|^2 L'^2)$, dominated by pairwise DTW computations for all $\frac{|\mathcal{D}'|(|\mathcal{D}'|-1)}{2}$ pairs. The space complexity is $O(|\mathcal{D}'|^2)$ for storing the distance matrix M, where L' is the reduced time series length. Average distance computation adds $O(|\mathcal{D}'|^2)$ time but does not affect the asymptotic bound.

4 Experiment

Experimental Settings. The system is Windows 11 and the hardware has an Intel(R) Core(TM) i7 - 9750H CPU @ 2.60 GHz, with 16 GB of memory.

Datasets. The dataset used is the UEA multivariate time series dataset [15], selecting 4 larger - scale ones for experiments as smaller ones don't need M4 preprocessing. The statistics of the datasets are summarized in Table 2.

Table 2. Statistics summary of datasets (L denotes series length and $|\mathcal{D}|$ denotes set size following Definition 1).

Dataset	L	$\|\mathcal{D}\|$	Total records
EigenWorms	107904	300	32,371,200
Heartbeat	24705	205	5,064,525
MotorImagery	192000	100	19,200,000
PEMS-SF	138672	173	23,990,256

Evaluation. We quantify diversity by dividing the minimum DTW distance in the representative set by the maximum DTW distance in the whole dataset.

$$S_{\text{div}} = \frac{\min_{X_i, X_j \in \mathcal{D}_r} \text{DTW}(X_i, X_j)}{\max_{X_m, X_n \in \mathcal{D}} \text{DTW}(X_m, X_n)} \tag{9}$$

The normalization in Eq. 10 bounds $S_{\text{cov}} \in [0, 1]$ by comparing the representative set's coverage (numerator) against the worst-coverage scenario where $\max_{X_n \in \mathcal{D}} \sum_{X_j \in \mathcal{D}} \text{DTW}(X_n, X_j)$ represents the most dispersed sequence. This guarantees: (i) $S_{\text{cov}} = 1$ when $\mathcal{D}_r$ perfectly covers $\mathcal{D}$, and (ii) $S_{\text{cov}} \approx 0$ when $\mathcal{D}_r$ contains only the most distant sequence.

$$S_{\text{cov}} = 1 - \frac{\sum_{X_m \in \mathcal{D}} \min_{X_i \in \mathcal{D}_r} \text{DTW}(X_m, X_i)}{\max_{X_n \in \mathcal{D}} \sum_{X_j \in \mathcal{D}} \text{DTW}(X_n, X_j)} \tag{10}$$

Methods For Comparison. The K-Shape method [7] requires all time series to be of equal length, but large-scale time series data often vary, so we exclude it. As shown in Fig. 4, the M4-preprocessed greedy algorithm has significant speed advantage. For example, with the 192k - length MotorImagery dataset, M4 acceleration achieves an average speedup exceeding 10× across five repeated runs. It requires little extra storage for preprocessed data, negligible for modern devices.

Diversity-Coverage Trade-Off Comparison. As illustrated in Fig. 5, the greedy algorithm with simultaneous optimization of diversity and coverage achieves balanced performance between these two metrics.

Simple Random and Binning methods show unstable performance. The Maxmin method has a slight edge in diversity on some datasets but is poor in coverage. Both greedy algorithms perform well. The M4-preprocessed greedy algorithm is affected by dataset features and aligns well with inherent patterns in datasets like MotorImagery.

Visualization Case Study. As presented in Fig. 6, gray lines are original data, colored lines are selected ones. For three selections, blue is for coverage first, then green and orange for diversity, giving a clear data view. The visualization uses M4-reduced data because the full dataset is too large to show directly.

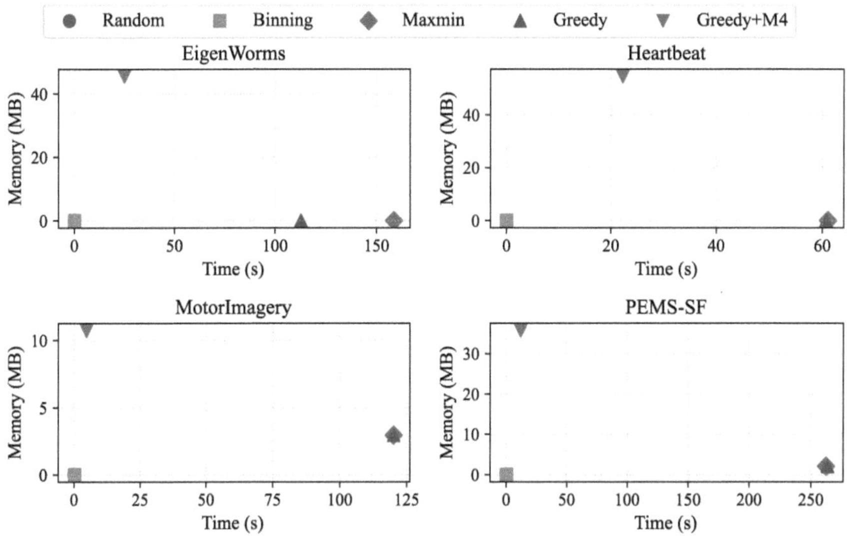

Fig. 4. Trade-off of time and memory of different methods.

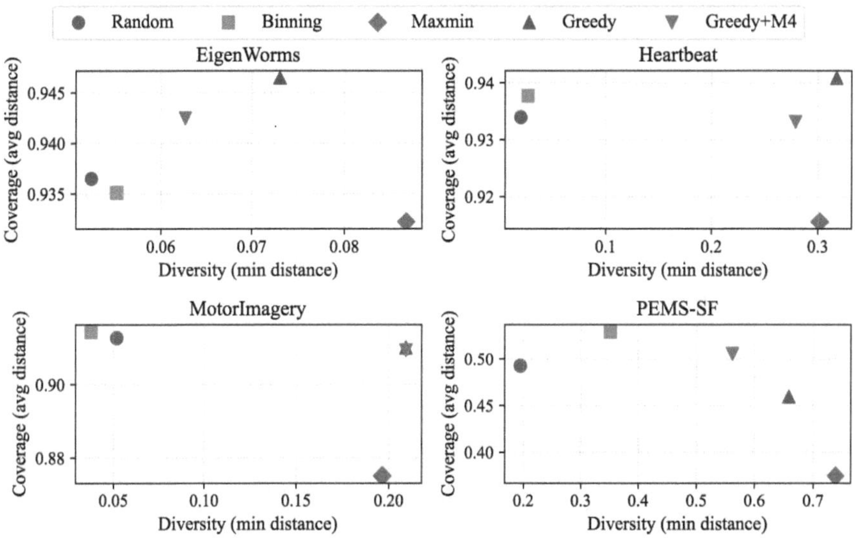

Fig. 5. Trade-off of diversity and coverage of different methods.

5 Related Work

For high - volume time series data, direct rendering has issues. Jugel et al. [10] proposed M4 for dimensionality reduction. M4 selects the first, last, min, and max values within each time span corresponding to screen pixels, ensuring error-free rasterization without visual distortion. OM3 [11] is a multi-level min-max

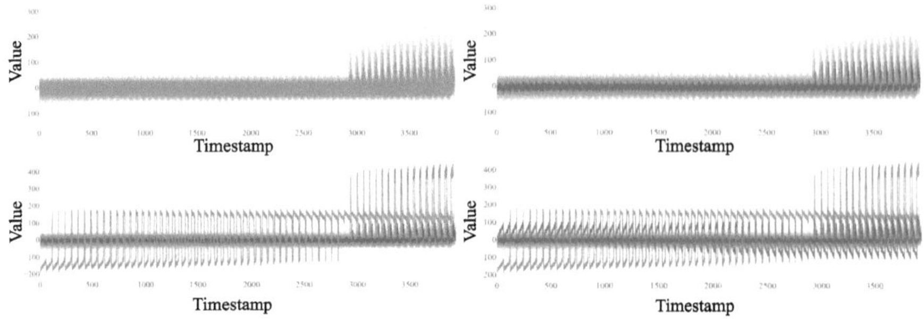

Fig. 6. Three representative time series from the MotorImagery dataset.

representation designed for large-scale time series visualization. It enables progressive, error-free interactive visualization by supporting incremental tree-based queries that reduce latency during operations such as resizing, panning, and zooming. Rui et al. [12] introduced M4-LSM, an optimized implementation of M4 tailored for LSM-Tree-based databases (e.g., Apache IoTDB). This optimization retains error-free visualization accuracy while improving query performance.

In information theory representative sample selection identifies a representative subset by maximizing mutual information while minimizing redundancy (relative entropy). This ensures that the selected samples fully cover the body of information in the original dataset while keeping redundancy low. Pan.et al. [13] used greedy and heuristic algorithms to design an objective function to iteratively select the optimal representative samples. Guo.et al. [14]. perform representative selection on a large number of points of interest (POIs) of interest in the spatiotemporal map. Since this problem can be reduced to the independent set problem (an NP-hard problem), they approximately select representative POIs by means of a greedy method.

6 Conclusion

This paper focuses on representative selection for visualization-oriented time series data, presenting a comprehensive workflow for large-scale representation, from M4-based preprocessing to greedy selection. It balances diversity and coverage while accelerating both selection and visualization. Ablation studies confirm that the space-for-time strategy using M4 preserves visualization quality while enhancing efficiency. Comparative experiments show the greedy algorithm outperforms Random, Binning, and Maxmin selection. Further analysis reveals how diversity-coverage trade-offs vary with different parameter settings across data distributions. In future work, we aim to support dynamic updates of representatives during zooming to improve real-time visualization capabilities.

References

1. Walker, J., Borgo, R., Jones, M.W.: Timenotes: a study on effective chart visualization and interaction techniques for time-series data. IEEE Trans. Visual Comput. Graph. **22**(1), 549–558 (2015)
2. Zhao, Y., Wang, Y., Zhang, J.: KD-Box: line-segment-based KD-tree for interactive exploration of large-scale time-series data. IEEE Trans. Visual Comput. Graph. **28**(1), 890–900 (2021)
3. Yang, C., Chen, L., Wang, H., Shang, S.: Towards efficient selection of activity trajectories based on diversity and coverage. In: Proceedings of the AAAI Conference on Artificial Intelligence, vol. 35, no. 1, pp. 689–696 (2021). https://doi.org/10.1609/aaai.v35i1.16149
4. Ben Abdennour, G., Gasmi, K., Ejbali, R.: An optimal model for medical text classification based on adaptive genetic algorithm. Data Sci. Eng. **9**, 378–392 (2024)
5. Berndt, D.J., Clifford, J.: Using dynamic time warping to find patterns in time series. In: Proceedings of the 3rd International Conference on Knowledge Discovery and Data Mining, pp. 359–370. AAAI Press (1994). https://doi.org/10.5555/3000850.3000887
6. Lee, G., Huang, S., Bao, Z., Zhao, Y.: Representative time series discovery for data exploration. Proc. VLDB Endow. **18**(3), 915–928 (2024)
7. Paparrizos, J., Gravano, L.: k-Shape: efficient and accurate clustering of time series. In: Proceedings of the 2016 ACM SIGMOD International Conference on Management of Data (SIGMOD 2016), pp. 69–76. ACM, New York (2016). https://doi.org/10.1145/2949741.2949758
8. Ravi, S.S., Rosenkrantz, D.J., Tayi, G.K.: Heuristic and special case algorithms for dispersion problems. Oper. Res. **42**(2), 299–310 (1994)
9. Jain, A.K., Dubes, R.C.: Algorithms for clustering data. Prentice-Hall Inc, Englewood Cliffs (1988)
10. Jugel, U., Jerzak, Z., Hackenbroich, G., Markl, V.: M4: a visualization-oriented time series data aggregation. Proc. VLDB Endow. **7**(10), 797–808 (2014)
11. Wang, Y., et al.: OM3: an ordered multi-level min-max representation for interactive progressive visualization of time series. Proc. ACM Manage. Data **1**(2), 1–24 (2023)
12. Rui, L., Huang, X., Song, S., Kang, Y., Wang, C., Wang, J.: Time series representation for visualization in apache IoTDB. Proc. ACM Manage. Data **2**(1), 1–26 (2024)
13. Pan, F., Wang, W., Tung, A.K.H., Yang, J.: Finding representative set from massive data. In: Fifth IEEE International Conference on Data Mining (ICDM'05), pp. 8–1. IEEE Computer Society, Washington (2005). https://doi.org/10.1109/ICDM.2005.69
14. Guo, T., Feng, K., Cong, G., et al.: Efficient selection of geospatial data on maps for interactive and visualized exploration. In: Proceedings of the 2018 International Conference on Management of Data (SIGMOD 2018), pp. 567–582. ACM, New York (2018). https://doi.org/10.1145/3183713.3183738
15. Bagnall, A., et al.: The UEA multivariate time series classification archive, 2018. arXiv preprint arXiv:1811.00075 (2018)

MAD-GCN: Multi-attention Diffusion Graph Convolutional Network: A Novel Approach for Low-Altitude Traffic Flow Forecasting

Hao Wu[1], Hailian Yin[2(✉)], and Changjie Wang[3]

[1] College of Civil Aviation, Nanjing University of Aeronautics and Astronautics, Nanjing 210016, China

[2] Ministry of Industry and Information Technology Key Laboratory of Advanced Technology for Small and Medium UAVs, Nanjing University of Aeronautics and Astronautics, Nanjing 210016, China
yinhailian@nuaa.edu.cn

[3] College of Computer Science and Technology, Nanjing University of Aeronautics and Astronautics, Nanjing 211106, China

Abstract. Low-altitude traffic flow forecasting is the core task of urban management and Intelligent Transportation System (ITS). Due to the complexity and uncertainty of traffic patterns, capturing the spatial-temporal correlation of traffic patterns is a big challenge. Recent research methods mainly rely on the predefined adjacency matrix to simulate the traffic relationship, which leads to poor prediction effects. Meanwhile, if only the adaptive adjacency matrix is relied on, some inevitable relations between nodes may be ignored. To solve the above problems, this paper proposes a novel deep learning model: Multi-Attention Diffusion Graph Convo-lutional Network (MAD-GCN). The model consists of three main modules: 1) Spatial-Temporal Positional Embedding module, in which historical and predicted traffic flows are embedded in the form of spatial-temporal embedding vectors; 2) Spatial correlation module, in which dynamic node relationships are captured and spatial models are established; 3) Time correlation module, in which introduces multi-attention to capture a variety of temporal features in traffic data series. Our proposed model was compared with multiple baseline models on real datasets, and the results show that our MAD-GCN is optimal and can capture the spatial-temporal dependence well.

Keywords: Low-altitude traffic flow forecasting · Spatial-temporal · Graph convolution networks · Attention networks

1 Introduction

Currently, with the rapid development of the low-altitude economy, we are faced with many complex issues regarding low-altitude traffic flow. With the maturity of spatial-temporal big data, deep learning, and other technologies, intelligent transportation system (ITS) has been developed. Traffic flow forecasting is the focus of ITS development

J. Li et al. (Eds.): APWeb-WAIM 2025, CCIS 2660, pp. 324–341, 2026.
https://doi.org/10.1007/978-981-95-7996-9_28

and plays a vital role in low-altitude urban development and management. Figure 1 shows the modeling difficulties in low- altitude traffic flow prediction. From the perspective of time dimension, traffic conditions at different historical moments will affect traffic conditions at other moments. From the perspective of space dimension, different traffic nodes will also affect each other. Even if two traffic nodes are far apart, they may have a spatial correlation with each other (this spatial correlation is represented by a dashed line). These complex spatial temporal relationships make low-altitude traffic flow forecasting face many challenges.

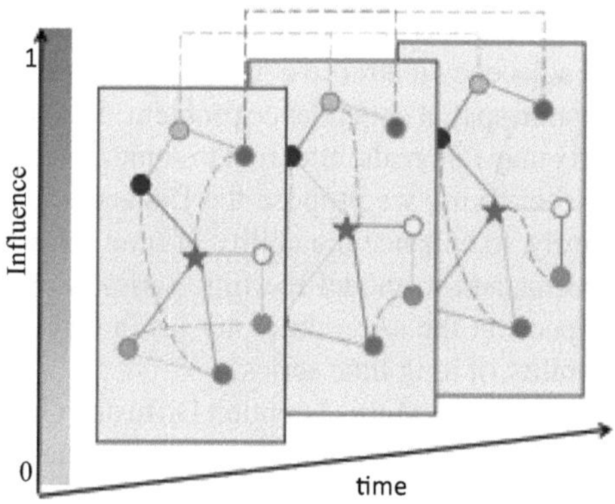

Fig. 1. Spatial-Tempotral correlation.

In previous studies, low-altitude traffic flow forecasting methods primarily depended on mathematical statistical models, with prominent examples like the Historical Average method (HA), Auto-Regressive Integrated Moving Average (ARIMA), and Vector Auto-Regression (VAR). These models initially delivered favorable results, effectively meeting basic forecasting needs in early-stage low-altitude traffic management.

However, their limitations soon became evident: they struggle to handle nonlinear problems, failing to capture complex, non-linear patterns in traffic flow, and they cannot account for critical spatial-temporal correlations—how traffic in one area affects nearby regions over time. As data volumes expanded dramatically, driven by advancements in sensing technologies, these methods' performance deteriorated significantly, unable to cope with the growing complexity.

In recent years, deep learning has gained traction due to its strong ability to model nonlinear relationships. Convolutional Neural Networks (CNN) [1] emerged as a new solution, processing historical traffic information in matrix form to extract hidden features. Yet, a key issue remains: sensor-based traffic grids do not form a classical Euclidean space, meaning the data lacks a regular grid structure. This mismatch undermines CNNs' inherent strength in processing grid-like data, greatly reducing their modeling capability in low-altitude traffic flow forecasting.

Graph convolutional neural network (GCN) was proposed for solving this problem, which has good applicability to non-Euclidean spatial structures. Depending on the graph structure, the adjacency matrix can realize the aggregation function of node and neighbor information, which is similar to the function of convolution. Graph neural network can also be used in different graph structures to achieve feature extraction and vector embedding functions for graph nodes. However, GCN still faces many challenges when dealing with node space modeling. Firstly, existing studies mainly rely on the predefined adjacency matrix as the adjacency matrix of graph convolution, which cannot truly reflect the spatial correlation between nodes. For example, some implicit dependencies cannot be represented by the geographical adjacency relationship between nodes. Guo et al. [2] attach attention weights to nodes to represent the spatial correlation strength between nodes, but they still do not consider the implicit spatial correlation. Bai et al. [3] compute node associations based on point of Interest (POI). Genget al. [4] model spatial relationships using multi-graph convolutions. However, these methods are still based on the predefined adjacency matrix to construct the static spatial topology, which cannot simulate the dynamic spatial dependence problem. Secondly, relying only on the adaptive adjacency matrix may ignore the attention to some inevitable node relationships. To enhance the node associativity, we propose the Distance-Step correction matrix to solve this problem. Thirdly, to enhance the ability of time series modeling, we use the multi-head attention mechanism to model the time series, so that the model can pay attention to different aspects at the same time, not just a single attention pattern, and improve the modeling ability of long time series.

To address the above problems, Multi-Attention Diffusion Graph Convolutional Network (MAD-GCN) is proposed. Compared with previous studies, MAD-GCN can adaptively capture spatial associations of nodes and increase the focus on non-negligible dependencies. It can also better model time series correlation through multi-attention. The main contributions are as follows:

We propose a novel deep learning framework MAD-GCN for low-altitude traffic flow forecasting, which can dynamically capture the spatial-temporal dependence of low-altitude traffic flow.

The historical and predicted low-altitude traffic flows are embedded, and the integration of the Distance-Step correction matrix with the attention matrix is proposed to construct a dynamic node association matrix. GCN combined with diffusion convolution is used to capture spatial dependencies.

The multi-head attention mechanism is used to capture many different temporal features in traffic data series.

We conducted experiments on four real-world datasets and compared MAD-GCN with multiple baseline models. The experimental results show that our proposed model outperforms the baseline models and has good spatial-temporal modeling capability.

2 Related Works

2.1 Traditional Method

The field of low-altitude traffic flow forecasting has experienced decades of development, from the initial traditional statistical methods, gradually developed into methods based on traditional machine learning and the current popular prediction model methods based on deep learning. We take urban traffic prediction as an example. In traditional traffic flow forecasting tasks, statistical algorithm models, such as history average model (HA)

[5], are often used to perform weighted calculations based on historical moment data to predict traffic. AutoRegressive Integrated Moving Average (ARIMA) [6] establishes time series models for time series prediction. Ahmed et al. [7] introduced ARMIA model into the field of low-altitude traffic flow prediction. Stephanedes et al. [8] used HA for the first time to predict traffic conditions. In addition, traditional mathematical statistical models such as Kalman filtering, Support Vector Machine (SVM), linear regression, and other models are also widely used. These methods have achieved good prediction results in the early stage, but with the increase of data volume and types, these methods cannot deal with complex data, so the effect is not good.

2.2 Deep Learning-Based Methods

Neural networks have been fully applied in low-altitude traffic prediction currently. The task of low-altitude traffic flow forecasting is usually divided into two parts: spatial correlation modeling and temporal correlation modeling.

2.2.1 Spatial Correlation Modeling

In spatial correlation model, studies [9, 10] choose to use CNN on the node spatial correlation model. Wu et al. [11] proposed a CLTFP to predict low-altitude traffic flow based on CNN. CNN has achieved a good effect on grid data, but the traffic network is a non-European network, and its discrete characteristics and irregular connection topology make it unsuitable to use conventional convolution kernel for analysis in most cases. Graph neural network (GNN) can solve this problem well [12–16]. It has good applicability to non-European spatial structures and can capture the spatial correlation among them. Since the current low-altitude traffic flow network can be described as a network form, in which the low-altitude traffic intersections can be represented as nodes, and the roads connecting each intersection can be represented as edges, GNN is very suitable for low-altitude traffic flow prediction. Geng et al. [4] proposed Spatiotemporal Multi-Graph Convolution Network ST-MGCN, which encodes traffic conditions in multiple different regions to generate multiple graph structures, and uses graph convolution to model multi-graph correlations. Li et al. [17] proposed the Diffusion convolutional recurrent neural network DCRNN to capture the implicit node relationships by simulating the diffusion process on a graph. Zheng et al. [18] proposed a multi-head graph attention neural network GMAN, based on the encoder-decoder framework, to simulate traffic data in historical time and future prediction time. When the graph structure is unknown, Wu et al. [19] proposed GraphWaveNet. Based on the random walk strategy, it uses two node embedding vectors to represent road information and further constructs an adaptive adjacency matrix.

2.2.2 Temporal Correlation Modeling

In temporal correlation modeling, Recurrent Neural Networks (RNN) [20] stand as a classic neural network architecture, boasting robust capabilities in processing sequential data. A multitude of RNN variants have been developed to address its limitations, including Long Short-Term Memory (LSTM) networks, Gate Recurrent Units (GRU), and others. These variants introduce specialized gating mechanisms to enhance the modeling of long-term time series data, effectively mitigating the original RNN's struggle with retaining distant temporal dependencies. For instance, Cui et al. [21] proposed SBU-LSTM, an LSTM variant designed to identify, fill in, and correct missing or abnormal data points— a critical feature in real-world scenarios—yielding impressive results in low-altitude traffic flow prediction tasks. Similarly, Bai et al. [22] introduced the Adaptive Graph Convolutional Recurrent Neural Network (AGCRN), which integrates an adaptive adjacency matrix with GRU to dynamically capture spatial-temporal interactions, further improving prediction accuracy.

However, RNN and its variants suffer from notable drawbacks: they involve a large number of parameters, leading to low training efficiency, and are plagued by the vanishing gradient problem, which hinders their ability to learn long-range temporal correlations. Convolutional Neural Networks (CNN) have also been adapted for temporal processing. Yu et al. [23] combined a gating mechanism with 1-D convolutions, allowing the network to slide across time slices and extract localized temporal features efficiently. Additionally, the attention mechanism has gained widespread use in time series modeling. Analogous to spatial relationship modeling, it constructs an attention matrix to assign varying weights to different time slices, emphasizing more relevant temporal segments. Li et al. [24] employed a transformer model, which relies heavily on self-attention, to encode temporal information in a parallelizable manner, overcoming the sequential processing bottleneck of RNNs. Meanwhile, Jiang et al. [25] integrated both attention mechanisms and convolutional components, leveraging the strengths of convolution for local feature extraction and attention for global dependency modeling to handle complex time sequences more comprehensively.

3 Methodology

3.1 Preliminaries

In this paper, we denote the transportation network as $G = (V, E, A)$, where, $V \in R^{N}$represents the edges between nodes, $E \in R^{N\times N}$ represents the nodes, and $A \in R^{N\times N}$is the adjacency matrix between nodes. Where each node has F eigenvalues at different time instants. At time t, the feature of point i is represented by xit, and the feature matrix of the road network is donated as $Xt = (x1t, x2t, ..., xNt)^{T}$, the historical traffic flow information is expressed as $H = (X1, X2,..., XT)$, where there are T time steps. Our task is to predict the traffic flow condition P time steps into the future. Traffic flow forecasting task can be expressed as follows:

$$X(T+1):(T+p) = F\theta(H; G) \quad (1)$$

where θ denotes all the parameters in the model, and F denotes the model function.

3.2 Model Architecture

To capture spatial-temporal dependency, we propose a novel deep learning model MAD-GCN. Illustrated in Fig. 2, our MAD-GCN comprises three modules: Spatial- Temporal Positional Embedding module, spatial correlation module, and temporal correlation module.

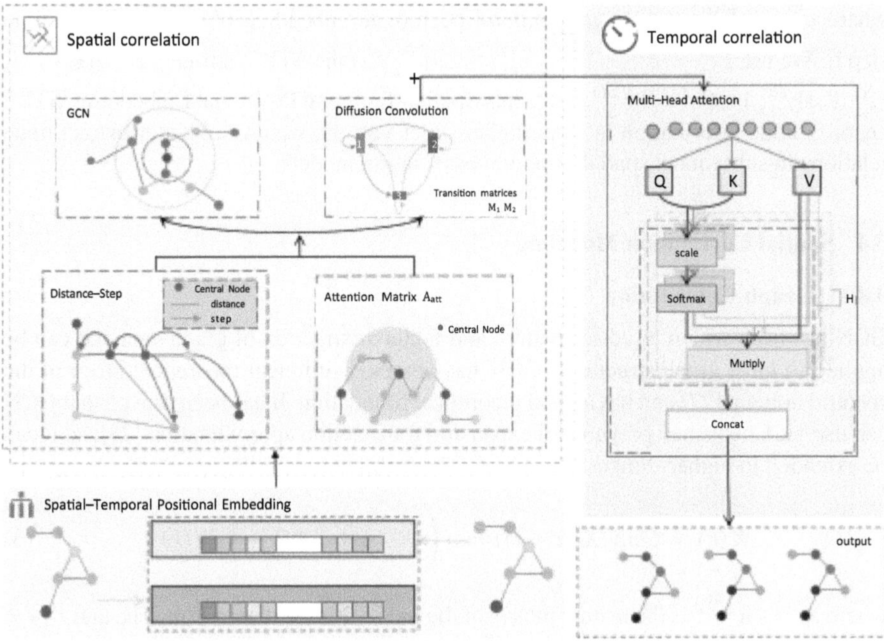

Fig. 2. MAD-GCN architecture.

3.3 Spatial-Temporal Positional Embedding

The condition of the low-altitude traffic flow is shown by the data recorded at the monitoring point. The data dimension of historical observation value is low, which means it can be easily understood by people, but it is not conducive to exploring the implied spatial-temporal relationship. Similar to [18], we propose a spatial-temporal positional embedding (STPE). Based on node2vec [26], STPE aims to embed the historical and predicted traffic flows, and extend low-dimensional data into high-dimensional vectors to obtain deep node relationships. Spatial-Temporal Positional Embedding is divided into two parts, spatial embedding and temporal embedding. In terms of spatial embedding, $e^{s}_{vi} \in R^{D}$ represents the spatial embedding of node i, where vi $\in$ V, D is the dimension of the feature. This means that there is only one spatial representation per node at all times. Spatial embedding can only represent fixed path coding information of nodes. When traffic changes on the time axis, temporal information of nodes will be generated accordingly. So we thought about adding temporal embedding. In terms of temporal embedding, H time slices in the past and P time slices in the future are embedded. $e^{T}_{tj} \in R^{D}$is the temporal embedding of time slice *tj*, where *tj* $=t_1, t_2, t_3, ..., t_H, ..., t_{H+P},$ *D* is the dimension of feature.

After obtaining the spatial and temporal embedding vectors respectively, we further construct the spatial-temporal positional embedding vectors. The spatial- temporal positional embedding is as follows:

$$e^{ST}_{vi,t_j} = e^{S}_{vi} + e^{T}_{t_j} \tag{2}$$

where $e^{ST}_{vi,tj} \in R^{D}$is the spatial-temporal positional embedding of node vi in the time step tj. We use $E^{ST}_{tj} =(e^{ST}_{v1,tj},, e^{ST}_{v2,tj}, e^{ST}_{v3,tj}, ..., e^{ST}_{vN,tj})$ as the STPE in time *j*. E^{ST}=($E^{ST}_{t_1}$, ..., $E^{ST}_{t_H}$, ..., $E^{ST}_{t_{H+P}}$) $\in R^{(H+P)\times N\times D}$ represents Spatial-Temporal Positional Embedding. STPE encodes node information to form higher-order vectors, which can better extract node relations in subsequent spatial-temporal association models.

3.4 Spatial Correlation Modeling

3.4.1 Graph Convolution

GCN has been widely used for pattern and feature extraction of graph data and can be applied to most graph structures. GCN has obtained sufficient theoretical proof in the spectral domain [27] and has a good theoretical foundation. It has been proved that GCN can use 1st Chebyshev polynomial expansion for function approximation [28], so it can be extended to higher-dimensional GCN. The formula is as follows:

$$X(h) = Conv(X(h-1)) = \sigma\left(\tilde{D}^{-\frac{1}{2}}\tilde{A}\tilde{D}^{-\frac{1}{2}}X(h-1)W(h)\right) \quad (3)$$

where $X^{(0)} \in R^{N\times d}$ is the feature matrix of the input. D is the diagonal matrix, and $\tilde{D}_{i,i} = \sum_j \tilde{A}_{i,j}$. $\tilde{A} = A + IN \in R^{N\times N}$, where A is the adjacency matrix, IN is the identity matrix. The matrix $W \in R^{N\times d}$ is a learnable parameter. $\sigma(\cdot)$ is the activation function. $\tilde{A}^{-\frac{1}{2}}\tilde{A}\tilde{D}^{-\frac{1}{2}}$ is the normalized adjacency matrix. h is the number of convolutions, a higher value of h means that the central node can aggregate information of further nodes, but too high h may lead to problems such as over-smooth. The core function of graph convolutional neural network is to propagate between nodes through the adjacency matrix, which is suitable for non-Euclidean topology.

3.4.2 Dynamic Node Association Matrix

However, previous graph convolutions usually use a predefined adjacency matrix, but in many practical applications, traffic region data is a special form of traffic point data when it is regularly distributed in spatial dimensions [29], the relationship between nodes of the graph may change over time or according to different environments. The spatial relationship represented by the predefined adjacency matrix is the static spatial topology of the traffic graph, which cannot simulate the dynamic spatial dependence. Therefore, we propose a dynamic node association matrix to replace the predefined adjacency matrix and construct a dynamic spatial topology. Our proposed dynamic node association matrix consists of two parts, namely, the Node Attention matrix and the Distance- Step correction matrix. The Node Attention matrix is used to adaptively simulate the dynamic association of nodes, and the Distance-Step correction matrix is used to assist in enhancing the ability to pay attention to some nodes in the graph.

Node Attention Matrix

In terms of spatial association, there may be dependencies between any nodes in the graph, which cannot be simply represented by the predefined adjacency matrix. The attention mechanism is used to dynamically and adaptively calculate the dependency between nodes. The formula of the node attention matrix is as follows:

$$Aatt = Vs \cdot \sigma((\chi h - 1W1)W2(W3\chi h - 1)T + bs) \tag{4}$$

$$A_{atti,j'} = SoftMax(A_{atti,j}) = \frac{\exp(A_{atti,j})}{\sum_{j=0}^{Tr} \exp(A_{atti,j})} \tag{5}$$

In the hth layer, we have $\chi^{h-1} = (X_1, X_2, X_3, ..., X_{Tr}) \in R^{N \times C \times Tr}$ as the input. $V_s, b_s \in R^{N \times N}$, $W_1 \in R^{Tr}$, $W_2 \in R^{C \times Tr}$, $W_3 \in R^C$ are the parameters to be learned. The weight matrix $A_{att} \in R^{N \times N}$ signifies attention between nodes, where $A_{atti,j}$ denotes the correlation between nodes. Where the larger value of $A_{atti,j}$ represents the stronger mutual influence between node V_i and node V_j. we apply the SoftMax function to normalize the attention matrix and get the normalized attention matrix A_{att}.

Distance-Step Correction Matrix

In the real transportation network, a certain transportation network node is most susceptible to the influence of its neighboring nodes, but this does not mean that it is only affected by its neighboring nodes. It may still be affected by nodes farther away from it. To model the spatial correlation between nodes, we use the Distance correction matrix and Step correction matrix to simulate this spatial relationship. Firstly, the distance between nodes in the road network directly determines the strength of the association between the two nodes. For example, when unexpected traffic conditions such as traffic accidents, traffic jams, etc. occur at a certain traffic node on the road network, this will directly affect the traffic conditions of adjacent nodes of that node, which may cause connected traffic jams and hinder traffic. However, nodes that are further away from this node are greatly affected after passing through the buffering between these two nodes. Considering this impact, we propose a Distance correction matrix, whose calculation is as follows:

$$D(i,j) = e^{\frac{-d(v_i,v_j)}{d_{\max}}} \tag{6}$$

where d(vi, vj) represents the distance between node i and node j, dmax represents the maximum value of the shortest path between any two nodes in the road network. Secondly, the number of transfer steps between nodes also affects the spatial association between nodes. Two points that are far apart may be directly connected, without the need to experience other nodes. In this case, there may be a strong spatial correlation between these two road network nodes, which cannot be captured solely by the Distance correction matrix. Due to this relationship, we propose a Step correction matrix with the following formula:

$$P(i,j) = e^{\frac{-p(v_i,v_j)}{p_{\max}}} \tag{7}$$

where p(vi, vj)represents the shortest step between node i and node j, pmax represents the maximum value of the shortest step size between any two nodes in the road network. Moreover, because two nodes with longer adjacent steps may have shorter distances, using only the Step correction matrix cannot effectively describe the spatial association between nodes. To combine Distance correction and Step correction matrix, we propose a Distance-Step correction matrix DP. The formula is as follows:

$$DP = D \odot P \tag{8}$$

Following the acquisition of the Distance-Step correction matrix, we combined it with the Node Attention matrix to generate the Dynamic node association matrix ADA. The formulation is outlined below:

$$ADA = DP \odot Aatt \tag{9}$$

$$X(h) = Conv_A P(X(h-1)) = \sigma\left(\tilde{\mathrm{D}}^{-\frac{1}{2}} \mathrm{A_{DA}} \tilde{\mathrm{D}}^{-\frac{1}{2}} X(h-1) W(h)\right) \tag{10}$$

3.4.3 Diffusion Convolution

After obtaining the dynamic node association matrix, a diffusion convolutional layer is used as a complementary process to graph convolution. Diffusion convolution expands the receptive field of the convolution kernel on the graph by dilating the kernel. The graph convolution formula fused with diffusion convolution is as follows:

$$X(h) = ConvAD(X(h-1)) = \sigma(D - AACD - X(h-1)W(h) + QD) \tag{11}$$

$$QD = \sum_{k=0}^{R} (\theta_0 M_0^k X^{(h)} W_0 + \theta_1 M_1^k X^{(h)} W_1) \tag{12}$$

$$M_0^k = A / \sum_j A_{i,j}, M_1^k = A^T / \sum_j A_{i,j}^T \tag{13}$$

where M_0^k and M_1^k are the forward and backward transition matrices, *θ0, θ 1, W0,W1* are the parameters matrices to learn. $M_0^2 = M0 \cdot M0$. The function of M_0^k and M_2^k is to represent the transition probability between nodes, and K is the number of diffusion steps. The forward transition matrix and the backward transition matrix are used to simulate the diffusion process on the graph, which can capture the implicit node space dependence while reducing the computational cost. At the same time, it can enhance the attention to the nearby nodes in the convolution process and improve the aggregation ability.

3.5 Temporal Correlation Modeling

The multi-head attention mechanism is effective when dealing with many different feature patterns, and can be used for parallel computation. The input sequence can be divided into multiple heads for weight and calculation and modeling to capture different feature relationships in multiple input sequences. Therefore, we use multi- head attention mechanism to model the temporal characteristics of low-altitude traffic flow data. The formula for long head attention for head H is as follows:

$$X^{(h)} W_q^{(h)} = Q^{(h)}, X^{(h)} W_k^{(h)} = K^{(h)}, X^{(h)} W_v^{(h)} = V^{(h)} \tag{14}$$

Table 1. Detailed information on datasets.

Dataset	PeMSD3	PeMSD4	PeMSD7	PeMSD8
Data type	Traffic flow	Traffic flow	Traffic flow	Traffic flow
Nodes(Sensors)	358	307	883	170
Edges	547	340	866	295
Time steps	26,208	16,992	28,224	17,856
Features	3	3	3	3
Data frequency	5 min	5 min	5 min	5 min

$$Attention(Q(h), K(h), V(h)) = Softmax\left(\frac{Q^{(h)}K(h)^T}{\sqrt{d_k}}\right)V^{(h)} \tag{15}$$

$$head_{\mathrm{i}} = \mathrm{Attention}\left(QW_i^Q, KW_i^K, VW_i^V\right) \tag{16}$$

where $Wq^{(i)}$, $W_k{}^{(i)}$, and $Wv^{(i)}$ are projections on the matrix and learnable parameters, and $\mathrm{Q}^{(h)}$, $\mathrm{K}^{(h)}$, $\mathrm{V}^{(h)}$ is the query, key, and value matrix to be constructed. After the attention matrix of a single head is obtained, the results of H head are concatenated and connected with the residual through the normalization layer to obtain the final prediction result. The formula is as follows:

$$S = Concat(head1, head2, \ldots, headH) \tag{17}$$

$$X'^{(h)} = W'\mathrm{Reshape}(S) + b' \tag{18}$$

4 Experiments

4.1 Datasets

We first referred to a low-altitude urban network simulation experiment, as shown in the figure. The figure illustrates the time histories of the corresponding tracking error, control input, and unknown equivalent external disturbance signal for a specific region.

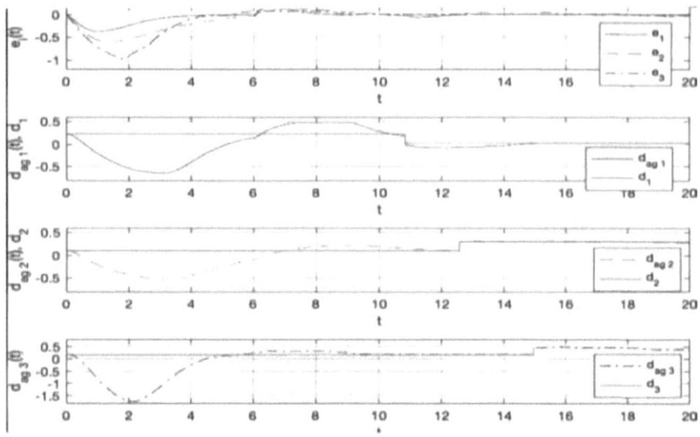

Fig. 3. Time histories of different variables

To assess the impact of our novel MAD-GCN model, we choose to employ highway datasets (specifically, PEMSD3, PEMSD4, PEMSD7, and PEMSD8) acquired from California as the foundation for our experimental examination. These data sets were obtained from California Department of Transportation's Performance Measurement System (PeMS), offering information regarding the state's road networks. Table 1 shows the detailed descriptions of PEMS. In our approach to data preprocessing, we employ Z-Score normalization to standardize the datasets, guaranteeing uniform orders of magnitude for the input data. The formula for the Z-Score is as follows:

$$\hat{X} = \frac{\mathrm{X - mean(X_{train})}}{\mathrm{std(X_{train})}} \tag{19}$$

4.2 Settings

We partitioned the datasets into three distinct subsets: the training set, the validation set, and the test set, maintaining a balanced distribution with a ratio of 6:2:2. Each consecutive time interval spans 5 min, effectively breaking down a day into 288 individual time segments. The prediction step is 12. In MAD-GCN, we set the hidden dimension of graph convolution as 64, the number of diffusion hops R = 2, and the number of attention heads H is 4. During the training phase, a batch size of 64 is utilized, with a learning rate set at 0.001. The adamoptimizer is employed, and the epoch count is configured to be 100. In the experiment, MAE, MAPE, and RMSE were chosen as the error evaluation standards. The three evaluation metrics are formulated as follows:

$$MAE = \frac{1}{N \times T}\sum_{t=1}^{T}\sum_{n=1}^{N}\left|Y_t^n - \vec{Y}_t^n\right| \tag{20}$$

$$RMSE = \sqrt{\frac{1}{N \times T}\sum_{t=1}^{T}\sum_{n=1}^{N}(Y_t^n - \vec{Y}_t^n)^2} \tag{21}$$

$$MAPE = \frac{1}{N \times T} \sum_{t=1}^{T} \sum_{n=1}^{N} \frac{\left| Y_t^n - \vec{Y}_t^n \right|}{Y_t^n} \quad (22)$$

4.3 Baseline Methods

We used the following seven baselines to compare with our proposed MAD-GCN model.

HA [5]: Historical Average method, we use traffic data from the past 12 time steps a s input.

VAR [30]: Vector Auto-Regression, a classical approach used for modeling time series data.

ARIMA [31]: Autoregressive Integrated Moving Average, an established method in time series forecasting and analysis.

LSTM [32]: Long Short-Term Memory networks, designed as an extension of Recurrent Neural Networks to capture temporal dependencies.

GRU [33]: Gate Recurrent Unit, a variant of an RNN that uses a gating mechanism to control how much information is retained before and after.

STGCN [23]: Spatial-temporal Graph Convolution Network, a framework that combines Graph Convolutional Networks with convolutional layers to model spatial and temporal dependencies.

DCRNN [17]: Diffusion Convolutional Recurrent Neural Network, introducing dilated convolutions to capture spatial-temporal correlations effectively.

ASTGCN [2]: Attention Based Spatial-Temporal Graph Convolutional Networks, based on attention mechanism, attach weight to spatial-temporal association, and use graph convolution to aggregate information.

4.3.1 Results on the PEMS Dataset

In the experiment, we choose eight traffic flow prediction models as the baseline model. We conduct experiments on four data sets: PEMSD3, PEMSD4, PEMSD7, PEMSD8. Here, we use the average results of the next twelve time steps as the performance of the model. From the experimental results in Table 2, it can be seen that our proposed

Table 2. Prediction Errors on the PeMSD3, PeMSD4, PeMSD7 and PeMSD8 Datasets.

Dataset	Metric	HA	VAR	ARIMA	LSTM	GRU	STGCN	DCRNN	ASTGCN	**Ours**
PEMSD3	MAE	28.15	23.65	33.51	20.62	21.33	17.49	17.99	17.31	**16.85**
	MAPE (%)	30.58	24.51	33.78	20.85	23.33	17.15	18.34	17.21	**16.57**
	RMSE	46.13	38.56	47.59	30.94	35.11	30.12	30.31	29.56	**29.34**
PEMSD4	MAE	37.92	33.85	32.54	28.88	27.25	24.92	22.95	23.36	**21.14**
	MAPE (%)	27.89	18.09	24.18	18.74	18.20	14.59	14.17	14.96	**12.85**

(*continued*)

Table 2. (*continued*)

Dataset	Metric	HA	VAR	ARIMA	LSTM	GRU	STGCN	DCRNN	ASTGCN	**Ours**
	RMSE	54.48	52.43	56.94	46.32	45.53	38.05	34.56	35.13	**33.25**
PEMSD7	MAE	55.14	53.93	38.17	29.71	29.98	25.38	25.22	25.61	**24.57**
	MAPE (%)	39.78	32.22	19.46	13.51	13.20	11.08	11.82	10.99	**10.79**
	RMSE	90.37	87.34	59.27	45.14	45.94	38.78	38.61	39.56	**37.91**
PEMSD8	MAE	28.76	21.23	23.49	22.18	22.04	18.07	17.72	18.17	**17.18**
	MAPE (%)	34.86	15.42	22.73	14.21	14.20	11.40	10.92	11.64	**10.48**
	RMSE	43.29	30.29	42.11	36.44	35.22	26.95	24.34	25.01	**23.61**

MAD-GCN is superior to the baseline model and has good spatial-temporal modeling ability. In addition, it can be seen from the evaluation results that HA, VAR, and ARIMA have poor prediction effects, among which PEMSD7 has the worst performance, and the predicted results have a large error. This is because these two models belong to classical statistical models, which do not have the ability to deal with time and space together. Other deep learning models, such as LSTM and GRU, have better modeling ability than the first two models, but still only have the ability to model one of the attributes of time or space, and cannot combine the two well. STGCN, DCRNN, and ASTGCN achieve good prediction results because they can well deal with the spatial-temporal dependency in low-altitude traffic flow. However, our proposed MAD-GCN has better results than these three baseline models. Compared with STGCN, this model uses a predefined adjacency matrix, which is based on the adjacency relationship of traffic nodes as the graph convolution kernel, and cannot dynamically model complex traffic conditions. DCRNN uses diffusion convolution kernel to simulate the spatial relationship of nodes, but it still relies on predefined adjacency matrix, and it cannot simulate implicit traffic conditions.

ASTGCN uses the attention mechanism to simulate the spatial relationship of nodes, but it still relies on predefined adjacency matrix, and it cannot simulate implicit traffic conditions. ASTGCN uses the attention mechanism to model the spatial-temporal correlation, and the attention weight matrix between nodes can dynamically simulate the node association, but it uses the Laplacian operator to calculate the graph convolution, which has high complexity. The use of 1-D CNN in the time dimension leads to its insufficient temporal modeling ability. Our proposed MAD-GCN can handle the spatial relationship well by using the adaptive node correlation matrix and combining with diffusion convolution. The multi-head attention mechanism is used to deal with temporal dependency, so it has good spatial-temporal modeling ability. In addition, Fig. 3 shows the specific error values between our proposed MAD-GCN and baseline model in two datasets, PEMSD4 and PEMSD8, among the 12 predicted time steps. The results show that when the predicted step size is larger, models perform worse. The error values of HA, VAR, and ARIMA are higher. The effect of LSTM and GRU, these two neural network variants of RNN is also not good. Although the prediction error in the first few time steps is similar to that of the spatial-temporal model, the error increases rapidly with the increase of the prediction time due to the lack of consideration of the dependency on the spatial correlation of traffic.

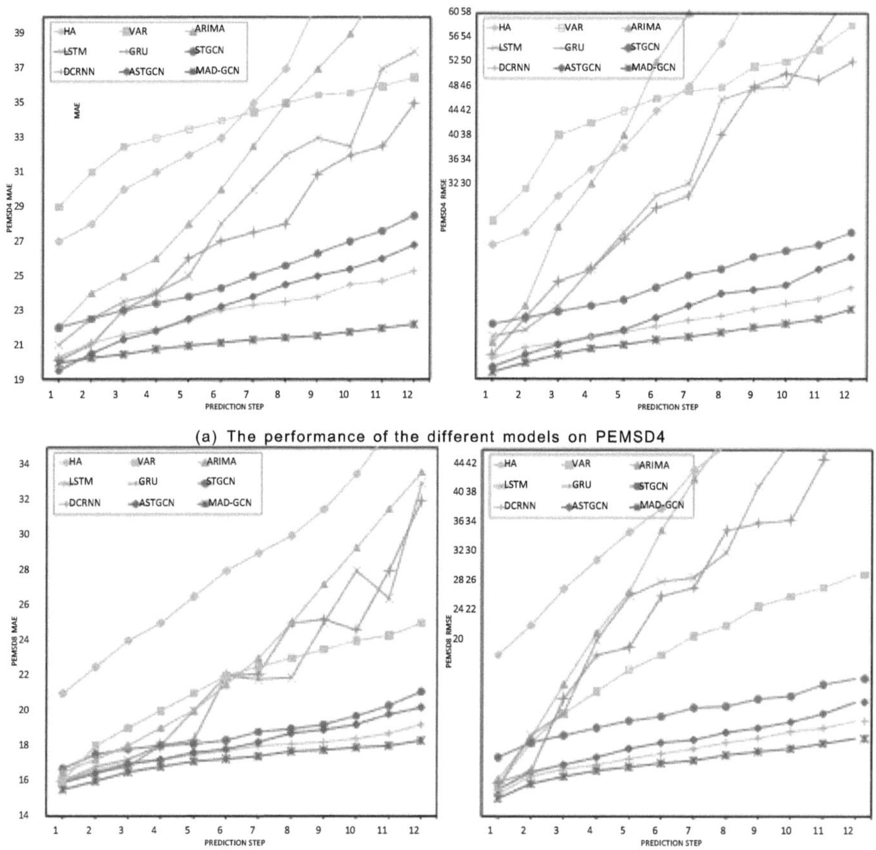

(a) The performance of the different models on PEMSD4

(b) The performance of the different models on PEMSD8

Fig. 4. Predictive performance visualization for different periods.

Compared with the spatial-temporal model, the effect of our proposed MAD- GCN is better than that of STGCN, DCRNN, and ASTGCN. In general, the effect of our proposed MAD-GCN is better than the baseline model at different time steps, which can well capture the spatial-temporal association.

4.3.2 Ablation Experiment

To verify the effectiveness of the modules in our proposed MAD-GCN, we designed the model as follows. (1) RemSTEP: It removes the Spatial-Temporal Positional Embedding in the MAD-GCN. (2) RemDNA: It removes the Dynamic node association matrix in the MAD-GCN. (3) RemDP: It removes the Distance-Step correction matrix in the MAD-GCN. (4) RemDC: It removes the Diffusion Convolution in the MAD-GCN. The experiments performed by the above model without modules are the same as the evaluation functions adopted by MAD-GCN, which are MAP, MAPE, and RMSE. Figure 4 depicts the prediction results of each model in detail.

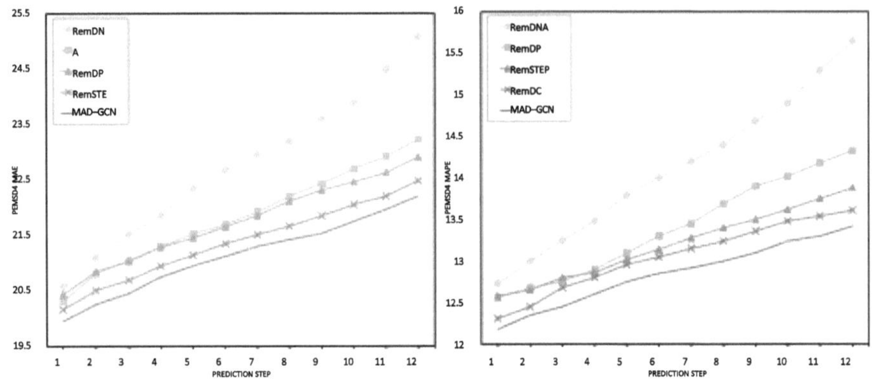

Fig. 5. Detailed framework of MAD-GCN.

The results in Fig. 4 show that the prediction results of several variant models designed by us are inferior to our proposed MAD-GCN model. Among them, RemDNA has the worst prediction effect, which proves the effectiveness of our proposed dynamic node correlation matrix in spatial-temporal modeling. We can also find that the effect of RemDC is better than that of RemDP, indicating that the effect of Distance-Step correction matrix in simulating the hidden relationship of nodes is better than that of the diffusion convolution through the transfer matrix. The prediction results of RemSTEP also demonstrate the effect of the Spatial-Tempor Positional Embedding (Fig. 6).

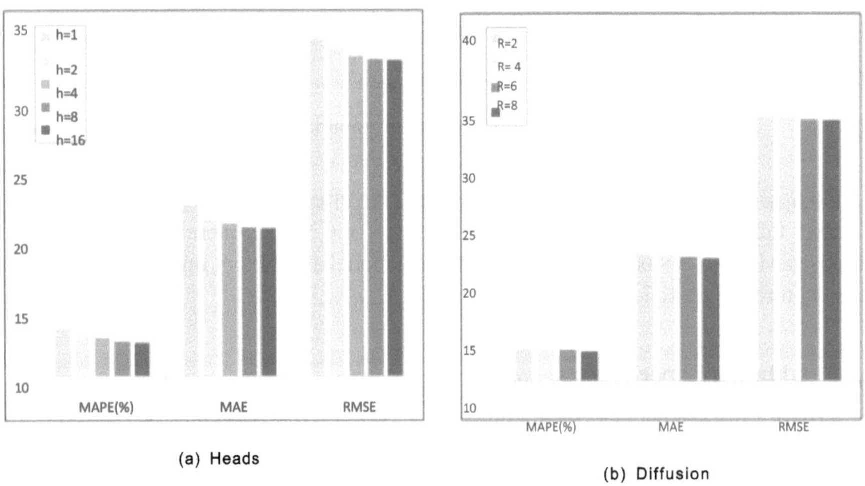

Fig. 6. Network configuration analysis.

4.3.3 Effect of Different Network Configurations

Figure 5 shows the prediction results of our proposed MAD-GCN in PEMSD4 under different hyperparameter values in network training. The left figure represents the number of heads in multi- head attention, and the right figure represents the number of diffusion hops R. It can be found that when the number of diffusion hops R increases, the improvement of prediction accuracy is small. The increase in the number of attention heads will slightly improve the prediction accuracy, but it brings a large computational overhead and may lead to problems such as overfitting.

5 Conclusion

In this paper, a novel deep learning model MAD-GCN is proposed. The model optimizes the predefined neural network in the graph convolutional network, combines the Distance-Step correction matrix and the attention mechanism, constructs the node correlation matrix, captures the dynamic node spatial relationship, and introduces diffusion convolution to enhance the modeling ability of the implicit dependence of nodes. The multi-head attention mechanism is introduced to explore the temporal correlation and enhance the processing ability of long time series. We conduct experiments on multiple datasets and compare it with baseline models, which shows that our proposed MAD-GCN performs better than the baseline model and has good spatial- temporal modeling ability. In the future, we will consider adding other variables to assist the forecast and enhance the accuracy of the forecast.

References

1. Dong, C., Loy, C.C., He, K., Tang, X.: Learning a deep convolutional network for image super- resolution. In: Computer Vision–ECCV 2014: 13th European Conference, Zurich, Switzerland, September 6–12, 2014, Proceedings, Part IV 13, pp. 184–199. Springer, Cham (2014)
2. Guo, S., Lin, Y., Feng, N., Song, C., Wan, H.: Attention based spatial-temporal graph convolutional networks for traffic flow forecasting. In: Proceedings of the AAAI Conference on Artificial Intelligence, vol. 33, pp. 922–929 (2019)
3. Bai, L., Yao, L., Kanhere, S.S., Yang, Z., Chu, J., Wang, X.: Passenger demand forecasting with multi-task convolutional recurrent neural networks. In: Pacific-Asia Conference on Knowledge Discovery and Data Mining, pp. 29–42. Springer, Cham (2019)
4. Geng, X., et al.: Spatiotemporal multi-graph convolution network for ride-hailing demand forecasting. In: Proceedings of the AAAI Conference on Artificial Intelligence, vol. 33, pp. 3656–3663 (2019)
5. Liu, J., Guan, W.: A summary of traffic flow forecasting methods. J. Highw. Transp. Res. Dev. **21**(3), 82–85 (2004)
6. TO, B.-J.A.: Analysis of freeway traffic time-series data by using box-jenkins techniques
7. Ahmed, M.S., Cook, A.R.: Analysis of Freeway Traffic Time-series Data by UsingBox-Jen kins Techniques, vol. 722 (1979)
8. Edes, Y.J.S., Michalopoulos, P.G., Plum, R.A.: Improved estimation of traffic flow for real-time control. Transp. Res. Rec. **95**, 28–39 (1980)

9. Ma, X., Dai, Z., He, Z., Ma, J., Wang, Y., Wang, Y.: Learning traffic as images:a deep convolutional neural network for large-scale transportation network speed prediction. Sensors **17**(4), 818 (2017)
10. Zhang, J., Zheng, Y., Qi, D.: Deep spatio-temporal residual networks for citywide crowd flows prediction. In: Thirty-first AAAI Conference on Artificial Intelligence (2017)
11. Wu, Y., Tan, H.: Short-term traffic flow forecasting with spatial-temporal cor-relation in a hybrid deep learning framework. arXiv preprint arXiv:1612.01022 (2016)
12. Kong, X., Zhang, J., Wei, X., Xing, W., Lu, W.: Adaptive spatialtemporal graph attentio n networks for traffic flow forecasting. Appl. Intell. **52**(4), 4300–4316 (2022)
13. Zhang, C., et al.: Augmented multi-component recurrent graph convolutional network for traffic flow forecasting. IS PRS Int. J. Geo-Inf. **11**(2), 88 (2022)
14. Wang, Y., Jing, C., Xu, S., Guo, T.: Attention based spatiotemporal graph attention networks for traffic flow forecasting. Inf. Sci. **607**, 869–883 (2022)
15. Zhang, W., Zhu, K., Zhang, S., Chen, Q., Xu, J.: Dynamic graph convolutional networks based on spatiotemporal data embedding for traffic flow forecasting. Knowl. Based Syst. **250**, 109028 (2022)
16. Zhang, S., Guo, Y., Zhao, P., Zheng, C., Chen, X.: A graph-based temporal attention framework for multi-sensor traffic flow forecasting. IEEE Trans. Intell. Transp. Syst. **23**(7), 7743–7758 (2021)
17. Li, Y., Yu, R., Shahabi, C., Liu, Y.: Diffusion convolutional recurrent neural network: Data-driven traffic forecasting. arXiv preprint arXiv:1707.01926 (2017)
18. Zheng, C., Fan, X., Wang, C., Qi, J.: GMAN: A graph multi-attention net-work for traffic prediction. In: Proceedings of the AAAI Conference on Artificial Intelligence, vol. 34, pp. 1234–1241 (2020)
19. Wu, Z., Pan, S., Long, G., Jiang, J., Zhang, C.: Graph wavenet for deep spatial- temporal graph modeling. arXiv preprint arXiv:1906.00121 (2019)
20. Laptev, N., Yosinski, J., Li, L.E., Smyl, S.: Timeseries extreme event forecasting with neural networks at uber. In: International Conference on Machine Learning, vol. 34, pp. 1–5 (2017)
21. Cui, Z., Ke, R., Pu, Z., Wang, Y.: Stacked bidirectional and unidirectional LSTM recurrent neural network for forecasting network-wide traffic state with missing values. Transp. Res. Part C: Emerg. Technol. **118**, 102674 (2020)
22. Bai, L., Yao, L., Li, C., Wang, X., Wang, C.: Adaptive graph convolutional recurrent network for traffic forecasting. Adv. Neural. Inf. Process. Syst. **33**, 17804–17815 (2020)
23. Yu, B., Yin, H., Zhu, Z.: Spatio-temporal graph convolutional networks: A deep learning framework for traffic forecasting. arXiv preprint arXiv:1709.04875 (2017)
24. Li, Y., Moura, J.M.: Forecaster: a graph transformer for forecasting spatial and time-dependent data. arXiv preprint arXiv:1909.04019 (2019)
25. Jiang, S., Zhu, M., Li, J.: Traffic flow forecasting using a spatial-temporal attention graph convolutional network predictor. In: International Conference on Spatial Data and Intelligence, pp. 107–121. Springer, Cham (2021)
26. Grover, A., Leskovec, J.: node2vec: scalable feature learning for networks. In: Proceedings of the 22nd ACM SIGKDD International Conference on Knowledge Discovery and Data Mining, pp. 855–864 (2016)
27. Defferrard, M., Bresson, X., Vandergheynst, P.: Convolutional neural networks on graphs with fast localized spectral filtering. In: Advances in Neural Information Processing Systems, vol. 29 (2016)
28. Kipf, T.N., Welling, M.: Semi-supervised classification with graph convolutional networks. arXiv preprint Hamilton, J.D.: Time Series Analysis. Princeton university press (2020)
29. Peng, L., Liao, X., Li, T. et al.: An overview based on the overall architecture of traffic forecasting. Data Sci. Eng. (2024)

30. Hamilton, J.D.: Time Series Analysis. Princeton University Press, Princeton (2020)
31. Williams, B.M., Hoel, L.A.: Modeling and forecasting vehicular traffic flow as a seasonal arima process: theoretical basis and empirical results. J. Transp. Eng. **129**(6), 664–672 (2003)
32. Hochreiter, S., Schmidhuber, J.: Long short-term memory. Neural Comput. **9**(8), 1735–1780 (1997)
33. Chung, J., Gulcehre, C., Cho, K., Bengio, Y.: Empirical evaluation of gated recurrent neural networks on sequence modeling. arXiv preprint arXiv:1412.3555 (2014)

Author Index

J. Li et al. (Eds.): APWeb-WAIM 2025, CCIS 2660, pp. 343–344, 2026.
https://doi.org/10.1007/978-981-95-7996-9

MIX
Papier aus verantwortungsvollen Quellen
Paper from responsible sources
FSC® C105338

If you have any concerns about our products,
you can contact us on
ProductSafety@springernature.com

In case Publisher is established outside the EU,
the EU authorized representative is:
Springer Nature Customer Service Center GmbH
Europaplatz 3, 69115 Heidelberg, Germany

Printed by Libri Plureos GmbH
in Hamburg, Germany